IMPROVING SCHOOLS THROUGH TEACHER DEVELOPMENT

CONTEXTS OF LEARNING
Classrooms, Schools and Society

IMPROVING SCHOOLS THROUGH TEACHER DEVELOPMENT: Case Studies of the Aga Khan Foundation Projects in East Africa

EDITED BY

Stephen E. Anderson
The Ontario Institute for Studies in Education, University of Toronto, Ontario, Canada

CRC Press
Taylor & Francis Group
Boca Raton London New York

CRC Press is an imprint of the
Taylor & Francis Group, an **informa** business

CRC Press
Taylor & Francis Group
6000 Broken Sound Parkway NW, Suite 300
Boca Raton, FL 33487-2742

First issued in paperback 2020

ISBN-13: 978-90-265-1936-9 (hbk)
ISBN-13: 978-0-367-60479-0 (pbk)
ISSN 1384-1181

Visit the Taylor & Francis Web site at
http://www.taylorandfrancis.com

and the CRC Press Web site at
http://www.crcpress.com

Library of Congress Cataloging-in-Publication Data

Contents

Acknowledgements

The idea for this book grew out of my work as a consultant to the Aga Khan Foundation (AKF) school improvement efforts in Tanzania, Kenya, and Uganda between 1994 and 1999. I participated in the evaluation of two of the school improvement projects (SIPs) described in this volume and visited others at varying stages of their mobilization, implementation, and continuation. Through these contacts, I realized that the SIPs were not isolated efforts. The AKF experience with school improvement in East Africa was a cumulative story of a non-governmental organization over time learning to plan, mobilize, implement, and sustain school-based improvement initiatives in partnership with local education providers. As with many such initiatives in developing countries, access to findings from these efforts is difficult. Evaluation reports are distributed on a limited basis, and project briefs may appear in the sponsoring agencies' publicity materials. The idea for this book was to assemble in a single volume the evaluations of the Aga Khan Foundation's East African SIPs from 1985 to 2000 and to provide a historical and analytical commentary in relation to their evolution and to school improvement efforts world wide.

I presented this idea to the Aga Khan Foundation - Geneva in the fall of 1999 and received an endorsement of the publishing project with an agreement to assist in various ways (e.g., facilitating communication with the authors of the SIP evaluations, providing access to photographs, assisting with the costs of preparing the manuscript). I am thankful in particular for the encouragement provided by Jeremy Greenland (Chief Education Officer, Aga Khan Foundation-Geneva), who also contributed the historical perspective on AKF's school improvement projects in Chapter 2. Without his interest and support, the book would have remained no more than a fleeting inspiration. Katherine Hinckley, formerly with Aga Khan Foundation, was instrumental in the inclusion of photographs from the AKF photograph files. Kathy Bartlett (Education Programme Officer, Aga Khan Foundation - Geneva) and Seth Ong'uti (Education Programme Officer, Aga Khan Foundation - East Africa) gave useful feedback on the original proposal and on my overview of the SIP efforts in Chapter 1.

I am deeply appreciative of the willingness, thought, and time committed by the authors of the original SIP evaluations to rework their reports into case studies for this book. Joe Farrell and David Hopkins, who had no prior association with the projects, kindly agreed to provide external perspectives on the Aga Khan Foundation SIP experiences in the concluding chapters. In preparing the manuscript, I was assisted initially by Anastasia Meletopoulous. Final copyediting was accomplished with meticulous professionalism by Margaret Oldfield and her colleague Zafar Shamsi, who did the layout. I would also like to acknowledge the interest, guidance, and patience of the editorial team at Swets & Zeitlinger Publishers.

Stephen E. Anderson
November 2001

Preface

The school improvement programs highlighted in this book were all initiated and carried out under the auspices of the Aga Khan Foundation (AKF). The Aga Khan Foundation is a private, nondenominational development agency established by His Highness the Aga Khan in Switzerland in 1967. AKF is a member of the Aga Khan Development Network, a group of institutions working in social and economic development and in the field of culture to improve living conditions and opportunities in specific regions of the developing world. Their common goal is to help poor people achieve a level of self-reliance sufficient to plan their own livelihoods and help those even poorer than themselves. It seeks sustainable solutions to long-term problems of poverty, with special emphasis on the needs of communities in resource-poor areas. The Foundation's program priorities include education, health, rural development, and NGO enhancement, with an emphasis on community participation, gender, the environment and human resource development. The AKF Education Program has been instrumental in the development and implementation of the school improvement projects featured in this book. The objectives of that program are to attain a quality education for all children, youth, and adults (by increasing student access, retention, and completion rates, and by ensuring active participation in the learning process); to increase gender equality; and to strengthen local institutions that can provide leadership and build capacity in early childhood, primary, and post-primary education. Expected outcomes of the program include whole-school development, linked to effective parent and community involvement (in curriculum, management, and financing); quality, community-based, early childhood development services that build on local culture and resources; local and regional institutional development to ensure relevant and effective curriculum, training and leadership programs; and well-tested programing approaches to quality, sustainability, parent and community mobilization, and private-public partnerships. Although formally a funding agency, AKF involves itself actively in the planning and execution of its projects. Grants are made to AKF's sister organization, Aga Khan Education Services, to the Aga Khan University Institute for Educational Development, to governments, and to nongovernmental and private-sector institutions that share the Foundation's goals. The Foundation's programs are funded with the assistance of more than 60 national and international development agencies and of many individual and corporate donors. The geographical focus of the Foundation's work is East Africa, and South and Central Asia. AKF's head office is in Geneva, Switzerland. Its other offices are located in Bangladesh, Canada, India, Kenya, Mozambique, Pakistan, Portugal, Tajikistan, Tanzania, Uganda, the United Kingdom, and the United States of America.

1

The Double Mirrors of School Improvement: The Aga Khan Foundation in East Africa

Stephen E. Anderson

This book presents a remarkable story of school improvement activity over the past decade and a half in the East African countries of Tanzania, Kenya, and Uganda. The school improvement projects (SIPs) that provide the focus for inquiry and discussion in the chapters that follow were all initiated and supported by the Aga Khan Foundation and Aga Khan Education Services, in collaboration with various other international donors and with local education authorities within the participating education jurisdictions. The core of the book consists of a series of six case study evaluations of the implementation and outcomes of these Aga Khan Foundation (AKF) sponsored school improvement projects.

The timing of this book is propitious. First, there has not been a recent collection of case studies of school improvement initiatives in the developing world (For previous collections, see Rust & Dalin, 1990; Lewin & Stuart, 1991; Levin & Lockheed, 1993), although interest in this topic remains high among donor agencies, governments, and practitioners (e.g., Craig, Kraft, & du Plessis, 1998). Second, dramatic changes in the context of elementary and secondary education in many developing countries over the past decade (e.g., reduced government funding, decentralization of school governance, more community involvement in school funding and management, privatization) have created a different political, economic, and organizational context for efforts to universalize and improve basic educational provisions. The pioneering experiences of AKF in creating effective, sustainable, field-based support systems for continuous school and teacher development within these changing contexts can contribute usefully to discussions in this area. Third, these shifts reflect similar trends worldwide (e.g., increased local control and responsibility for school development and improvement within a context of centralized curriculum policy, test-driven external accountability, and policies promoting greater competition and choice among schools). As a result, school improvement experiences in the developing world have increasing theoretical and policy relevance to contemporary education movements in non-Third-World contexts.

The Aga Khan Foundation's Approach to School Improvement

The story this book tells is remarkable in several ways. First, it is a long-term story of international donor and local educator commitment and efforts to improve the quality of education in African schools under very challenging circumstances. Individually, each project highlighted in the case studies extended over multi-year (3-10 years) funding, mobilization, and implementation cycles. Within each project, it is possible to discern an evolutionary progression of strategic actions and results. The local knowledge that emerged about more and less effective strategies for improving the quality of education in the targeted schools is embedded in the cases. Collectively, the story encompasses an interconnected series of AKF-engendered school improvement initiatives in East Africa over a period of about 15 years.[1] The knowledge from each project fed into the planning and implementation of successive projects across the region. While the evaluations of the projects reported in the case studies do not conclude that a miraculous solution to improving schools in developing countries has been found, the case studies provide a rich and cumulative body of knowledge about school improvement issues, strategies, and outcomes in the developing world.

The individual and collective school improvement experiences told in these case studies are also remarkable because of what can be learned from them about the potential and the limitations of an approach to school improvement grounded in a common set of research-based strategic principles that have been variously enacted in diverse settings subject to changing political and economic contexts over time. Six key principles underlie the design of each of the projects studied. The chances for improving the quality of teaching and learning in schools are greater when improvement efforts (a) are school-based, (b) involve the whole school as the unit of change, (c) emphasize the ongoing professional development of teachers, (d) attend to school management and other organizational conditions affecting the capacity of teachers to implement change at the school level, (e) prepare for the institutionalization of organizational structures and processes that enable continuous school development beyond the school improvement project phase, and (f) evolve through genuine partnerships among relevant education stakeholders.

'School-based' implies that targeted needs for improvement reflect identified concerns at the school level, and that multi-school interventions are tailored to school-specific needs as well as to common needs among schools within a system. School-based also implies that on-site assistance is provided to head teachers, teachers, and other stakeholders involved in planning and implementing improvement initiatives at the school level, and that access to external resources, support, and help is provided.

School improvement efforts that target whole-school change are based on the presumption that the overall quality of teaching and learning in schools is more likely to increase when all teachers in a school participate and cooperate in the pursuit of shared goals for school-wide development and change. Strategically, this implies that school improvement initiatives are designed to maximize the probability that all or most teachers will participate in and benefit from the resources, professional development, decision-making, and other forms of school-improvement-related support provided through a project.

Teacher development activities are a cornerstone of all the projects. At a minimum, this includes in-service learning to improve the quality of teachers' instructional practices. The forms of in-service learning invoked are variable across the projects, and include such interventions as access to teacher-center- and school-based workshops; in-class coaching by consultants, supervisors, or peers, or all three; team planning and problem-solving by collegial work groups within the school; action research; teacher inter-visitation within and between schools; and professional study groups. In-class coaching by master teachers and peers is a common in-service denominator across the projects. Most projects have also included interventions to develop local teacher leadership for teacher and school development at the school and district levels.

The project designs have all addressed teachers' in-service learning as one component in a systemic approach to whole-school improvement that includes simultaneous attention to various focuses of school development. The range of organizational conditions addressed grew in comprehensiveness over successive project experiences. The range includes capacity building for basic resource acquisition and development (e.g., low-cost teaching and learning aids), head teacher training in school and change management, mobilization of school-level parental involvement and funding, and development of district leadership and support for school improvement. Planned interventions to improve school management became more prominent in response to project evaluations that highlighted the centrality of effective school management in initiating and sustaining school improvement, and in the context of persistent state and international pressures toward decentralization of school funding and governance. Another common feature across projects has been an investment in the development and improvement of teacher resource centers at the district, school cluster, or school level that have the capacity to support teachers' professional learning needs (e.g., workshops, consultants, libraries) and the production of teaching aids.

Each of the school improvement projects highlighted in the case studies was initiated with explicit intention to prepare for sustainability once the special project status and funding ceased. Sustainability applies not only to the use of innovative teaching practices and to changes in student learning, but also to the organizational structures and processes created at the school and district levels to support the ongoing development of teachers and schools. The establishment of innovative teacher leadership positions, for example, has been accompanied by overt problem-solving about the challenges of continuing those positions under normal budgetary, staffing, and operational conditions (e.g., compatibility with traditional teacher and supervisory roles, funding, professional development for new roles). The introduction of new ways of supporting teacher learning, such as collaborative work groups, in-class coaching, and incentive grants, has been followed by reflection on both the impact of these methods on teacher development and on local capacity for continuation. Parental contributions to school governance and funding as a result of initial project interventions have turned into challenging scenarios for renewal of ongoing parental support for school improvement. All the project evaluations address the prospects for local sustainability of project strategies that might enable continuous teacher development and school improvement.

Efforts to forge partnerships among different stakeholder groups in the pursuit of education quality improvement have been integral to all the Aga Khan Foundation school improvement projects. Initial conceptions of partnership were invitational and couched in terms of dissemination of project interventions and outcomes, that is, encouraging 'nonproject' schools to join in on 'project school' professional development activities, offering to share project developed resources with other schools. As the projects grew in scope and comprehensiveness, the concept of partnership evolved, and included interventions to foster joint responsibility for school improvement planning, resource provision, and implementation among the Aga Khan Foundation, Aga Khan Education Services, local school district authorities, parents and communities served by targeted schools, both public and private school sectors, and other NGOs and education quality improvement initiatives at play in the school improvement project sites. The mobilization of parent and community involvement in school management and development emerged as a focus of need and intervention in the initial projects in Kenya and Uganda. Community development for school improvement became a formal component of the school improvement model with the 1994 Mombasa project, drawing upon AKF's experiences with community-based Madrasa nursery schools.

The case studies and commentaries in this book offer a comparative perspective on a series of long-term school improvement initiatives in East Africa. The projects illustrate the enactment, results, and challenges of school improvement designs that are grounded in a cluster of strategic principles for educational change: change that is school-based, school-wide, participatory and collaborative among key stakeholder groups (e.g., education personnel in different roles and levels in the education system, school parents), focused on teacher development, systemic in addressing school management and organizational conditions impinging on teacher development, grounded in local inquiry, connected to local teacher centers, and attentive to the sustainability of school improvement structures and processes, as well as of teacher and pupil outcomes.

The Road to School Improvement: A Work-in-Progress

This book is relevant to anyone with a genuine interest in learning about school improvement policies and practices. Its potential readers include education policy-makers, education leaders at local district and school levels, education researchers and consultants, and external agencies that invest in the promotion of quality improvement in education. It offers a challenging set of stories and findings; however, it is not a recipe for change. The efforts of AKF and its partners in school improvement in Tanzania, Kenya, and Uganda remain a work-in-progress. The case studies open a window on that work-in-progress and invite readers to engage in, learn from, and reflect upon the experiences of the projects, just as the contributing authors have done.

Neither the sponsors and developers nor the evaluators of the case study projects would claim that the strategies for school improvement embodied in these initiatives represent the only route to improving the quality of education in African schools. Indeed, one might argue, as some of the evaluators do in their critiques, that other strategies are worth considering in light of the issues and challenges presented in the implementation and outcomes of the projects at the school and school district levels. Four such debatable strategies common to most if not all the school improvement projects are (a) an investment in district-level teacher development personnel, (b) an emphasis on enhancing teachers' general pedagogical expertise, (c) a focus on the adoption and use of child-centered activity-oriented methods of teaching and learning, and (d) the promotion of data-based decision-making for school development.

The district-wide school improvement projects rely heavily on the establishment and in-service interventions of cadres of district-level teacher development consultants. In recent years, many school districts in North America and other areas of the developed world have drastically reduced the availability of support services for teachers at the school-system level, partly as a cost-saving measure, but also because of the absence of compelling evidence that investment in curriculum and professional development consultants affects the overall quality of teaching and learning (regardless of benefits for individual teachers in need of assistance). The emphasis in the Aga Khan school improvement projects on developing teams of teacher development agents within the school system but external to schools, could be said to run counter to trends elsewhere in the world. It may well be, however, that lack of access to opportunities for professional development and low awareness of possibilities for classroom improvement among teachers in many areas of the developing world creates a fundamentally different set of conditions for ongoing professional learning, thereby creating a much stronger rationale for continuous, locally based external assistance from master teachers who are linked to a broader research- and practical-knowledge base on effective teaching. The case studies do not argue that local consultant-centered support systems for in-service teacher development are the only way to go. They do offer insight into the experiences of school districts that have pursued that route to improvement, often in combination with more participatory forms of teacher learning at the school level (Shaeffer, 1990).

One of the continuing challenges in the AKF school improvement projects has been to find cost-effective ways for district consultants to provide assistance to teachers in large numbers of schools at a sufficiently intensive level to make a real contribution to teachers' efforts to master new methods of instruction. Project results suggest that periods of 6 to 12 months of in-school coaching by one or more external consultants can significantly enhance teacher learning of new teaching methods. They further suggest that it is feasible to organize that level of intensive service on a district-wide basis using cyclical schedules, whereby all schools get access to external professional development activities on an ongoing basis and subsets of schools get intensive aid at the school level during particular terms or years. While there are practical problems in such systems related to such things as role definition, financing, and transport to and from schools, the major difficulty arises in creating conditions and expertise at the school level for teachers to carry on with their own development without dependence on persistently high levels of external assistance. Efforts in the Aga Khan projects to create school-level cadres and networks of teacher leaders to facilitate continuous improvement have been difficult to put into practice and to sustain. Another finding from the projects and evaluations is the importance of training for people acting in teacher development roles at the system and school levels concerning both the content (e.g., desired teaching methods) and the process of change (e.g., peer coaching, action research, assessing teacher needs).

Most of the school improvement projects target teachers' instructional knowledge and skill as a prime focus for development. This contrasts with strategies that emphasize the development, in-service training, and implementation of specific programs that contain packaged teaching and learning materials, lessons and teaching guides, assessment procedures, and so on. In the former approach targeting teachers' instructional knowledge and skill, the long-term aim is to develop teacher understanding and skill in lesson planning and a broad repertoire of teaching strategies, with the expectation that teachers will gain the capacity to independently decide how to plan and apply these strategies to targeted subject matter objectives, content, and student needs. In the latter approach involving development, testing and implementation of specific programs, similar teaching methods might be employed; however, they are learned and mastered as tools associated with specific lessons and expectations for program delivery. The evaluations of teaching reported in the SIP case studies reveal the complex and long-term challenge of raising the level of teachers' instructional expertise in settings where the baseline level of teachers' basic literacy and professional knowledge can be quite low. Critics might conclude that well developed programs combined with best-practice in-service designs are a more efficient and effective way to raise teaching standards on a large scale (Thiagarajan, 1990), notwithstanding the all-too-frequent tales of failed or superficial implementation of external programs (e.g., Olorundare, 1990; Ibrahim, 1991; Prophet, 1995). The findings from the case study project evaluations add empirical substance, but do not bring closure to this debate. Others might argue that prior or equal attention should be given to developing teachers' subject matter expertise, and there is certainly evidence from past research to suggest that the quality of student learning in developing countries is correlated with the level of teachers' subject knowledge (Fuller & Clarke, 1994). While consideration of teachers' subject knowledge has not been ignored within the AKF school improvement projects, the issue has not been tackled head on, and the projects have tended to leave that focus of teacher development to upgrading programs provided through local teacher training institutions.

The advocacy and support for activity-based discovery-oriented teaching methods within the Aga Khan Foundation school improvement projects provides a further focus for reflection and debate. This pedagogical orientation and its associated practices are promoted as an alternative or complement to the teacher-centered direct teaching methods typically reported in studies of teaching practice in African schools (e.g., Fuller & Snyder, 1991). Reviews of research on effective teaching practices in developing countries offer modest but mixed support for the claim that more student-centered, interactive, and open-ended teaching and learning methods are associated with higher levels of student achievement (e.g., Avalos, 1980; Fuller & Clarke, 1994). Past studies of efforts to introduce and support the implementation of such pedagogical methods in Africa are often disappointing in terms of the degree and quality of teacher change (e.g., Arthur, 1998; Kunje & Stuart, 1999; Prophet, 1995; Rowell & Prophet, 1990). This has led some scholars of educational change in developing countries to express doubts about promoting child-centered pedagogical methods in curriculum, cultural, organizational, and professional contexts that have long favored teacher-centered didactic methods of instruction (Guthrie, 1990). One alternative proposed is to invest in helping teachers become more skilled at direct teaching methods that may be more practical in the resource-poor systems typical of so many developing countries, and perhaps more 'effective', as long as education quality is judged on the basis of curriculum content coverage and student performance on examination systems is closely aligned with content expectations. The projects and evaluations presented in the cases do not resolve the question of what images of pedagogy should be advocated and supported in African schools. They do add to knowledge about the circumstances affecting efforts to introduce progressive teaching methods in developing countries, including the difficulties of developing widespread teacher expertise, of accessing low-cost teaching and learning resources, and of adapting those methods to classrooms with high pupil-teacher ratios (e.g., 50-

100 pupils per teacher in some settings). The significance of these projects, however, has less to do with the choice of teaching methods than with what can be learned about creating infrastructures and organizational cultures that can support ongoing school and teacher development focused on locally determined student learning needs.

Who would argue against the intent to get local educators at the district and school levels to engage in ongoing assessment and evaluation of student learning and teacher development needs, and to utilize that information in deciding on school improvement goals, plans, and outcomes? Most of the district-wide Aga Khan SIPs included proposals for systematic baseline data collection and tracking of student performance, teacher beliefs and practices, school management, parent involvement, school resources, school finances, and so on. A few have required project personnel or external evaluators to create independent standardized tests of student performance and to conduct comparisons of student results in schools affected by but not targeted by project interventions. While some of the projects successfully grounded their teacher development efforts in project-generated assessments of teacher and student needs, efforts to set up new information systems for tracking and evaluating student, teacher, and other outcomes associated with SIP interventions have stumbled and stalled in practice. Should externally supported school improvement initiatives invest in creating separate monitoring and evaluation systems, including the testing of students, that focus uniquely on project initiatives and outcomes and that ultimately live and die with the projects? Or should they invest more in developing local district and school capacity to study and determine school improvement needs, actions, and outcomes, including but not limited to specific project initiatives? To date, given AKF's own accountability concerns and those of other funders, the Aga Khan Foundation projects have placed more emphasis on project-specific monitoring and evaluation processes than on building local capacity to generate and effectively use good information on school-improvement-related conditions and actions.

About the Case Study Evaluations

Each of the school improvement program (SIP) evaluations reported in the case studies was commissioned by the Aga Khan Foundation as an independent summative evaluation at the end of one or more phases of project funding. The results of these evaluations fed into decisions about the design and funding of further cycles of project activity. In the cases of the Kisumu and Mzizima SIPs, the evaluation studies reported here were conducted toward the end of multiple phases of project funding and were preceded by evaluations at the end of earlier phases, the findings of which were incorporated into the final reports. The other evaluations were undertaken toward the conclusion of initial 3- to 5-year cycles of project implementation and funding. Some common features and limitations of the evaluations and their representation in this book are important to consider in advance.

The evaluations were all conducted under contract to the Aga Khan Foundation and were designed to respond to terms of reference stipulated in those contracts. So, while they were 'independent' evaluations, the questions they were designed to answer were guided by the terms of reference of the evaluation contracts. Contractual relations notwithstanding, the evaluation sponsors have been open and receptive to the evaluation findings, including those that are critical of the projects. The Foundation has publicly disseminated the evaluation reports in their entirety.

The Aga Khan Foundation has insisted that evaluations of all its school improvement projects be designed and conducted by evaluation teams that include host-country as well as international education researcher-consultants. This has helped ensure that the evaluations are sensitive to local education contexts in their design, in their interpretation of findings, and in the recommendations arising from those findings.

The evaluation studies of each of the school improvement projects were essentially one-shot case study evaluations, though some of them incorporated post-intervention comparisons of teaching practice and student learning outcomes in project and nonproject schools in their evaluation designs. Data gathering in each study typically included a 2- or 3-week period of on-site interviews with project stakeholders (teachers, head teachers, parents, project personnel, etc.); observation of classroom practices and project activities (e.g., in-service interventions, meetings with parents, teacher center activities); and review of past project evaluations, monitoring reports, in-house evaluation data, and project files (e.g., records of meetings, professional development events, budgets). Inspection of project-related curriculum, teaching, and learning materials figured into some of the evaluations, as well as independent testing of student learning outcomes. The timing and scale of the studies limited the evaluators' capacity to analyze the longitudinal effects on student learning and the cost-benefit ratios of project inputs and outcomes. The case studies are based on the original evaluation reports. In the interest of space, the editor asked the case study writers to provide enough information about the evaluation methods employed for readers to appreciate the sources of information for the findings and recommendations, but to refer interested readers to the original reports for more detailed descriptions of methods and instruments.

One of the unique features of the case studies and cross-case commentaries presented in this book is that the writers of the case studies were invited to report not only their evaluation findings and judgments, but also to highlight key recommendations included in the original evaluation reports. All of the evaluation teams consisted of education researchers-consultants with recognized expertise in one or more of the areas addressed in the evaluations, such as educational change processes, teacher learning, education accountability processes (e.g., student evaluation and large-scale testing), curriculum content and pedagogy (e.g., English-language education, mathematics education). The case studies contain not only information about the school improvement projects, their implementation, and impact, but also reflections and suggestions for further development grounded in the evaluators' special areas of expertise and interest. The same invitation was extended to the cross-case commentators in this book, as part of their comparative analyses of the case studies vis-à-vis the literature on school improvement in the developing world and beyond. This input from the contributing authors pushes the discussion beyond academic critique into research-based proposals for some of the unresolved challenges encountered in the school improvement projects.

The case studies and cross-case commentaries are not framed within a common theoretical framework for describing and interpreting school improvement processes and outcomes. Each connects with the literature on educational change in its own way, consistent with the expertise that the authors bring to their respective analyses. The unifying thread across the evaluations and the supplementary cross-case chapters is inherent in the conceptual, strategic, and practical linkages among the school improvement projects themselves. The cases are presented chronologically, based on the initiation dates of the various projects. This is deliberate, because the coherence in school improvement design strategies across the cases is best understood as an emergent and evolutionary phenomenon, as the experiences and learning from one project feed into others and as all the projects adapt over time to the changing political and socioeconomic contexts in the countries and local jurisdictions in which they are set.

Introduction to the Case Studies and Cross-Case Analyses

Following this introduction, the book begins with an historical account of the Aga Khan Foundation's teacher development and school improvement initiatives in the developing world. While this book concentrates on AKF's school improvement activities throughout East Africa, AKF's support for teacher development and school improvement includes field-based teacher development initiatives in the Northern Areas, Pakistan (Bude, 1992; Sales, 1999); the development of teacher resource centers in Karachi, Pakistan (Ali et al., 1993); projects specifically targeted at improving school management in India (Simpkins & Garrett, 1995); and school improvement projects in Bombay (Black et al., 1992), the Pakistani province of Sindh (Heneveld & Hasan, 1989), and Andhra Pradesh, India (Menon & Sandeep, 1994). AKF is one of the founding partners in the establishment of the Institute for Educational Development (IED) and its in-service education programs at the Aga Khan University in Karachi. AKF has also been deeply involved in the development of community-based-and-supported early childhood education programs through Madrasa preschools associated with local mosques in Zanzibar, Mombasa, and Kampala (Brown, Brown & Sumra, 1999). In Chapter 2, Jeremy Greenland, Director of Education Programmes at the Aga Khan Foundation's headquarters in Geneva, tells the story of AKF's school improvement initiatives, with an emphasis on the East African projects highlighted in the cases. Greenland's story provides a historical context for the case studies that follow, and reveals AKF's strategic thinking about school improvement and its role in promoting school improvement, as it evolved from project to project within the changing social and political contexts of Tanzania, Kenya, and Uganda in the last quarter of the 20th century.

Chapters 3 to 8 present the case study evaluations of six primary and secondary school improvement projects initiated by AKF. All but two were managed in partnership with Aga Khan Education Services in East Africa. In Chapter 3, Anderson and Sumra describe and evaluate one of AKF's earliest school improvement ventures, which was based at Aga Khan Mzizima Secondary School in Dar-es-Salaam, Tanzania. The Mzizima SIP is one of two secondary school improvement projects among the cases. It is unique in that the project was focused mainly on school-wide improvement in a single private school within the Aga Khan Education Services system in Tanzania, over an extended (10-year) period of project funding. Efforts were made to disseminate the professional learning experiences and products of curriculum development activity to other secondary schools in Dar-es-Salaam and other regions of the country, though the main focus of development was within the school. In its various phases of development and implementation, a variety of school-based in-service strategies were utilized with varying degrees of effectiveness: curriculum workshops, peer observation and coaching, action research, team problem-solving, and mentoring of teachers new to the school. Other innovative features of the project included the creation of a deputy head (professional development) position and of an in-school teacher resource center to support school in-service activities and teachers' needs for professional resources (e.g., library, materials for production of teaching and learning aids). The evaluation confirmed the development and institutionalization of a truly collaborative school culture oriented toward continuous improvement in teaching and learning. It also revealed the limitations of a demonstration-school model for disseminating the pedagogy and organizational conditions associated with school improvement within Mzizima Secondary School and elsewhere.

In Chapter 4, Capper, Nderitu, and Ogula report on their summative evaluation at the end of the second 3-year phase of the AKF school improvement project in Kisumu, Kenya (1990-1996). This was the first of the SIPs to attempt to provide school-based in-service assistance to teachers across an entire primary school district using a cadre of teacher trainers employed by the project (project officers), as well as key teacher leaders in targeted schools. The project officers delivered workshops for teachers at the project headquarters and visited successive clusters of schools to provide school-based assistance at the classroom level with the help of the teacher leaders, called 'school coordinators'. A system was also introduced through the project whereby local communities contributed funds through their schools to aid in the acquisition of low-cost teaching materials. Efforts were made to work in partnership with the Municipal Education Officer, district inspectors, and local Teacher Advisory Center (TAC) tutors, particularly during the latter stages of the project, in the interests of sustainability of the project model for school improvement. While the project succeeded in delivering workshops and school-based in-service assistance to teachers in 42 schools as specified in the project design, the impacts on teaching practice and on student learning were disappointing. Teacher understanding and skill in the use of child-centered teaching methods was often superficial and varied within and between schools. The evaluators concluded that the in-service design, as implemented, was biased toward helping individual teachers, not whole schools as intended. They made several recommendations for the creation of professional communities of teachers who would jointly study their craft within their schools and through teacher networks among schools. The evaluators also identified weaknesses in the project model. They made recommendations such as ongoing access to change-facilitator training for district teacher-development roles (e.g., project officers, inspectors, TAC tutors), intensive instructional leadership training for head teachers, more systematic support for parental involvement, earlier attention to district policies and structural issues impeding sustainability, and improvements in the student examination system driving classroom instruction.

In Chapter 5, Siraj-Blatchford, Odada, and Omagor report on their evaluation of the first phase of the Kampala SIP (1994-1997) in Uganda. The Kampala SIP design was similar to the Kisumu project model. A team of local master teachers (project assistants) were recruited and trained by an expatriate project director to provide workshops and intensive school-based assistance to successive cohorts of government- and private-sector primary schools in urban Kampala. The team worked with sets of three to eight schools for one academic term (about 3 months) before moving on to a new cohort of schools. By the end of Phase 1, the in-service design had been implemented in 15 schools. Unlike the Kisumu project, the Kampala SIP relied primarily on the project assistants to support the in-service learning of classroom teachers. Initial plans to train teacher leaders in the project schools to cascade and sustain in-service learning about child-centered pedagogy were set aside in response to teacher opposition. One unique feature of the Kampala SIP was the establishment of a central Teacher Resource Center (TRC) as a site for workshops, consultation, a professional library, and teacher production of low-cost teaching and learning materials. Several project schools created in-school resource centers modeled on the project-level TRC. The evaluators found a high level of enthusiasm for the project and a growing ethos of improvement among educators in the project schools and elsewhere across the district. They reported dramatic changes in the material features of the classroom environment in project schools, including teacher- and student-made teaching and learning materials, wall charts, activity centers, and the organization of desks and students into learning groups. As in Kisumu, however, observations of teaching practice suggested that the quality of teacher understanding and use of child-centered activity-oriented teaching methods remained highly variable within and across the project schools, suggesting that teachers were still in the novice stage of mastery of that pedagogy. The Government of Uganda introduced a Universal Primary Education policy midway through the project. Class sizes typically doubled, which created an unexpectedly daunting context for teaching and learning and for the implementation of child-centered teaching methods. The evaluators were led to wonder about the practicality of the kinds of child-centered methods advocated by the project in situations characterized by enormous classes (e.g., 50-100 students), few resources, inadequately trained teachers, and low teacher remuneration. They concluded, nonetheless, that the results of the first phase of the project warranted continuation. They made recommendations for strengthening the project in-service design, including lengthening the period of in-school assistance to schools, more head teacher management training, efforts to mobilize increased parental contributions and support, linking central and school-based teacher resource center operations, and planning for sustainability of the project's support system for school improvement under the auspices of the school district.

The project experiences and evaluations from Kisumu and Kampala significantly influenced the Mombasa SIP design reported by Anderson and Nderitu in Chapter 6. The Mombasa SIP plan was more ambitious than its predecessors. The project was designed to reach and provide a combination of external workshops and intensive in-school assistance to all primary schools across the district over a 5-year period. (At the outset of the project in 1994, there were 112 public and private primary schools.) The project was planned and implemented from the start in partnership with the Municipal Education Office, in order to strengthen local capacity for sustainability beyond the project. The project officers in Mombasa were hired and paid by the school district, rather than with project funds. The Mombasa SIP was the first project to employ a local educationalist, rather than an expatriate, as the project director. Drawing from the Kisumu and Kampala experiences, the project design specifically included management training for head teachers concurrent with in-service assistance to teachers, a community development officer to help mobilize parental involvement (funding, materials, interest) in school and teacher development, and collaboration with district supervisory officers and Teacher Advisory Center consultants (TAC tutors). The project also provided material support and technical assistance to activate an existing network of 10 teacher centers as sites for the decentralized delivery of in-service support to school clusters across the district. The SIP community development officer played a critical role in mobilizing parental involvement in the governance and funding of schools and teacher centers. Sustaining school-based funding for school development initiatives and for TAC services emerged as a major challenge for continuation of the decentralized in-service support system. The evaluation occurred toward the end of the fourth year of the project (1998). By that time, 50 schools had benefited from workshops and in-school assistance from the 10 project officers (POs), with 40 more schools targeted for the final year. Each PO worked with three schools at a time for up to a year, in collaboration with the TAC tutor and inspectors from one of the district's 10 zones. The evaluation of teaching revealed that teachers who benefited from the workshops on child-centered teaching methods were less likely to put those methods into practice than teachers who received in-class coaching by POs in addition to workshops. As in Kisumu and Kampala, the evaluators concluded that even teachers coached in the classroom required longer term support to achieve a high level of expertise in the use of child-centered pedagogy. They recommended that more attention be paid to developing the internal capacity of each school for self-sustaining teacher development and school improvement, and to reducing the dependence on one-to-one assistance from external helpers.

The first phase of the Dar-es-Salaam Primary Schools Project (DPSP) ran from 1996 to 2000. As reported by Welford and Mosha in Chapter 7, the project was set up to serve an entire primary school district, though only 15 of 73 primary schools benefited from the full range of project support during Phase 1. The project team worked intensively with teachers, head teachers, and school committees from five schools at a time for a full year before moving on to another cohort of schools. The project was based in an existing district-level teacher center. Key features of the DPSP project design included the establishment of two teams of project consultants, one focusing on school management and the other on teacher development; an emphasis on teacher development in the core subjects (English, mathematics, science) in the upper primary grades; intensive in-service training in external workshop settings for cadres of up to four subject specialists from each project school (100 contact hours over the full year); project-facilitated school-based workshops delivered by those key teachers to their peers; the provision of school development incentive grants based on indicators of school performance and on school capacity to generate matching funds raised by local school committees; the development and use of replicable program modules for head teacher and school committee management training; and processes for monitoring of in-service and school management activities, and outcomes at the teacher-center and school levels. The evaluators reported positive effects on basic management practices of head teachers and school committees. They recommended more attention to curriculum management expectations for head teachers and the need to plan for ongoing management training due to turnover of head teachers and school committees. The incentive grants proved a powerful motivating force and support for school development, especially in terms of funding for pedagogical materials and for school facilities; however, it remained to be determined how the grant mechanism could be sustained without external project funds. The evaluators found evidence of material change in classrooms (e.g., rearrangement of desks to facilitate small-group learning, use of teaching and learning aids) and noted teacher efforts to engage students in more interactive questioning. Overall, they concluded that teacher understanding of the approach to pedagogy supported by the project remained quite variable, and that teacher skill in the use of those methods remained at the beginning level. The practicality of these methods, as opposed to more teacher-centered, didactic teaching methods, was severely challenged by class sizes that often exceeded 100 pupils. The evaluators argued that the potential for team teaching in large classes was not being exploited and recommended that more attention be given to teachers' assessment literacy (understanding and use of student evaluation for programing and for school improvement), in light of the pervasive testing of students in the exam-driven education system. They also identified a need for change-facilitator training for people acting in change agent roles, such as the project subject consultants and lead teachers at the school level. They addressed the technical challenges of evaluating project effects on student learning.

Brumfit and Hikmany present their evaluation of the Secondary Language Orientation Project (SELOP) on the Islands of Zanzibar (Tanzania) in Chapter 8. This project involved the development and implementation of a program to support the English-language development of students in their first year of secondary school (Form I). While it was not a whole-school improvement initiative, the teacher-development model employed was similar to those used in the East African SIPs. Throughout Tanzania (including Zanzibar, which has its own Ministry of Education), the official language of instruction shifts from Kiswahili to English when students enter secondary school. English is only taught as a primary grade subject in government schools. The change to English across the curriculum presents an extremely challenging teaching and learning situation for both students and teachers at the secondary level. The SELOP project had two major components. The first consisted of the production of curriculum-based materials to support the development of students' English-language proficiency and comprehension across the curriculum during the first year of secondary school. Using these materials, teachers in all core subject areas (English, social science, mathematics, general science) act as teachers of English through the medium of their subject specialties. The second major component was the establishment of an in-service training strategy to support implementation of the program by Form I teachers in all 92 of Zanzibar's secondary schools. Brumfit and Hikmany conducted their evaluation of the curriculum development process, program materials, in-service strategy, and teacher and student outcomes near the end of the project's initial phase (1994-1997). By that time, the program materials had gone through several cycles of locally managed development, field testing, and revision. The program had been trialed in 64 schools, with the remaining 28 scheduled for implementation in the final months of the project. In addition to their English content, the program materials were designed to introduce more interactive teaching and learning processes. All lessons included paired activities for students that were supposed to exemplify 'active learning' in English. The in-service training structure had several components, including four half-day workshops per year for all teachers, delivered through a decentralized network of existing teacher centers, and individual classroom visits by a specialized team of part-time district program consultants, head teachers, and other users of the program within and between schools (i.e., peer coaching). Overall, the evaluators were positive in their judgments of the program materials (clarity, simplicity, curriculum fit), though they questioned the quality of 'active learning' embodied in some of the pupil activities built into the lessons. They reported a high degree of teacher ownership and commitment to the program, instilled in part by a highly participatory approach to developing, piloting, and refining the program materials. The in-service training strategy appeared to be working efficiently, and had resulted in a common facilitative teaching style in the context of program implementation. Teachers' lack of confidence in their own English-language proficiency was identified as a major constraint on the effectiveness of implementation. The evaluators proposed the establishment of English-language improvement programs for secondary school teachers in all subject areas, at both the in-service and pre-service levels. There was little evidence that the SELOP teachers transferred teaching methods embedded in the program to other classes, or that non-program teachers visiting SELOP classes adopted those methods without any supporting curriculum materials. The evaluators raised concerns about local capacity for sustainability, such as transportation for district consultants to visit schools, incentive payments for teacher-workshop participation, and the need to develop internal strategies for evaluating curriculum materials, training, teaching, and student achievement.

The authors of Chapters 9 and 10 have many years' experience studying and working with teacher development and school improvement internationally, though neither has been directly involved in the Aga Khan Foundation projects in East Africa or elsewhere in the developing world. They were invited to provide independent cross-case commentaries on the SIP case studies and to connect the experiences and findings of the Aga Khan Foundation school improvement efforts to the literature on school improvement more generally. To assist with this purpose, they were provided with copies of the original evaluation reports in addition to the chapter-length versions contained in this book. Farrell's task was to draw comparisons between the Aga Khan Foundation projects and the literature and practice of school improvement in developing countries. Hopkins was asked to comment on the Aga Khan SIP projects and research from the perspective of the literature on school improvement and teacher development that has evolved largely through the North American and European traditions of policy, practice, and research.

Farrell, in Chapter 9, puts forth a challenging critique of the East African SIP experiences described in this book. The essence of this critique is that these SIP initiatives, like most planned school improvement efforts worldwide, are set up to preserve a taken-for-granted model of formal education that is basically antilearning, both for the pupils and for the teachers who work in those schools. Some core features of this standardized, antilearning model of schooling include placing children in buildings called schools for 3 to 6 hours a day, in age-graded groups of 20 to 60 pupils, to work with a single adult for discrete periods of 40 to 60 minutes devoted to separate subjects, in order to learn and be tested on standardized curriculum content prescribed by external authorities. Farrell notes that educational change efforts that do not radically question or alter these taken-for-granted assumptions about schools routinely yield disappointing results in terms of teacher and student learning. This is the 'bad news,' according to Farrell. The 'good news' is that some alternative models of education (mainly at the primary level) have emerged and spread over the past 30 years in various regions of the developing world. Some common features of these alternative models are that (a) teaching and learning is more child-centered and active than teacher-driven and passive, (b) classrooms are multi-graded and pupil progress is continuous rather than age graded, (c) teachers are assisted by other adults (often from the community) in the classroom, (d) older students tutor younger students, (e) learning materials are self-guided and allow pupils to work at their own pace, and (f) locally directed ongoing training and peer mentoring occurs for teachers. Studies of these alternative models of education are not well generally well known but show promising, sustained, positive results for student learning and teaching quality in regions of the world where conventional forms of schooling have failed dismally, despite best efforts at school improvement. Farrell compares the Aga Khan Foundation SIP initiatives to characteristics of the standard and alternative approaches to education and schools and concludes that, while these SIP initiatives have some positive features (e.g., the emphasis on child-centered active learning), they are basically designed to support and sustain traditional forms of education and schools that hold scant hope for large-scale improvement in the quality of education for the people they are intended to serve. His thesis defies the potential for educational reform taken for granted within the standard model of schools in these contexts.

Hopkins compares the principles and practices exemplified in the Aga Khan Foundation projects to international trends in school reform and to the knowledge base on school improvement, particularly with reference to whole-school approaches to change. He suggests that the emphasis on school capacity building through instructionally focused in-service training of teachers, local development of teaching and learning materials, and management training for head teachers is consistent with findings from past research on the implementation of new programs and practices in the classroom. At the same time, however, this approach to school improvement does not take advantage of, nor fulfill, the results-oriented expectations of current performance- or standards-based reform initiatives around the world. A better balance between capacity building and accountability measures is needed at the local level. With the knowledge base about school improvement as a foreground, Hopkins discusses the successes and challenges of the Aga Khan Foundation's East African SIPs in relation to six 'issues,' as follows: (a) the commitment to child-centered learning, (b) curriculum versus pedagogic development, (c) the focus on teacher and leadership development, (d) the school as the unit of change and capacity building, (e) the local support infrastructure, and (f) sustainability. He notes, for example, that findings from the evaluations portray limited teacher understanding of, and use of, child-centered teaching methods, and by extension, little evidence of positive impact on student learning. Hopkins argues that teacher learning would be enhanced, and the potential for improvement in student learning increased, if the in-service teacher development practices used in these projects were more embedded in programs aligned with the local curriculum context and performance standards. He further suggests that steps already taken to create sustainable support systems for school and teacher development could benefit from increased attention to peer support and networking among teachers and across schools. Drawing upon the literature and findings from the East African SIP studies, Hopkins offers a policy framework for educational change in his conclusion.

The Double Mirrors of School Improvement

In the introductory sections of this chapter, I described several design principles and strategies for school improvement that have been enacted in various ways across the Aga Khan SIP initiatives. These principles and strategies offer an experienced-based strategic framework for thinking about and planning how to approach school improvement. They make good sense and are consistent with the research on school improvement and in-service teacher development. The principles are pitched at a level of abstraction, however, that likely falls short of the more concrete specification of organizational structures and processes that school improvement agents would find useful in creating or enhancing sustainable infrastructures for ongoing school and classroom development. Drawing from the experiences of the Aga Khan SIPs in East Africa, the graphic organizer depicted in Figure 1 represents an array of specific organizational conditions that need to be addressed in order to achieve school-wide changes in classroom teaching aimed at improving the general quality of teaching and learning: teacher learning (external and in-school), teaching and learning resources, school and district management, parental and community involvement, funding for school and teacher development, assessment and evaluation, change agent development, teacher working conditions, and coordination.

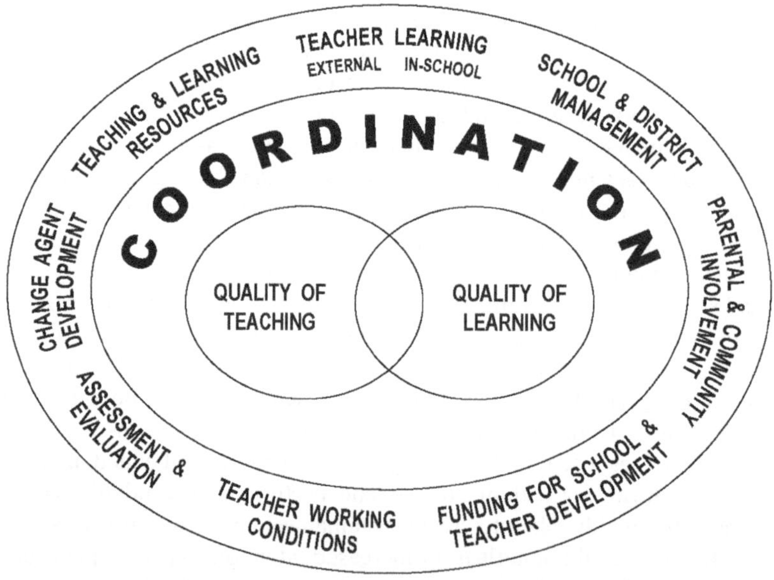

Figure 1. Organizational conditions for teacher development for school improvement

The overlapping circles in the center of the model portray the relationship between the quality of teaching and the quality of student learning. As depicted, the quality of teaching is important, but is not the only determinant of the quality of student learning. (The model does not attempt to specify all the factors that impinge upon student learning.) This is significant to keep in perspective, given that the school improvement initiatives on which the model is based have focused fundamentally on teacher development and on the conditions that facilitate or hinder teachers' professional learning and change on the job. Each of the elements in the outer circle highlights a focus for school improvement-related planning and intervention. The earliest AKF school improvement initiatives focused mainly on teacher learning and on classroom resources. The range of organizational conditions addressed, however, evolved through the experiences and evaluations of the projects to include those specified in the model depicted in Figure 1. 'Teacher Learning' directs attention to teacher-learning opportunities both inside and outside the school, and to the linkages between those two contexts for knowledge acquisition and growth.

In developing this graphic organizer, I struggled to create a representation that would depict the relationships among different elements of the model. In the end, this proved hopelessly complicated, overly deterministic, and questionably valid on the basis of the existing project and evaluation data. The resolution I came to borrows from a model of classroom interaction and change developed by Rachel Hertz-Lazarowitz (1992). Hertz-Lazarowitz describes a circle of six mirrors in the classroom that affect the social and academic behavior of children: classroom organization, learning tasks, teachers' instruction, teachers' communication, pupils' academic behavior, and pupils' social behavior. The mirrors reflect each other, like double mirrors in a 1950s barber shop. In theory, change in one mirror should project into and result in changes in the other mirrors. The problem with change, as explained by Hertz-Lazarowitz, is that teachers often get caught in one mirror and do not modify the others accordingly. In the graphic organizer for school improvement presented in Figure 1, the elements specified in the outer circle can be thought of as mutually reflecting mirrors akin to Hertz-Lazarowitz's model of classroom dynamics and change. In principle, the more cohesive the interface between those organizational elements that impinge on the quality of teaching, the greater the likelihood of significant change and improvement. 'Coordination' floats in the field between the double mirrors of school improvement. It, like the other elements, requires specific attention as part of an overall school improvement initiative. Without it, images in the mirrors are less likely to be reflected in complementary actions in the other mirrors. In the school improvement initiatives reported in the Aga Khan SIP evaluation studies, the project management structures have typically played an important, albeit temporary, coordinating function. A major challenge for these and future school improvement initiatives is to enhance the capacity of local schools and districts to effectively coordinate changes in one part of the teacher development system with concurrent actions in other components of that system.

2

Evolution in School Improvement with the Aga Khan Foundation, from 1984 to 2000

Jeremy Greenland

In 2000, the Aga Khan Foundation (AKF) was funding 20 education programs, in South and Central Asia and in East Africa, with a combined annual budget of roughly US $11 million. The grants fell under one of two programs: school improvement and early childhood development, and were made to governments, NGOs, and private agencies such as the Aga Khan Education Service. A major grantee is the Aga Khan University Institute for Educational Development in Karachi (AKU-IED), which trains teacher trainers to master's degree level and conducts research via an expanding cluster of Professional Development Centers (PDCs) in countries where the school improvement (SIP) and early childhood development (ECD) programs are located (AKU-IED, 2000). To monitor these programs and provide professional support, AKF currently has education program officers in India, Pakistan, Tajikistan, Kenya and Tanzania, as well as a director of education programs and an early childhood specialist at the Foundation's head office in Geneva.

This chapter charts the evolution of AKF's school improvement programs from the early 1980s to 2000 and places the East African programs in the wider context of similar activities that AKF has sponsored over the same period in South and Central Asia. Early childhood development programs constituted roughly half the Foundation's education portfolio, in terms of number of programs and budget, throughout this period. Many of the topics and trends in school improvement discussed in this chapter apply equally to the ECD programs but are reviewed elsewhere (Bartlett, 2000). The chapter describes how AKF's interpretation of the term 'school improvement' has evolved, how the size and geographical spread of the programs have grown, and how AKF is learning from the programs to influence national and international education policies in this field. The chapter concludes with a summary of 'work in progress'.

Initial Approaches to School Improvement

School-Based Teacher Training and Curriculum Development

AKF's first school improvement program started at the Aga Khan Mzizima Secondary School in Dar-es-Salaam, Tanzania in 1984. It focused on training teachers in their own school and involving them in developing locally relevant curriculum materials. These emphases were derived in part from an analysis of the school's predicament. Despite its being one of the best schools in the country in terms of exam results, many of the teachers were un- or under-qualified and did not have the means or the time (because of second jobs, notably private tutoring) to access conventional in-service courses, few of which were, in any case, available. Production of syllabuses and textbooks at the National Curriculum Development Center was slow and expensive. The emphasis on school-based training and curriculum development derived from AKF staff's conviction and experience that the AKF model was superior to the conventional 'cascade' type of training, in terms of effecting long-term change in teachers' skills and raising their sense of professionalism (Joyce and Showers, 1988).

Also in 1984, the Field-Based Teachers Development Program (FBTDP) was launched by the Aga Khan Education Service in Northern Areas of Pakistan, at two AKF schools and in two government schools, to address the problems of remoteness from training colleges and cultural obstacles to allowing female teachers to leave their villages. At each site, 10 unqualified teachers were given charge, in pairs, of the 5 primary grades. Two master trainers, one of whom doubled as head teacher, were posted to each school. During the 9-month school year, the master trainers supervised the trainees and gave model lessons, based on manuals produced by the core team, which were in Urdu and which followed closely the primary school syllabus. At the end of the year, the trainees took the government Primary Teachers' Certificate exam as private students. This program encompassed at one time eight sites, and a similar course was added for middle school teachers (Classes 6 to 8). The course's content has been modified in response to three external evaluations (e.g., Bude, 1992). There is continuing demand; the program is currently also training 720 teachers from government schools over a 5-year period, as part of the Northern Education Program of the Government of Pakistan.

The Contributions of Teacher Resource Centers

At roughly the same time as it launched the FBTDP, AKF made its first school improvement grant in India, to the Parisar Asha Environmental Studies Center in Bombay, enabling it to expand its library of audiovisual and print materials, which gave local color to the formal state curriculum, and to make these materials available to teachers in some of the Municipal Corporation's poorest schools. Parisar Asha was thus the first of AKF's grantees to address the dilemma (still a feature of the portfolio today) of developing human and material resources at a central location and then taking them into the realities of teachers' classrooms. For example, distributing pencils and worksheets on the topic of trees in Bombay to a class of sixty six-year-olds sitting four to a sloping desk, is not easy.

Parisar Asha, an NGO, existed prior to AKF's grant. In Karachi, Pakistan, a group of teachers, nearly all women and working in private-sector schools, requested a grant to establish from scratch a Teacher Resource Center (TRC) that, along the lines of centers then flourishing in the United Kingdom and elsewhere, would make a wider range of books and other resources available than any individual school could afford and, at the same time, empower ordinary teachers to define their own professional development needs and the content, timing and style of the appropriate training (Ali et al., 1993). Defining the Resource Center's governance structure, in particular the respective roles of teachers, schools and other interested parties, setting levels of fees, and raising funds toward an eventual endowment proved more complex and time consuming than planning the curriculum and training services the Center would provide.

AKF soon came to see that the Resource Center's emphasis on individual teachers (despite the fact that schools as well as teachers could become members) had the effect of ignoring 'whole-school' development, and indeed often had the negative effect of discouraging collaboration among teachers and among schools. Teachers who were not selected for training were envious of those who were. Although the Center's certificates of attendance carried no formal status, they quickly gained currency and were used by teachers to bargain for salary increases and to switch to competing schools if they did not get salary increases. Most employers of teachers in Asia and Africa, in both the government and private-school sectors, are extremely authoritarian and secretive in their recruitment, deployment, appraisal, and promotion procedures. In this context of little transparency and often-fierce competition, AKF's modest contribution has been to persuade its school improvement grantee partners to publish (and actually use) teachers' salary scales and set criteria for appraisal and promotion.

An important element of the original TRC program in Dar-es-Salaam was the intention to use teachers and resources from an already good, improving school to reach out to neighboring government and private schools. This notion was formalized in the planning of the program at Kisumu in Kenya that started in 1990. Memoranda were signed with the Municipal Education Office, detailing the arrangements for including a progressively wider circle of schools and rotating the best teachers after training within and between schools to mentor their less experienced colleagues. The 'ripples in the pond' idea had had little effect in Dar-es-Salaam and, despite the more careful planning, only had modest success in Kisumu. The reasons in both locations were the same: little attention in the program design to securing ownership and commitment from the head teachers of either the 'home' or the 'outreach' schools; logistical difficulties in getting mentors to these schools, even in urban and peri-urban settings, even when transport costs were included in the program budget; no formal accreditation of teachers after training; low levels of mentors' mastery of subject-content knowledge and methodology, even after intensive training at the original program school; and difficulties in negotiating the transfer of mentors and their resources from program schools to the government's Teacher Resource Center, to ensure some measure of sustainability, as the grant period moved towards its end.

The SCRE Report

A comparative analysis of the programs in Kisumu and Bombay was published by Harry Black of the Scottish Council for Research in Education (SCRE) in 1993. The individual case studies had been written by Florence Kiragu (Kisumu) and R. Govinda (Bombay), following a conference in London the previous year to discuss the findings with representatives of the international donor and academic community (Black et al., 1993).

The SCRE report, and AKF staff's involvement in its production, had the valuable function of endorsing certain features of the existing school improvement programs and of recommending areas for further work and new directions. The policy of focusing on individual schools was found to be "appropriate ... though insufficient attention may be being paid to the role of head teachers ... teachers and senior school staff should have [more] opportunity to develop their professionalism by networking with colleagues in other schools and indeed in other school systems." The report concluded that the programs' adherence to a 'clinical' method of teacher training, that is, addressing teachers' needs in the environment of their own classroom rather than via a centralized, 'cascade' type of training, had resulted in "substantial changes in classroom practice". However, the report encouraged the programs "to explore the relative advantages of different modes of clinical support." In a similar vein, an evaluation of the Field-Based Teachers Development program in Northern Pakistan (Bude, 1992) noted that the trainees' acquisition of new skills tapered off after initially being rapid and that consolidation and extension of their skills during the last part of the nine-month training cycle was minimal. Also, because of the high level of teacher turnover, there is a constant need for new teachers in a given school to be inducted. The notion, therefore, of a school improvement team 'bringing a school up to speed' in a prespecified period of time and then moving on to repeat the process in a new school breaks down in practice: The ripples in the pond may spread out, but those near the center require ongoing attention.

The 'Whole-School' Approach

AKF's commitment to the whole-school approach to improving quality derives in large part from staff's contacts with the Schenley School experiment in Pittsburgh (Wallace et al., 1990) and the Michigan School Partnership in the US, and with the Oxford Internship Scheme in the UK (Benton, 1990). These contacts were made in the early 1990s when the task force planning the Aga Khan University Institute for Educational Development (AKU-IED) was investigating potential academic and professional links with schools and universities in the West. The key notions of the whole-school approach are that schools will not improve unless every teacher is exposed to in-service training (as opposed to the 'cascade' method, where a 'representative' teacher is sent away to 'bring back the message') and that every school is encouraged to define its distinctive ethos, culture and objectives. AKF has found that the whole-school approach is also a more logical accompaniment to its emphasis on community involvement than more centralized forms of teacher training and curriculum development.

The whole-school approach has been developed and strengthened in the subsequent school improvement programs in Kampala, Mombasa and Dar-es-Salaam in the 1990s and has become a guiding principle at AKU-IED from its establishment in 1993 onwards. The AKU-IED's Professional Development Center (PDC) in Karachi works intensively with a cluster of schools (currently 35) in its vicinity. AKU-IED and the programs in East Africa are now, admittedly somewhat belatedly, instigating research to show whether the attention to the needs of head teachers and managers as well as teachers in individual schools has an impact, superior in terms of students' learning, access and completion rates, to conventional cascade training. A major snag with the whole-school approach is that it gives its administrators the political headache of explaining why one school was included and not another. The scattered, mountainous terrain of Northern Pakistan has led the PDC in Gilgit, established in 1999, to initially cluster its 'whole schools' within a short jeep ride of Gilgit, to inevitable and loud protests from schools located further away. The Kisumu program administrators tried to defuse this pressure by organizing periodic workshops for all teachers in a given zone of the town, but this diverted resources from the main task and, if the research eventually confirms the program staff's hunch, had little effect on classroom practice. The response of the programs in Kampala and Mombasa to this equity issue has been to agree that all schools in the respective municipal area will be covered before the end of the grant period. This begs the question of what 'cover' means (see section Work in Progress, below), and the SIPs in East Africa are constantly seeking a compromise between raising the 'graduation' criteria for schools in the program and reducing the time and resources that have to be deployed at each school.

Training Head Teachers and School Managers

The SCRE report credits the programs, Bombay somewhat more than Kisumu, for clarifying respective roles and responsibilities of staff, head teachers, school committees and local officials, but it calls for much more attention to the area of management training. This recommendation prompted AKF to make two new grants in India in the mid-1990s. The Narsee Monjee Institute for Management Sciences, a private university in Bombay focusing on business studies, was keen to adapt some of its structures and curriculums to the training of school principals and managers in clusters of schools in Bombay and Ahmedabad. At the same time, a group of senior female teachers in New Delhi set up the Center for Education Management and succeeded in obtaining local university accreditation for their head-teacher training program. Formal courses in educational management were not part of AKU-IED's original 'menu', but the need for such training was recognized very early, and the Certificate in Educational Management course, run on a modular basis in Karachi, has consistently been oversubscribed. It is currently being introduced by the PDC in Gilgit and by the growing cluster of schools in East Africa who want to consolidate their school improvement programs into one or more local PDCs. These efforts have resulted in publications that tend to be project specific, though four manuals for head teachers produced by the Dar-es-Salaam program can be potentially disseminated and used throughout Tanzania. Another feature of the increased attention to school leadership has been the use of program funds to promote women to senior jobs. Progress to date has been mixed. AKU-IED has built an impressive record in this regard. In Northern Pakistan, by contrast, cultural factors are a strong impediment to change. In urban Africa, the wives of 'big men' get posted to classroom teaching jobs but most head teachers are men, and AKF has little leverage over the formidable bureaucracies that determine postings and promotions. Work on determining the range of skills involved at the various program sites is still very much 'in progress', notably whether it is appropriate to try and turn head teachers into professional leaders as well as better managers and administrators.

Child-Centered Methods

The SCRE report commends the programs for trying to effect a shift from teacher-centered to child-centered methods (defined by the SIPs as attending to the needs of small groups and individual children, in an enquiry- and activity-oriented manner). But, as subsequent evaluations of the Kisumu program (Chapter 4) and of the Mombasa program (Chapter 6) discuss at greater length, this shift implies a much bigger change in teachers' understanding than program resources can usually afford. It also begs the question of whether the shift is desirable in all teaching situations. Enabling a class of 60 to identify countries of Africa on a map may best be done by the teacher in conventional didactic mode, from the front of the room, with the aid of one large map (or, more probably, by eliciting chorus responses to a map which she or he has tried to draw on the scratchy, shiny blackboard).

Locating Resources Centrally or Within Schools

Most of the countries in which AKF works have, in the last two decades, gone through spells of enthusiasm for Teacher Resource Centers (TRCs) and Teacher

Advisory Centers (TACs) of varying shapes and sizes, followed by disenchantment (Knamiller et al., 2000). AKF's response to the dilemma of how best to allocate resources (human and material) at successive points between the center and the periphery of a school system has been to encourage experimentation at both ends of the spectrum. On one hand, the Foundation has supported AKU in establishing the AKU-IED in Karachi as a national and international resource in teacher education of university-level quality. To date, 120 teachers have completed an intensive, two-year master's degree in teacher education and now divide their time between teaching at the AKU-IED's PDCs and working on the professional development of their fellow teachers in their 'home' schools.

At the other end of the spectrum, the current programs in East Africa, particularly in Kampala, are experimenting with the allocation of materials (limited often to recycled 'waste' materials) between zonal centers and 'learning centers' in each teacher's classroom. Interviews with school management committees in Dar-es-Salaam and Mombasa show a rapid tailing off of parents' willingness to pay the many fees and levies demanded of them, once that money leaves the strongbox in their own school and is 'lost' at the nearest teachers center, let alone at district or national headquarters.

Parent and Community Involvement

Devolving Authority from Central to School Level

Parents' and communities' cofinancing of public education systems has reached dramatic levels in many parts of the Third World (as much as 65% of the per-student cost, according to a recent estimate in Kenya). Concurrently, many governments have introduced 'decentralization' policies, by which they devolve responsibility for decisions from a central ministry to district, subdistrict and school levels. The difficulties in both cofinancing and decentralization have been three-fold: first, defining what aspects of school life should come under local control; second, providing local officials and volunteer managers with the knowledge and skills to exercise these new responsibilities; and third, overcoming the authoritarian forces in many local government offices that militate against the intention to share power with individual schools and communities.

Clearly, community involvement differs in important respects from decentralization of administrative and professional responsibility, and the two processes may combine very differently in government-, NGO- and private-school systems. One might ask to what extent the huge network of schools run by the Bangladesh Rural Advancement Committee (BRAC) is a centralized or a decentralized system and question the assumption that NGO-run systems are necessarily better at promoting decentralization than government- or private-sector systems are. If school management committees are required by law, should they be viewed as citizen or governmental organizations?

Since the early 1990s, AKF has become convinced that involving communities is essential to attaining the objectives of Education For All and Gender Equity and has made community involvement an increasingly important component of its school improvement programs, both in East Africa and South Asia. This growing preoccupation, however, has a number of antecedents. First, the network of Aga Khan schools in East Africa and South Asia, one of the major constituencies with which AKF works, has resulted from a combination of the philanthropy of the present Aga Khan and his grandfather and the spirit of self-help and volunteerism that the Ismaili community has consistently practiced. The 175 Aga Khan schools in Northern Pakistan are still often called 'DJ' schools because many of them were started with funds collected to mark Sultan Mohamed Shah Aga Khan's Diamond Jubilee in 1947. The self-help tradition was drawn upon in designing the Self-Help School Construction Program (SHSCP) in Northern Pakistan, which has been operating since 1984 and has to date resulted in nearly 500 new classrooms and 100 toilet blocks (outhouses). The basic formula is that the community donates land, unskilled labor and locally available materials (gravel, sand, water), while program funds supply skilled labor and 'imported' materials (cement, metal rods). Extensive dialog is needed between program staff and community leaders to clarify these and other terms of the funding contracts (for example, that a given percentage of school enrollment must be female).

AKF's first foray into involving parents in cofinancing improvements in training and curriculum came in Kisumu, with an attempt to levy the equivalent of US $0.50 per child in each participating school to ensure that teacher centers could be sustained independently of AKF's grant. However, the scheme had not been part of AKF's original agreement with the municipal authorities, and the results were very patchy. Since that project, matching grants have been introduced, with much more success, at the Integrated Pre-Primary and Primary Program in the southern Pakistan province of Sindh: The fees paid by parents are banked, teachers' salaries are paid with an equivalent portion of AKF's grant, and the school management committees have access to their bank investment plus accrued interest when their schools achieve an agreed set of quality criteria. A similar scheme is getting under way in northern Pakistan, where communities can apply for matching funds on a 4:1 basis to open coaching centers for girls, as a prelude to establishing full schools, and to add secondary classes to existing primary schools. In the Gorno-Badakhshan province of eastern Tajikistan, AKF is promoting a number of schemes at school and district levels, such as a revolving textbook fund, to offset the dramatic decline in central government funding since the collapse of the Soviet Union. The Madrasa program in Mombasa, Zanzibar and Kampala allows individual, community-run preschools that meet agreed quality standards to build an endowment fund, with contributions from AKF and the community. The proceeds from the fund can be reinvested or used to supplement teachers' salaries. The Dar-es-Salaam Primary Schools Project offers incentive grants to participating schools, grants that are based on performance measures associated with financial management, institutional development (e.g., school management, school-community links), and curriculum improvement (e.g., resources, teacher development), with a matching award for local fundraising.

In attempting to publicize these experiments and take them to scale, AKF is finding that many government policy-makers and donor agencies are ambivalent about whether greater community involvement and devolved authority are intrinsically desirable or just 'stop-gap' measures dictated by the decline in levels of public funding of social services. Governments distrust NGO- and community-led education systems as potential seedbeds of political dissent, while donor agencies worry that promoting community involvement favors the already well-to-do at the expense of the poor. They point out that nowhere in the world is there an example of a substantially sized, poor, rural or urban population financing 'quality education' entirely on its own.

Defining the Limits of Community Involvement

Even if only envisaged as a short-term measure, community involvement requires detailed definition and decisions about whether it includes responsibility for resources (a school's income and expenditure on teachers' salaries, buildings, equipment, consumables); for personnel (teachers and their selection, deployment, promotion and dismissal, head teachers, teacher center staff); and for curriculum and the teaching and learning process (Bray, 2000). The weakest version of community involvement in curriculum limits it to the contribution of information on local history and culture, traditional crafts and medicines; however this risks patronizing community members and associating the notion of community involvement with a view of the past that is static and no longer relevant to today's needs. Deciding whether to give parents and communities a voice in determining whether their children should be taught about HIV/AIDS, sex, and women's and children's rights is, at the same time, much more controversial and urgent.

Teachers as Community Development Agents

Countries without a cadre of community development officers (CDOs) within their Ministries of Education are reluctant to add such officers, especially when international donor agencies, in so-called structural adjustment discussions, are demanding a reduction in the number of teachers and other government employees. But the evidence from the Mombasa program is that some form of 'social marketing' is essential to increasing enrollment and achievement rates. Its cost must therefore be included in budgets. The cheaper alternative, where all teachers and head teachers 'double up' as CDOs and spend a proportion of their time on community-related tasks, is a matter for ongoing experiment and dialogue. Yet another option is to adapt and retrain an existing cadre of personnel, for example, Tanzania's ward education coordinators. Whatever the formula, the SIPs are finding that communities and parents need to be visited several times for a particular message to be understood and accepted.

Community-Specific Approaches to School Improvement

Another lesson common to the construction and school improvement programs has been the need to allow for significant differences within a superficially homogenous environment. One size does not fit all. The same mountainous terrain in Northern Pakistan is home to communities from different traditions of Islam, and, although AKF's grant has been open to requests from all communities, the Ismailis' greater and older commitment to girls' education means that they applied for and were awarded most of the grants in the early years of this initiative. In the last few years there has been a dramatic surge in the level of interest from non-Ismaili communities, resulting in a major shift in the distribution of grants, but these communities still find the self-help conditions much more onerous. The dilemma, with which AKF is gradually coming to terms in its community-focused approach to improving schools in East Africa, is to decide how far to relax the terms of grant agreements with 'weak' communities, just to get them started. A recent study of poverty indicators such as water supply and housing conditions in the Kinondoni district of Dar-es-Salaam, where the school improvement program is located, shows extremes just as great within this one district, as those that exist between this allegedly favored district and the underprivileged rural parts of the country.

Capacity Building

Agencies such as AKF that advocate a strong emphasis on training (as opposed to the dissemination of knowledge) as the key element in capacity building face several difficulties. Not least is the first difficulty of finding a politically correct term for what they want to do: 'Mentoring' seems somehow more acceptable than 'coaching' (despite the fact that world-class football players and musicians have coaches). Officials above a certain level are invited to meetings where 'information is exchanged': It seems that they can no longer be trainees and be taught. Ironically, the Self-Help School Construction Program (SHSCP) suffers less from this syndrome because master-builders are used to the concept of apprenticeships and do not mind being shown how to make reinforced steel joists and lintels. The second difficulty is that such training and coaching needs to be embedded in the local social and political reality, and personalized. The third difficulty, a lesson from SHSCP for school improvement in East Africa, is that responsibility for community mobilization needs to be in the hands of staff who enjoy the respect of the local communities, a factor with formidable political implications.

The largest space in the 'black box' of uncertainty about capacity-building programs is the time (and hence cost) that is required for them to be effective. At the start of SHSCP in Gilgit in 1984, the first day was spent getting the prospective managers to figure out how to compute percentage increases on the pocket calculators that had been proudly distributed. Capacity-building programs are rarely able to make similar unscheduled adjustments. Few organizers of such programs take Joyce's and Shower's (1988) description of the coaching or mentoring cycle seriously; instead, they rely on variations of the so-called cascade method (whose main feature usually seems to be that it soon becomes too shallow and diffuse to be recognized as a cascade). Then there is the choice between (potentially cheap) computer-based 'distance' methods and (certainly expensive) face-to-face tutoring, with some experts claiming that for every $10 invested in the former, six more dollars need to be spent on the latter. Finally, there is the feasibility of even motivated individuals following a sustained course of training. Most of the trainees have large families to care for, have second (if not third) jobs economic survival, and live in houses that have no private space for study and an electricity supply that, if it exists at all, frequently breaks down in the evening. Such are the realities of building the capacity of teachers and community leaders.

Teachers and Community Leaders as Development Managers
AKF's NGO Enhancement Program is developing a course and a curriculum, called Development Management Qualifications (DMQ), that will attempt to steer a middle path in capacity building between the notion of a 'generic' curriculum for all types of 'development managers' and the sector-specific needs of managers in education, health, rural development, and other sectors. It will also steer a middle path between providing 'customized' or 'personalized' capacity-building programs to individuals and programs catering to conventionally sized groups of students. A further compromise is between the needs of 'development managers' in the public and civil-society sectors (where civil society is itself an umbrella term used to include community-, NGO- and privately led initiatives), and whether the needs of public- and civil-society managers are sufficiently similar to make a single course and curriculum for both sectors.

Any capacity-building program is faced with compromise in reconciling employee and employer interests. A key feature of AKF's DMQ proposal is the assumption that the program must be driven by effective demand from its prospective clients and their willingness to contribute at least part of the cost. However, in reality, employers will be asked to pay much of the bill, directly or indirectly (by agreeing to an aid budget) and by granting employees some sort of study leave. This was the case with the 'distance' master's degree that the Kisumu and Mombasa programs conducted for roughly 40 senior school staff, with the Universities of Exeter and Leeds in the United Kingdom. Most employers are lukewarm toward any kind of professional development or capacity building because they fear that their employees will acquire new skills and then leave for a better paid job elsewhere. When the IED master's degree graduates return from Karachi to their classroom teaching jobs, IED is finding it difficult to insist that their employers respect agreements to deploy them in the school-based mentoring role for which they have been trained. The tendency, especially with respect to IED graduates formerly employed in government schools, is to dub them 'inspectors', if not 'administrators', and shut them away in offices behind large desks.

AKF is finding that the curriculum of these capacity-building programs for teachers and community leaders needs to go beyond 'school' issues if it is to attract clients' commitment. The program in rural Sindh is showing strong demand for an adult literacy component that should include information on and discussion of health issues, especially women's reproductive health, and sensitive sociopolitical issues such as land-holding rights and migration to Karachi and the Persian Gulf for employment.

The consensus among donor agencies on how to build capacity 'to get things done' (for example, implementing the recommendations of the Conference on Education For All held in Dakar in March 2000) falters on defining the relative importance of the knowledge base and the training to turn that knowledge into skills. Those who favor increasing the knowledge base are placing more confidence in information and computer technology (ICT) and the proliferation of web-sites sponsored by donor agencies to deliver knowledge. Community leaders in Northern Pakistan, in theory, have access to a web-site that describes alternative ways of organizing early childhood development programs to replace the current unsatisfactory practice of having four-year-olds chanting numbers and letters of the alphabet on the verandah outside the Class 1 classroom. But can community leaders access such information in practice, and do they? Is there a comparable web-site that illustrates alternative classroom-building techniques? AKF has not yet found an education management information system that is user friendly enough to enable local-government and NGO officials with modest levels of formal education to calculate and understand student dropout rates and male-female enrollment ratios.

One challenge for capacity-building programs is that implementation of any element of the school improvement formula involves actions that lie beyond the competence of primary stakeholders (which often include government, at least at the local level) to resolve. Understanding that sanitation facilities are essential to retaining adolescent girls in schools is one thing: Building them is another. Bringing water mains to the site and satisfying safety and sanitation standards requires the collaboration of three separate government departments that are unused to talking to each other and are certainly unused to being put under pressure by an NGO to collaborate. Frustrated donor agencies are increasingly resorting to the term 'lack of political will' to describe these difficulties, a phrase that implies more personal responsibility and more individual ability to effect change than is often the case.

Expanding the Scope of School Improvement Programs

Changes in National Education Systems

Starting from just one school in Dar-es-Salaam in 1984, AKF's school improvement programs now directly impact several hundred schools (well over 100 in Mombasa, 90 in Kampala and 15 in Dar-es-Salaam; 35 in Karachi and 175 in Northern Pakistan; 150 in the Bombay-Ahmedabad-Jaipur-New Delhi arc of northwest India; 330 in the Gorno-Badakhshan province of Tajikistan); and, if one includes participation in larger consortia, the 34,000 schools in Bangladesh run by the Bangladesh Rural Advancement Committee (BRAC), of which AKF is one of several funding partners.

AKF's decision to keep its experimental agenda but give more emphasis to wider scale implementation has been influenced by dramatic changes in the educational environment in East Africa and elsewhere. The region has seen huge increases in the number of students and the average size of classes and schools since the early 1980s. Municipal education officers no longer know, if they ever did, how many private schools come under their purview. The number of public and private schools in Mombasa has virtually doubled since AKF's program started in the city in 1994. The Government of Uganda's announcement of Universal Primary Education half-way through the first phase of AKF's Kampala SIP doubled class size in the city (in many cases doubling classes from 50 to 100 pupils) and necessitated an urgent reassessment of some of the program's assumptions about classroom organization and the viability of 'child-centered' teaching methods. Levels of government funding in East Africa continue to decline and various 'cost-sharing' schemes have been introduced. The hemorrhaging from the school system of qualified teachers in search of better paid jobs and, more recently, due to AIDS, has not been balanced by an infusion of new recruits. Government mechanisms for the training, deployment, appraisal and promotion of teachers, largely inherited and unchanged from colonial times, has become increasingly inefficient and ponderous. Teaching has become a mass profession, accounting for the biggest block of government employees in most countries in Africa and Asia. Teachers are progressively demoralized and steadily losing any claim to professional autonomy and the setting of rules for professional membership and advancement.

Frustrated by the inequitable, patchy results of funding bodies striking local 'deals' with small circles of schools, governments and major international aid agencies have recently become increasingly attracted to a 'sector-wide approach to educational planning', by which they mean persuading donors to pool their resources and draw on them to finance nationwide improvement schemes. The jury is still out on whether this approach will speed attainment of the objectives of Education For All and Gender Equity (Swift, 2000). The early evidence in Tanzania, for example, is that complex coordination arrangements have led to a slowdown in the rate of aid disbursement. AKF, in common with other foundations and NGOs, would argue that the sector-wide approach tends to marginalize ongoing experimentation and make it more difficult for foundations and NGOs to ensure their own adequate representation in national policy forums.

Grantees

Until the early 1990s, all the SIP grantees were NGOs, in most cases the local branches of the Aga Khan Education Service (AKES), who were keen to see their own schools access resources for improvement and then serve as a base for reaching out to neighboring schools. One measure of success of this process is that AKES's 'flagship' schools have established their own links with high-quality schools at home and abroad and see less need to complicate their mission by leading wide-scale improvement programs in tough and often unresponsive environments.

In 1989, AKF made its first school improvement grant to government, to the Islands of Zanzibar, to implement an English-Language 'sandwich' course in Standard VIII and ease the transition from Swahili- to English-medium instruction in the Islands' entire network of 80-plus schools. Kisumu SIP, in 1990, was the first program with the explicit 'school improvement' label to involve detailed negotiations and memoranda of understanding with government. Government participation has become progressively stronger since then: For example, the formal grantee in Dar-es-Salaam is now the Ministry of Education. The underlying trend has been away from a 'classical' grantor-grantee relationship to one in which AKF has become an active partner in the implementation of programs as well as their design. At least in its school improvement work, AKF has never experienced the situation where ready-to-implement proposals have dropped through the mailslot. However, the attraction for AKF of being an active and long-term stakeholder in programs is offset by the difficulty of eventually disengaging and leaving technically and financially sustainable institutions in place. A corollary of this difficulty is that, whereas the first series of grants in the 1980s assumed a lifespan of between three and five years, there is now an expectation on both AKF's and the grantees' part that an 8- to 10-year partnership is required if experimentation is to be converted into sustainable and permanent practice.

As a generalization, it has proved easier to negotiate grants with NGOs and governments than with universities. Although attractive proposals have been submitted from time to time by entrepreneurial faculty from universities in East Africa, for example, to cofund a school-based Bachelor of Education degree for the many thousand teachers who do not have an undergraduate degree, the proposals' sponsors have been unable to prevent their university senate committees from stifling the experimental dimension. They have been unable to persuade their universities' bureaucracies to make potential students' financial and residency requirements flexible enough to attract teachers who have children to look after and homes to run.

Cofunding

From the outset, it has been AKF policy to seek donations from other agencies for proposals that the Foundation and its grantees have developed. In the 1980s, the bulk of this cofunding came in the form of project-specific or 'block' grants from the 'nongovernment window' of agencies such as the Canadian International Development Agency (CIDA), the United States Agency for International Development (USAID), the European Community (EC) and the UK's Department for International Development (DFID), formerly the Overseas Development Administration (ODA). The agencies' principal points of contact were AKF's offices in the respective Northern countries. At least in the 1980s, they allowed AKF's staff and grantees in Asia and Africa considerable latitude in the design, implementation and evaluation of programs. The pattern of agency cofunding has since then evolved in two significant ways. First, the proportion of funding allocated by agencies to the 'NGO window' has not kept pace with the huge increase in the number of requests, and, to sift applicants, agencies have imposed more stringent and standardized conditions such as LogFrame Analysis, a tool for quality control that was not dreamed of in the 1980s. Second, in part because of their frustration with the slowness with which governments in Asia and Africa disburse bilateral funds, Northern donor agencies have opened their 'bilateral' window to NGOs. AKF's school improvement programs in India and Pakistan, AKU-IED, and the recently approved Kenya Basic Education Program in Mombasa have been beneficiaries of this new funding channel. The benefit has been larger grants that have enabled AKF to take experimental level work to scale. There are less congenial implications, however. First, there is a much more protracted and complex process of program design. Even if the grantee is an NGO, the host government becomes a key stakeholder and exerts political influence on the choice of schools and the selection of program staff. Second, the government needs convincing that their grant is 'additional' and does not prejudice the rest of bilateral aid from the agency in question. Third and most important, despite resolve not to be 'donor-driven', there is the inevitable tendency for AKF to become a subcontractor and trim its program objectives and content to appeal to the international donor agency's latest policy and strategy for the country concerned. Since the 1980s, a major change in the wider donor environment has been the growing dominance of the World Bank, and although its legal *modus operandi* remains loans on concessional terms to governments, AKF is benefiting, particularly in the thematic field of early childhood development, from the Bank's increasing pressure on its government partners to involve NGOs and the private sector in implementing programs.

Integration with National Institutions

Once the early stages of experimentation are past, the dilemma arises of whether to concentrate on retaining 'model' programs and institutions for inspection and replication by others, or whether to risk losing the model programs' character by merging them with existing or new institutions owned by third parties and sharing responsibility for managing that process. Thus in its next phase (2001-2006), AKU-IED has to decide how to divide its energies between setting up a cluster of Professional Development Centers on its own terms and supporting government-run provincial institutes of teacher education in Pakistan. The PDC in Gilgit has to decide how far to support the government's pre-service teacher training colleges on the other side of town. A three-person team is currently 'testing the market' in East Africa for the types of in-service training programs that an AKU-sponsored PDC could offer and assessing how such a PDC would fare if owned and operated by Aga Khan agencies alone or by a consortium that includes local stakeholders. The former option allows for greater freedom in selecting staff and determining program content, but the latter promises quicker national accreditation and, perhaps, better prospects for access to donor funding and long-term financial sustainability.

The first implication of shifting the balance in the AKF's school improvement programs from experimentation to large-scale implementation has been an effort to secure the long-term future of project personnel in the wider government- or private-school system, ideally before the programs start, as opposed to the Kisumu program of the early 1990s, where the future of project personnel was left to the closing stages of the grant period and was only partly in place at its end. If the new Mombasa program has obtained more solid guarantees from the outset of the future of project staff, the price has been considerable staff time and delays in reaching the starting line. A second implication has been resisting the head-hunting of 'prima donna' program leaders from abroad, who must perforce depart when the grants end and, instead, selecting the best person available in the local system.

Dissemination of Results and Research

Dissemination of Results

In the resolutions of the World Education Forum at Dakar in March 2000 (King, 2000*)*, in the follow-up plans of governments and donor agencies (UNICEF, 2000) and NGOs, and in earlier influential documents such as Heneveld and Craig's World Bank paper (1996), there appears to be widespread consensus on the factors that contribute to increasing quality in education and agreement that the key challenges are filling gaps in the knowledge base and building the capacity to turn these factors into reality. This part of the chapter describes what AKF has done in this regard.

The 'Voice' of the SIP

The choice facing any donor organization is where to situate itself on the dissemination continuum: letting the programs it funds speak for themselves, disseminating results in the form of evaluation reports and conferences, commissioning research, encouraging 'independent' studies and articles by third parties, getting involved in direct advocacy. AKF has invested substantial effort and resources along this continuum, except in the last channel, where AKF's approach has until now been much more cautious.

Visits to program sites by ministry officials, diplomats, aid agency staff and other interested parties are becoming increasingly frequent. The Karachi Teacher Resource Center, for example, organizes an annual Open House with an exhibition on a particular theme in which literally thousands of children and their teachers from local schools as well as dignitaries from the Sindh Government participate.

Dissemination via Evaluations, Publications and Conferences

It is AKF policy to commission an external program evaluation at the end of each funding phase. These reports usually run from 70 to 100 pages and result from a 2- to 3-week site visit by a foreign and a local consultant. The selection of consultants, defining of terms of reference, organization of logistics, debriefing of consultants, review of the draft report, selection of photographs, and finalization and dissemination of the report are all done in consultation with grantee and project teams, but overall responsibility rests with AKF's program officers. It forms a substantial proportion of their workload. Some 70 copies of each report are produced and distributed within the country concerned and internationally to government officials, donor agencies, NGOs, universities and individual 'friends of AKF Education'. The school improvement case studies in this book are abridged versions of a selection of the East Africa external evaluation reports.

The number of studies and articles by SIP staff and third parties is steadily increasing. Examples include the book-length report on the Computers in Education program in Kenya (Makau, 1988), the manual accompanying the *Teachers of the Karakorums* video (Holbrook & Greenland, 1990), the summary descriptions of the school improvement and Madrasa programs in AKF's Project Briefs series, and an article on gender-related aspects of the teacher training program in Chitral (Sales, 1999). Those institutions that produce an illustrated annual report, for example, the Karachi Teacher Resource Center, find that annual reports are a particularly effective channel for communicating with their own constituency and with a wider audience.

The main international event that AKF has organized to date was a conference in London in 1993 to publicize the previously discussed SCRE report. There have been more conferences at national levels; for example, in Bombay (1994) and Hyderabad (1995). Starting with the participation of Kisumu program leaders at an International Congress for School Effectiveness and School Improvement (ICSEI) in Vancouver, AKF staff and its grantee partners have made an increasing number of presentations on SIPs at conferences and workshops organized by international agencies such as the World Bank (Greenland, Kanji & Dean, 1998), UNESCO and the Comparative and International Education Society (Anderson, 2000; Bartlett, 2000). A paper describing the Kampala program was read at a conference of the National Institute for Educational Planning and Administration in New Delhi in 1999. AKU-IED has become a regular contributor to meetings of Asian Network of Training and Research Institutions in Educational Planning (ANTRIEP), a regional Asian body convened by UNESCO's International Institute for Educational Planning (IIEP).

Research

The theses of the 120 AKU-IED master's degree graduates and those of the 40-plus staff from the SIPs in Kisumu and Mombasa who completed distance master's degrees through Exeter and Leeds Universities, and the dissertations of the 15 IED faculty who are currently doing doctorates (mainly at the AKU-linked Universities of Toronto and Oxford) constitute a large bank of data on change and improvement in the schools from which these graduates and faculty come.

In East Africa, each of the three current SIPs has a researcher on its team. In addition, AKF has recently appointed a regional research coordinator, whose mandate is first and foremost to ensure that information is shared among the three programs and that they have access to literature from relevant programs elsewhere in Africa and beyond. The research coordinator also helps standardize and improve the quality of the instruments for collecting baseline data so that cross-site comparisons can be made, and also extracts common lessons and disseminates them via conferences and publications to government officials and donor agencies.

The common goal of the individual SIPs' research is to progress on the key issue of impact. What elements of the SIP formula can be shown to make what sort of difference? Officials and the program team in Mombasa were delighted when the city moved up from its customarily lowly standing of 50th in the national Primary School Leaving Examination ranking to 37th, but how much of this 'success', if any, can be attributed to the SIP? The Aga Khan Schools' Academic Development Unit in Gilgit will investigate whether the progressive transfer to English-medium education, and administrative reforms such as decentralization and greater parental and community involvement, correlate with increased access to education and retention rates and with students obtaining higher grades. IED and the programs in Northern Pakistan and Tajikistan want to see whether 'distance technology' can be used to train teachers in remote locations and achieve standards comparable to those resulting from on-site instruction.

A dilemma facing all sponsors of school improvement programs (governments, NGOs, communities, the private-school sector) and their donor agency partners is how much money to invest in 'research', in hiring statisticians and economists to help program staff compare factors believed to influence quality and rank them in order of significance, and how much time to wait before the research findings of the 'experts' support a particular implementation strategy.

Advocacy and Policy Influence

Audience and Focus

AKF uses the results of research it commissions to influence policy by organizing discussions with different levels of stakeholders, ranging from local school systems and communities, to district and central government officials, to consortia of national governments and international donor agencies working together on sector-wide approaches to educational reform. With respect to any advocacy and policy work, there is a dilemma about far AKF should stray from what it thinks its programs can 'prove' before it enters the wider political arena. Leaders of AKU-IED, the Gilgit PDC and the Aga Khan Education Service in Pakistan have recently been asked to participate in national education policy forums. Should they use these opportunities to express their views on the size and role of government's participation in the system as a whole, or lobby on behalf of their own interests, or both? During early discussion in Pakistan of what became the Northern Education Program, there was hope that, once World Bank and DFID funding had 'primed the pump', the Pakistani government would increase its own funding via its Social Action Program, through block grants to community-run schools, through provision of more government-paid teachers, or (most ambitious of all) through a per-student subsidy paid directly to communities. Unfortunately, little has happened on this front in the last five years.

Influence on Teacher Training Programs

Influence is difficult to quantify, but AKF's interim assessment at the end of 2000 is that its school-focused training and curriculum development programs have so far had little *formal* effect on national policy and practice. One reason is that AKF must rely on its grantees to lead any advocacy campaign, and their status as NGOs or private-sector agencies has made them ambivalent about involvement in activities that may prejudice their often delicate and complex relationships with government ministries and institutions.

With respect to teacher training, a significant factor in the SIPs' limited impact on the structure of and curriculum of government primary teachers colleges is that the latter are bastions of conservatism that have scarcely changed since the 1960s. The colleges tend to be staffed by elderly teachers who have missed out on other forms of promotion and whose own expertise is limited to one secondary school subject each. In East Africa, many teachers colleges are very small, having been established in colonial times by Christian missions to service their local circle of grant-aided primary schools. The colleges' main clients are school-leavers who failed to get high enough grades for university entrance and who regard teaching as a last resort; hardly the best recipe for fostering commitment to a profession. National bodies for accrediting teacher training, even in countries where they exist, tend to be dominated by academics anxious to apply the degree structure of their university. Ministry officials are fearful of the organizational and financial implications of the improvement measures advocated by the SIPs. If head teachers receive management training, someone has to decide whether they deserve salary increases as a result. The balance between resources allocated to pre-service versus in-service training was not very favorable to the latter in the 1970s (Greenland, 1983); the gap has widened much further since then. Regular visiting of students during practice teaching has deteriorated at most colleges and universities to mere formality as a result of budget cuts. Older teachers in East Africa refer nostalgically to 'the good old days' at Makerere University when 'proper practice teaching' was done.

To address the failure of the colleges to supply the number of new teachers required, various 'emergency' schemes have been introduced, for example the USAID-supported Teacher Development Management System (TDMS) in Uganda and the DFID-funded School Support Systems Program in Malawi. They involve varying combinations of short spells of residence at a teachers college in the school vacations, training sessions on weekends during term time at zonal resource centers, correspondence assignments and, more rarely, visits to teachers' classrooms by a project tutor. The dream that pre-service colleges might be shaken up and some of their resources used to finance serious in-service programs like those described above has not yet materialized, and few emergency schemes have taken 'coaching' or 'mentoring' very seriously.

The claim that AKF's SIPs have exerted a definite but *informal* influence on national programs is much more substantial. The Kampala SIP team is helping to train the managers and tutors of the five Teacher Resource Centers set up by the above-mentioned TDMS program in Kampala. The SIP team contributes to the course for nursery school teachers at the Institute of Teacher Education Kyambogo (ITEK), where 20% of the latter course is actually taught at the SIP Teacher Resource Center. The team also writes a weekly newspaper column that is read throughout the country. In Karachi, course attendance certificates from AKU-IED and the Teacher Resource Center have increasing currency for employment and promotion, particularly in the huge network of privately and community-run schools. AKU-IED has trained several professional development teachers (i.e., its master's degree graduates) seconded from the Government of Balochistan's Primary Education Program, who in turn have run short courses for teachers from schools throughout the province. The manuals developed by the Field-Based Teachers Development (FBTD) Program in Northern Pakistan have also been adapted and used by the Balochistan program. AKF's assumption that investing in a 20-minute video and manual describing the FBTD program, entitled *Teachers of the Karakorums* (Holbrook & Greenland, 1990), would influence practice in Pakistan's training colleges proved misplaced. Subsequent enquiries showed that virtually none had used it (though by contrast the video has been put to extremely good use as an information tool for aid agency and general audiences).

Influence on Curriculums and Course Materials

As noted in the first section of this chapter, AKF's approach to the improvement of classroom instruction has been to concentrate on developing training programs rather than new syllabuses and course materials. In consequence, formal contacts with national curriculum development centers and textbook boards have been limited. The main exception to this approach has been AKF's work with English-medium instruction in Pakistan and Tanzania. AKF's first grant to the Aga Khan schools in Karachi (1989-90), to improve 'generic' teaching skills, coincided with pressure from parents and school managers for a transfer from Urdu- to English-medium instruction. The ensuing Language Improvement Program (1991-1994) introduced an early literacy scheme from the UK and used it as the basis for an experimental syllabus for the first five Classes. A similar program is now being implemented in Northern Pakistan, where government schools want to introduce English earlier than before and are taking up the syllabus and materials produced under AKF's grant. There has been a parallel process in Tanzania. Materials were developed at the Aga Khan Mzizima Secondary School in Dar-es-Salaam to ease the Form I students' transfer from their Swahili-medium primary schools to English. A church-based press printed the cyclostyled materials (stencils) as sets of sturdier brochures, which were popular for a time in the southern part of the country, but nothing came of AKF's approaches to commercial publishers. In Zanzibar, however, the Ministry of Education applied for a grant to adapt the Mzizima materials and turn them into a complete set of student coursebooks and teacher guides, which it then printed at the government's own press and distributed to all schools throughout Zanzibar. The Aga-Khan-supported Health Education program in Tajikistan recently produced a draft set of coursebooks for Grades 1-5, which is attracting national attention.

Influence on Management Training and Community Involvement

Since the SIPs' focused attention to management training and community involvement dates only from the mid-1990s, their influence on the wider school system and on national policy has been correspondingly late to develop. Also, in East Africa, a number of other agencies are already active in the education field, and a range of printed materials exists; for example, the series published by the Commonwealth Secretariat. Nevertheless, the four handbooks produced for head teachers by the SIP in Dar-es-Salaam are already proving a valuable addition to this literature. A 'niche' in this market that AKF and the SIPs are currently discussing with UNICEF and other agencies, is materials, possibly also a web-site, that make choices between alternative 'quality' investments accessible to zonal officials, school committee members and parents. School committees and parents think their scarce resources should first go to buy computers and construct science labs, but they do not know what a small primary health care clinic would cost to run and whether it would have a greater impact on student attendance rates. Documents produced by AKF's early childhood programs, for example the Madrasas Integrated Curriculum, have so far had comparatively greater influence within and outside East Africa than the corresponding SIP materials (described above), in part because, with the exception of Kenya, few national curriculum centers have invested substantially in the early childhood field.

Work in Progress

AKF's overall goal of improving the quality of basic education in the countries in which it works means persevering with the experimental work described earlier in this chapter. The primary focus remains improving teachers' mastery of classroom teaching skills and the 'institutionalization' of mentors and resource centers to strengthen that process. The second focus is increased attention to the management and financing of schools, and the involvement of parents and communities. Achieving this goal means testing different formulas for public and private partnerships to provide education, train head teachers, administrators and teachers; and using incentive grants and mini-endowment schemes to motivate communities and philanthropists to support their local education facilities over the longer term. A third focus is studying how the different types of improvement measures outlined above, individually or in combination, influence enrollment, retention and academic achievement rates, particularly for girls and particularly at preschool and primary school levels. The final focus is taking a much closer look at costs and benefits.

Mastery and Mentoring

Despite the male connotation of the phrase 'master trainer', the SIPs are reluctant to abandon it because 'mastery' retains the important notion of achieving a publicly recognized level of skill, as in 'master craftsman' or 'master violinist' and assumes a lengthy, if not continuous, process of apprenticeship. The SIPs in East Africa, supported now by a regional research coordinator and more adequate research funds, are reassessing their emphasis to date on developing teachers' skills and are seeking to define a better balance between this emphasis and teachers' need for more subject-content knowledge. One reason for reassessing the emphasis is that when a sample of teachers in the Northern Pakistan program were recently asked to take the same Class 5 mathematics exam as the students, their performance was alarmingly poor. A study of pre-service teacher training in Pakistan reached the depressing conclusion that "measured by their learning, primary-school students benefit greatly from each level of formal education that teachers complete and little from having their teachers certified" (Warwick and Reimers, 1995, p. 47). The problem, as an interim report from the first phase of the Kampala program puts it, is that the SIP approach cannot improve teachers' subject knowledge or their command of English substantially in a short period of time. Given this constraint, should the SIPs limit themselves to plugging specific gaps in teachers' knowledge of the primary mathematics syllabus, or should they retain the ambitious aim of enabling teachers to reach a Grade 12 (Form IV in East Africa) standard? The Aga Khan Education Service's decision in Northern Pakistan to teach mathematics and science in English rather than Urdu up to Class 5, introduced under strong pressure from parents, is necessitating a larger and more prolonged investment in teachers' own English-language competencies than was originally envisaged.

In terms of classroom teaching skills, work continues on exploring whether some skills are more 'powerful' than others, and whether there are more or less efficient paths to acquiring them. Given the highly arbitrary nature of appointing and promoting teachers and the teachers' extreme distrust of any appraisal process, it is also important for the SIPs to reach an understanding with teachers of how these skills can be objectively assessed.

The label 'school-based training' needs defining. Ideally, it means one mentor and one teacher in the teacher's own classroom, working together through the 'classical' sequence of pre-lesson conference, lesson by the teacher to the class, post-lesson conference between mentor and teacher, and demonstration by the mentor. More realistically, it means one mentor working with a group of teachers, drawn from a cluster of neighboring schools and grouped in a 'typical' classroom. In Kampala, the basic 'formula' has been for each of the four senior trainers to spend one day each week in a school and then to bring all the teachers together at the resource center on the fifth day. The SIP team concluded that each school required a minimum of one year's intensive training along the above lines, followed by a second year of reduced support, during which the school would create its own school-based program and help with new schools. As one 'tool' in the struggle to achieve some measure of long-term sustainability, AKU-IED is sponsoring the creation of teacher associations, in mathematics and science for example, to see how far this low-cost instrument can sustain improvements in classroom quality and spread at least some of the benefits of the strict whole-school approach beyond the original circle of client schools. The challenge facing AKF is to deal with critics who argue that the cost and logistics of organizing school-based support on an extended basis rule it out for anything except small circles of privileged, probably private, schools.

Centers of Excellence

The objective of AKF's SIP initiatives is to become more confident in defining where the components of the school improvement formula are best delivered. This means substantiating and adding detail to the 'interim' claim by the present SIPs that classroom teaching skills are best improved by a combination of resource-center- and school-based training, and that teachers' subject-content knowledge can be strengthened via a combination of self-study and distance education.

A corollary of the decline in governments' ability to maintain educational quality is the need to encourage the establishment of centers of excellence for training and research in the NGO/private/community sector. This is the role envisaged for AKU-IED and its proposed network of Professional Development Centers, the four core partner institutions in the Program for Enhancing School-Level Education (PESLE) in India, and the clusters of Madrasas Resource Centers and school improvement programs in East Africa.

Community Involvement

Promoting an emphasis on community involvement presupposes clearer definitions and more widely shared understanding of both the words 'community' and 'involvement'. 'Community' needs defining in terms of its spatial distribution in a given context (rural, peri-urban or urban); how it relates to parents and families (for example, nuclear families, extended families or clans, other bases of association such as religious affiliation); and how it relates to other existing patterns of community organization or 'civil society' in a given location or project area, such as parent-teacher associations and school management committees. Clarifying the concept of community will be necessary to determine whether the 'school community' represents the 'normal distribution' of gender, age, income, education, and so on in the wider society. It means determining whether the key relationship is between an individual community and an individual school (and, if so, what the implications are for the relationship between a community and individual classes within a school); or between a community and several schools in a district that its children may attend; or between a community and institutions such as Teacher Resource Centers. Should the members of school management committees be elected and, if so, by whom: all the members of a defined community or only those who have children actually in the local school at the time? Some large villages in Northern Pakistan already have several village organizations and women's organizations: Should each school have its own governing committee, or might there be gains in quality and efficiency through encouraging the formation of a 'supra-village' committee to manage several schools?

Promoting the involvement of individual communities is requiring AKF to take sides on a number of contentious issues. First, whose views count most: those of the expert, the trained head teacher and teachers actually running the school system, government, funding agencies, or the local community, and how can conflicting views be reconciled? In Northern Pakistan, communities have linked their willingness to pay higher school fees to the NGO sponsors' willingness to switch from Urdu- to English-medium education, and they want walls built around schoolyards before investing in toilet blocks (outhouses). Local values will take precedence over expert and donor agency opinion. The second issue is that of equity among communities in a given area or school system. Will poorer, remoter communities ever be able to pay their way, or is it a case of 'unto him who hath, more shall be given'? Can a measure of cross-subsidy between prosperous and less prosperous communities be envisaged? In encouraging communities to start their own schools or take over declining or defunct government schools, AKF is accepting a measure of responsibility for the process; and for the eventuality of the communities' failure to manage the schools effectively. Should AKF be working with governments to design a 'safety net' for communities who find that they cannot cope with their new responsibilities? A recent study of decentralized school management in New Zealand showed that, even in an economically prosperous society, up to 30% of schools were unable to execute the responsibilities thrust upon them (Fiske & Ladd, 2000).

Costs and Benefits

AKF recognizes that it has been slow to calculate the per-student, per-teacher, and per-school cost of quality improvement measures (as opposed to salaries and administration) and that this calculation must now receive much greater attention if experimental level programs are to be adopted and taken to scale. Also at the local level, a school management committee entrusted with its own budget needs to know how much it costs to have a teacher trainer visit its teachers two days each month, as opposed to buying a new set of science books for Grade 3. AKF needs to define the optimal duration of providing 'concentrated' improvement measures to a given school or school system and then work out how to reduce the volume and cost of these measures over time in such a way that benefits are sustained and any loss in quality is minimized. The further step of comparing the costs of campus- versus resource-center- and school-based programs is complicated by the often-arbitrary figure selected for amortizing the cost of buildings.

One consequence of AKF's support for community- and NGO-owned schools is that it is increasingly being drawn into the challenge facing school owners of fixing student-teacher ratios, levels of teacher qualification and teacher salaries. Salary scales and working conditions inherited from colonial times that were viable when government was virtually the sole employer and the number of teachers was small can no longer be sustained, and the phenomenon of the 'paraprofessional' teacher, who commands a significantly lower salary, is spreading everywhere. AKF's contribution is not only to encourage its grantees to experiment with matching grants, and income-generating and endowment schemes; but also to help the grantees develop teacher appraisal and remuneration policies that are professionally acceptable and work out a balance between these administrative costs and investments in 'quality' items such as training, curriculum materials and infrastructure. These are issues requiring not only experiment in the field but also the drafting and implementation of a new and more robust legal framework for teachers' employment than probably existed before.

In Tajikistan, AKF's program is facing a crisis situation where teachers' salaries have dropped to the equivalent of a mere US $5.00 per month (usually paid several months late) since the collapse of the Soviet Union. As a result, teachers are abandoning the profession for better paid jobs in other sectors, or, more often, are reverting to subsistence farming. Helping communities mount income-generating schemes and persuading parents to increase their financial contribution may alleviate the situation somewhat, but the alternative of lobbying agencies such as the World Bank and the Asian Development Bank to underwrite salary supplements now has to be addressed. This, in turn, forces a choice between arguing for 'across-the-board' supplements or much more contentious performance-related rewards.

In East Africa, these problems may not yet be as acute as they are in Central Asia, but the fact that a primary teacher's salary probably buys twice as much in Kenya as it does in Tanzania and Uganda has a powerful effect on teachers' willingness to forego the income of a second job (most frequently, private tutoring) and invest their time in acquiring further training and qualifications; or even visiting the TRC to collect materials to brighten up their lessons.

Conclusion

Today, as a result of AKF's school improvement programs over the last 15 years, many more children in Mombasa, Dar-es-Salaam and Kampala are enjoying school and learning more than they did before. Their teachers are more motivated and are learning new methods of instruction. Decentralization by governments is creating a context in which head teachers and school committees are able to manage staff and resources better. Parents and communities are making greater contributions to their local schools, in money and in kind. Meanwhile, the number of children in these three conurbations who are *not* benefiting from these improvements is growing much faster than the number who are – to say nothing of the rest of the children and teachers in the region. Another 15 years will take us to 2015, the deadline chosen by the international community at Dakar for achieving Basic Education for All. Why

2015 was chosen, what Basic Education for All means, and what justifies the expectation of reaching this target, is unclear and somewhat arbitrary. AKF's school improvement agenda, as outlined in this chapter, is long and complex. To avoid the judgment, 15 years from now, that progress in East Africa has been as erratic and incomplete as in the world at large, the immediate task must be to prune the list of presumed improvement measures and pursue only those that, from an early date, can be shown to make the most, the quickest and the cheapest difference. Easier said than done.

3

Building Professional Community at Mzizima Secondary School, Tanzania

Stephen E. Anderson and Suleman Sumra

From 1985 to 1994 the Aga Khan Foundation (AKF) supported a continuous school improvement program at the Aga Khan Mzizima Secondary School in Dar-es-Salaam, Tanzania.[2] The Mzizima SIP incorporated a variety of principles, strategies, and structures for school improvement and teacher development that were quite innovative at the time for schools worldwide, especially in developing countries like Tanzania. These included

- *a school-based and school-wide approach to improvement* in which the school is viewed as the 'unit of change' and the changes are based largely on needs defined by local school personnel (i.e., teachers, administrators);
- *commitment to a multi-year school improvement effort encompassing multiple focuses of school development*, including curriculum, instructional methods, teacher learning, teacher leadership, school climate, and organization;
- *an emphasis on continuous teacher development* as key to improvement in the quality of teaching and learning;
- *utilization of nontraditional strategies to support school-based teacher learning and change*, including teacher leadership roles, collegial work

groups in curriculum-oriented teacher workshops, an in-school teacher resource center, peer coaching, action research, and new teacher induction;
- *an emphasis on local capacity building* to enable school personnel to carry on school improvement without dependence on external leaders and expertise; and
- *attention to institutionalization* of organizational innovations created to mobilize, support, and sustain teacher and school development.

The school improvement program evolved through three phases. Phases 1 and 2 (1985-1990) focused on the development of teaching and learning resources and on curriculum planning for English, mathematics, and sciences. Specific attention was directed toward the development of curriculum materials and teaching strategies to ease the transition from Kiswahili to English as the medium of instruction during the early years of secondary school. In Phase 3 of the school improvement program (1991-1994), participation was extended to teachers in all subject areas, and the main emphasis shifted from curriculum development toward developing teaching skills and teacher leadership. Explicit efforts were made to institutionalize the organizational interventions piloted through the SIP project in order to sustain continuous improvement at the school. Outreach activities were undertaken in all phases to disseminate components of the Mzizima SIP (e.g., curriculum materials, teaching expertise, professional development processes) to other educational entities in Tanzania (e.g., secondary schools, primary schools, education ministry, teachers colleges). This chapter presents a history of the Mzizima SIP and findings from an evaluation of the SIP experience conducted by the authors at the end of Phase 3.[3]

Context

Tanzanian Education System

The Tanzanian education system operates on a 7-4-2 model of primary and secondary education. Primary education (Standards I-VII) is offered mainly through publicly funded government schools. According to figures cited in Lassibille et al. (2000), only 15 of 10,960 primary schools in 1992 were privately owned and managed. At the secondary level, however, there are more private (57%) than public schools. The medium of instruction in public primary schools is Kiswahili. A few private primary schools employ English. The medium of instruction in all secondary schools is English. The central government exercises considerable control over school fees, curriculum content and timetabling, textbook selection, teacher supervision, and student promotion in both public and private schools. While various sources quote different figures, government spending on education has remained fairly constant over the past 20 years, at around 14% to 15% of the national budget. According to Lassibille et al. (2000), primary school enrollment has declined nationwide from over 90% in 1980 to about 66% by 1996. A third of primary school students do not complete the 7-year primary school cycle. Promotion from one cycle to the next is regulated by national examinations. All students sit the primary school-leaving examination at the end of Standard VII. The government uses a quota system based on geography and gender to control the number and distribution of students whose results qualify them for promotion to Form I of the lower secondary cycle in public schools. Examinations also occur at the end of Forms II and IV. Examination results determine which students are tracked into academics and which into commercial-stream programs, and which have access to upper secondary (Forms VI-VII) education.

Secondary Education in Tanzania

Access to secondary education has always been a contentious issue in Tanzania.[4] The government's priority since independence has been to provide basic education at the primary level. As primary school enrollment increased, secondary school enrollment remained fairly stable. The proportion of primary school graduates obtaining admission to secondary schools plummeted. At independence in 1961, more than 30% of students finishing primary school were admitted to secondary school. This proportion fell to around 4% by 1984. Given the limited capacity of the government to build secondary schools to meet increased demand for secondary education, secondary education expansion has occurred through policies enabling private efforts, or through such measures as double-shift schools. Government statistics as of 1997 reported 386 religious and secular private schools (282 registered since 1987) versus 286 publicly funded government schools and government-subsidized community-built schools. About 55% of all secondary school students were enrolled in private schools.

Comparisons of teacher survey and student performance data from Tanzanian government and private schools reveal a number of patterns. All schools collect student fees, though private school tuition averages one and a half times that in public day schools. The vast majority of public and private secondary schools are coeducational and offer instruction from Forms I to IV. Mzizima Secondary School is among the 10% of schools across both sectors that offer instruction at the upper secondary level, Forms V and VI. While enrollment in public schools tends to be larger than in private schools, the pupil-teacher ratio is lower by 20%. Schools in both sectors suffer chronic resource deficiencies; for example, textbooks for students and the low quality of school physical facilities. Private schools enroll a lower proportion of children of peasant background (39.5%) and a higher proportion of children of self-employed parents (19.5%) than public schools (54% and 7.9%). Teachers in private schools are paid more on average than public school teachers, but they are less likely to hold permanent contracts (95% in public schools, 39% in private schools). Private school teachers tend to have fewer years of experience, formal education, and training. More men are employed in private (82%) than in public (68%) schools. According to a secondary school survey in the mid-1990s, only 47% of public and 40% of private school teachers reported any in-service training in the previous 5 years. With the expansion of secondary schooling and employment opportunities for teachers, teacher mobility is high. Given higher salaries, despite losses in job security, teacher mobility tends to be from the public to the private sector.

Lassibille et al. compared 1992 and 1994 public and private school results on the national Form II and Form IV examinations. Student results for all schools were discouraging (only a third of answers responded to correctly). Public school students, however, outperformed those in private schools by 25% (Form II) and 20% (Form IV) in 1992, and by 15% across both examinations in 1995.

Mzizima Secondary School

Aga Khan Mzizima Secondary School is a private, coeducational, nonsectarian, racially integrated day secondary school located in a middle- to low-income residential neighborhood of Dar-es-Salaam. The Mzizima School was founded in 1967 by the Aga Khan Education Services (AKES). Apart from donor project grants, the Mzizima School operates on revenue from government-regulated tuition fees, parent contributions, and fundraising. In 1994, tuition revenues provided for a third of the school's operating cost.

The school building at the time of our evaluation (1994) consisted of six two-storey blocks, including three teaching blocks, one administration block, and two blocks for science labs. A school library, a computer lab, a staff room and a Teacher Resource Center were all housed within the administration block. The classrooms had desks for all students and were ventilated by ceiling fans and louvered windows. The library had a large selection of books and work space for students. The computer lab housed six functioning computers and three printers. The science laboratories appeared to be well stocked with equipment and supplies. Textbook shortages typical of Tanzanian government schools were not a problem at Mzizima School. The school had a large staff room furnished with individual carrels for teachers and a conference table. The Teacher Resource Center next to the staff room housed a professional library and a conference area (tables, chairs, blackboard) for in-service activities. While the facilities were superior to public schools visited during our visit, we saw another private school with superior grounds, decor, labs, and library.

Mzizima School offered a 4-year lower secondary program (Forms I-IV) and a 2-year upper secondary program (Forms V-VI). Students were organized in four mixed-ability streams from Forms I to IV. They followed the same government curriculum the first 2 years (English, Kiswahili, physics, chemistry, biology, commerce, accounts, geography, history, civics, and mathematics). Students entering Form III chose a science or an arts and commerce stream. The upper secondary program offered both streams.

The school admits 160 Form I and 80 Form V students annually. Form I admission is based on school-set tests in English and mathematics, and on parent interviews. As a community school, the management ensures that 20 to 25% of the students are Ismaili Muslim. Form V admission is based on national Form IV examination scores and is not limited to Mzizima School students. Classes in 1994 consisted of about 40 students per stream in Forms I to IV and 20 students per stream in

Forms V to VI. Enrolment in October 1994 was 790 pupils: 456 boys and 334 girls. The student-teacher ratio was 22:1. The racial and religious makeup was 23% Ismaili, 28% African, and 49% non-Ismaili Asian. Students entering Form I came from several elementary schools. One was an English-medium school, the rest Kiswahili. Aga Khan Education Services opened an English-medium primary school near Mzizima School in 1994. The first primary graduates were eligible for entry to Mzizima School the year after our evaluation. Mzizima School students came from lower middle to upper income families, and were mostly Tanzanian nationals.

The professional staff at Mzizima School consisted of 36 teachers, including a head teacher, a deputy head teacher (administration), a deputy head teacher (professional development), and 33 classroom teachers. The school employed an accountant, a nurse, and support staff (e.g., secretaries, drivers, caretakers, lab assistants). A majority of teachers were male (80%), and 75% had 6 or more years of teaching experience. Contrary to many Tanzanian private and public schools, most of the teachers were highly qualified academically: 23 (64%) held university degrees, 11 (31%) held teachers college diplomas, and two had A-level certificates. Reports earlier than ours cited a high number of expatriate teachers and a high rate of teacher turnover. By 1994, there were only two expatriate teachers on staff (the head teacher and one other). The rate of annual teacher turnover had stabilized at about 22%. About 75% of the teachers were indigenous Africans. The rest were of South Asian heritage.

The teaching staff were organized into five faculties (science, languages, social sciences, mathematics, and commerce) and 13 subject departments. There were both faculty heads and department heads: sometimes one and the same person. The average teaching load was 22 periods per week (out of 44 scheduled periods). Everyone taught, though the head teacher and deputy head teachers had reduced loads. Teachers were employed by AKES on 3-year term contracts. The average salary was over double that reported in three public schools visited, yet lower than what teachers reportedly earned in some other private schools. Mzizima School teachers also received medical benefits, tuition waivers for their children, leave allowances, and housing assistance through the Aga Khan estate.

History of the Mzizima School Improvement Program

Origins of Change

The Mzizima school improvement program originated in the mid-1980s out of concerns within the Aga Khan Education Services about the quality of education in Tanzania, and more particularly, at Mzizima Secondary School. A generalized decline in educational standards was reflected in high failure rates in national secondary school Form IV and VI examinations. The decline was attributed to multiple causes, including an outdated national syllabus; the need for more and better quality Tanzanian teachers; overreliance on noncommitted expatriate teachers; ineffective teacher training programs; low pay and poor working conditions for teachers; shortages and variable quality of teaching and learning materials; decreasing English-language proficiency of secondary students and teachers; predominance of lecture and rote learning approaches to teaching and learning; steering effects of national examinations emphasizing knowledge recall; low motivation and limited professional incentives for teachers (e.g., support for professional development, career opportunities); and low motivation among students. The teaching and learning situation at Mzizima School was described by the SIP developers in similar terms. Within this context of educational malaise, AKES and AKF joined in a long-term program of school improvement.

SIP Phases 1 and 2

Phase 1

Phase 1 of the SIP began in March 1985 and ended in March 1988. Three programs of action were envisioned: development of the curriculum, beginning with English and mathematics; acquisition of resources to meet teacher and student needs; and attention to school policy issues (e.g., homework, discipline, student counseling, parent involvement). Teacher involvement in curriculum development activities was expected to result in improved teacher motivation, curriculum and lesson planning, and methods of teaching and assessment. The development of the SIP's Phase 1 was strongly influenced by a 1985 national report on English-language proficiency at the secondary level. The report identified a need for strategies to support English language learning as students switched from Kiswahili to English as the medium of instruction.

The key components of the Phase 1 SIP plan were as follows: hire expatriate English and mathematics specialists to teach part time and to lead weekly full-day work sessions with relevant subject teachers for the purpose of developing course materials (e.g., schemes of work, lesson plans), teaching resources (e.g., teacher guides, worksheets, teaching aids), and new teaching methods suited to these curriculum modifications. Access to expertise in other subjects was envisioned for a later phase. The plan included the purchase of textbooks, library books, and reference books for teachers. Outreach plans were limited to inviting teachers from nearby schools to attend the weekly workshops free of charge.[5]

Phase 1 began with the appointment of an expatriate English language specialist in March 1985. The first specialist resigned and was replaced by another expatriate in March 1986. An expatriate mathematics specialist joined the project in January 1986. He remained until the end of his contract in December 1987. Implementation of Phase 1 followed the original implementation plan, though curriculum specialists brought their own views and expertise to bear on project development. The first English specialist oversaw the acquisition of class libraries for all Form I and Form II classes (60 novels per class), managed weekly English teacher workshops, and organized the production of a bank of teaching materials (e.g., worksheets for specific topics). The usefulness of the materials was reportedly limited by a failure to align them with the English syllabus of the national curriculum. Under the leadership of the second English specialist, the English language component of the SIP took two major thrusts: refinement of the class library initiative and development of a Form I English language orientation program. The class library program was supposed to support English language learning among lower secondary students. English teachers classified the books by reading level, produced comprehension tests, and developed a class library program teachers' guide.

The new specialist initiated a strategy for English language learning across the curriculum in Form I. Teachers in all subjects provided the English language team with topics and course materials covered during the first 6 weeks of their Form I courses. The English consultant and teachers rewrote the course materials in simplified English, and incorporated suggested teaching and learning activities into the modified courses that were less teacher directed and more practically relevant, yet still aligned with the national curriculum requirements. The Form I English Language Orientation Course (FIELOC) was piloted in Mzizima School and five nearby public schools in January 1987. The program was well received by teachers. Based on the pilot, the FIELOC materials were revised and packaged in booklets by the project team for use in 1988. The revised program was introduced into twelve more schools in other areas of the country.

Curriculum development activity in mathematics began in January 1986. Following the pattern of English workshops, a day a week was timetabled for the mathematics specialist and teachers to prepare mathematics lessons and teaching materials. The quality of the lessons and materials was reportedly diminished by a failure to relate them to an overall framework of topics and learning objectives, by poor alignment with the national mathematics syllabus, and by lack of provision for replacement of teaching aids as they wore out or disappeared. The poor curriculum fit was a disincentive to teachers, who felt a keen responsibility to prepare students for syllabus-based national examinations. In addition to weekly workshops for English and mathematics teachers, annual whole-staff seminars covering a variety of topics were organized by the specialists. These took place in June when classes were not in session.

During Phase 1, teacher representatives from five neighboring government secondary schools attended the curriculum workshops, shared in the development of English and mathematics materials, and took those resources back to their home schools. Some attended the whole-school seminars as well.

The results of the SIP's Phase 1 were reported in internal progress reports and in an external evaluation by C. E. Beeby (1987). The main impact of the SIP was on curriculum content and materials. Beeby described the class libraries as the most successful program of the project, and recommended this program's continuation and expansion. He commended the FIELOC program and recommended extending its development and use across the curriculum beyond the first 6 weeks of Form I. The authors of the monitoring and evaluation reports agreed, however, that teachers needed more time, in-service training, and experience to assimilate the new teaching methods introduced in both the workshops and FIELOC. Teacher participation in SIP in-service activities (workshops, seminars) was also having a positive impact on teacher motivation, commitment, and collegiality. Improvements in teacher attitudes and collegiality were greater for teachers directly involved in workshop activities. Beeby recommended that weekly workshops be held in all subject areas.

The potential and challenges of disseminating the Mzizima SIP experience were apparent by the close of Phase 1. Participation of teachers from other schools in the curriculum workshops and seminars was inconsistent, and SIP benefits were diminished by lack of follow-up support in participants' schools. The FIELOC and class library programs, however, coincided with government interests and attracted national attention. Form I class libraries were being introduced in government secondary schools across the country as part of an English language development project sponsored by the British Council. In his evaluation, Beeby argued that teachers needed more than introductory workshops to effectively implement the FIELOC program and that widespread dissemination support was beyond the capacity of Mzizima School teachers. He suggested that links be made to the Ministry of Education and British Council project.

The Beeby evaluation report identified several issues relating to management and continuation of the SIP beyond Phase 1. These included a continuing lack of teacher continuity and expertise to lead project activities, and head teacher leadership problems. The head teacher was not on good terms with the faculty. This contributed to a lack of student and teacher discipline and engagement in the school. He had little interest in the SIP and left after the first year. A senior teacher was appointed acting head teacher in a caretaker role. Project specialists ran the SIP activities independently of the school administration. Beeby argued that SIP activities and specialists should be responsible to the head teacher. He also expressed concerns about a "lack of unity, spirit and identity" as a 'whole school,' and proposed the development of extracurricular programs for students, staff activities to promote *esprit de corps*, and attention to school-wide issues initially targeted in the Phase 1 plan (e.g., homework, discipline, and parent involvement).

Phase 2

A new head teacher from Kenya was appointed midway through the 1987 school year. He set a plan for dealing with a variety of whole-school issues. The plan included promoting teacher community through staff social activities; greater teacher engagement by assigning extracurricular or administrative duties to all teachers; improved communication through regular department head and staff meetings; high expectations for teacher professionalism (e.g., punctuality, attendance, lesson planning, and student evaluation); consistent student discipline based on self-respect and self-reliance; regular homework; development of extracurricular activities (e.g., sports, clubs); increased contact with parents (e.g., open house, awards, and sports days); and continued acquisition of library and textbook resources. The new head teacher communicated his support for the SIP and his desire to integrate SIP activities into the school. The head teacher's plan created a more positive context for the implementation of the SIP's Phase 2.

Funding for 2 more years (March 1988-March 1990) of the Mzizima SIP was granted in March 1988. The Phase 2 plan had several components, beginning with project management. The Phase 1 English language specialist was to assume the role of project coordinator under the head teacher, in order to create a more unified project identity and approach. The reliance on expatriate consultants was to continue with the English specialist and newly appointed specialists in mathematics and science. Weekly curriculum workshops for English and mathematics teachers were to continue as in Phase 1, with a greater emphasis on alignment of curriculum development activities in mathematics with the national syllabus. Workshop participation was to be extended to the science teachers. The Phase 2 plan called for continued development of the class library and the FIELOC programs. More books were to be added to class libraries, and this program was to be extended up to Form IV. The project staff proposed to extend the FIELOC program from 6 to 12 weeks by 1989, and then to a full year in 1990. Further dissemination of the FIELOC program was envisioned through regional English language consultants associated with a nationwide British Council initiative. No strategy was proposed to deal with the obstacles to dissemination of SIP workshop activities to other schools (e.g., lack of continuity in participation, of release time for teachers to attend, of follow-up in teachers' home schools, and funding for their workshop materials).

Implementation of Phase 2 proceeded more or less as planned. The English specialist served as both project coordinator and English language consultant. Rather than hiring an external consultant, the project relied on in-house expertise for mathematics. The project coordinator facilitated the work of the mathematics teachers, and the new head teacher, himself an experienced mathematics teacher, participated in the workshops. An expatriate science specialist was contracted and stayed with the project the full 2 years. Science workshop activity was complicated by the fact that there were three science subjects (biology, chemistry, and physics) taught by teachers with little experience collaborating across subject areas or using student-centered interactive methods of instruction. The class library program achieved a routine level of implementation. The FIELOC program continued in Mzizima School Form I classes and in about a dozen other schools in Dar-es-Salaam and some northern cities. The proposal to lengthen the FIELOC program to a full year was not enacted, but the FIELOC materials and teacher training were disseminated to 20 schools in southern Tanzania, in cooperation with the Ministry of Education and with donor funding for local publication.

A consensus picture of the cumulative outcomes of the SIP's Phases 1 and 2 emerged from project monitoring and evaluation documents reviewed for our study (a 5-year review by the English specialist, an external evaluation of Phase 2 by Terry Allsop (1990), an external evaluation of the FIELOC program, and the proposal for Phase 3). All agreed that significant gains had been made in the acquisition and school-based development of improved teaching and learning materials, teacher motivation and commitment, and school climate. Weekly workshops for English, mathematics, and science teachers had become an integral part of the school timetable. Key issues for continuing the workshops were to develop and sustain teacher leadership and to broaden workshop participation to all teachers in order to make the project a truly school-wide program of improvement. Citing a staff survey at the end of 1989, program reports attributed improvements in teacher stability and attitudes (e.g., motivation for professional growth, commitment to teaching and the school, job satisfaction) to SIP activity and to the head teacher's emphasis on teacher professionalism (e.g., punctuality, participation, conduct). Organizational problems related to school management and climate, which had hindered SIP progress, were no longer an issue by the end of Phase 2.

The various commentators on SIP progress at the end of Phase 2 agreed that the institutionalization of school improvement activity, and the resulting improvements in pedagogical resources and materials, curriculum development skills, and teacher attitudes, were significant accomplishments. They also agreed that intended changes in teaching methods remained elusive in practice. In his evaluation of Phase 2, Allsop reported, "Years of workshop activity in languages, mathematics and science should be resulting in a transformation of the nature of classroom interactions between teachers and students. I saw little evidence of this in the small number of classes seen during my stay." Similar judgments from other observers of SIP outcomes contributed to proposals for a greater emphasis on instructional development during the project's third and final phase.

While no evaluations of the SIP's impact on student learning were undertaken during Phases 1 or 2, project reports provided anecdotal accounts of improved English language proficiency among students and of improvements in aggregate results on Form IV examinations in English, mathematics, and sciences. The project leaders and evaluators were cautious about attributing these outcomes to SIP activities and products.

SIP influence beyond Mzizima School by the end of Phase 2 was mixed. Workshop participation by non-Mzizima School teachers continued to wane after the introduction of double shifts in government schools. Allsop concluded that dissemination of the FIELOC program was diverting attention from improvement within the school and that support for FIELOC outreach should be addressed separately from the SIP work.

SIP Phase 3

From the end of Phase 2 (June 1990) to the start of Phase 3, AKF provided a bridging grant to support planning for Phase 3 and the continuation of SIP staff development and outreach activities. Phase 3 funding began in March 1991 and continued to June 1994. Here we provide an overview of the plan and implementation of Mzizima SIP Phase 3.

Phase 3 Plan

Phase 3 of the Mzizima SIP had six broad aims: (a) to sustain the accomplishments of Phases 1 and 2, (b) to establish a mechanism for continuous teacher development through institutionalization of the school improvement structures and processes, (c) to involve all subject areas and teachers, (d) to develop the leadership capacity of teachers to carry the process on their own, (e) to shift the major emphasis of SIP activity from curriculum development to improvement in teachers' pedagogical knowledge and skills, and (f) to continue efforts to share the SIP experience with other schools and institutions.

Teacher development objectives included improving teachers' ability to identify needs for improvement, their commitment to professional growth, subject matter knowledge, understanding of how children learn, and lesson planning and teaching methods. The plans gave ambitious, but nonspecific, direction for intended pedagogical changes: less teacher-centered instruction, less lecturing and note giving, less emphasis on rote learning, more interaction between teachers and students in the learning process, more active participation of students, more practical learning activities, more inquiry-oriented activities, more small-group work, and increased use of methods to encourage higher order cognitive outcomes.

The Phase 3 plan proposed a restructuring of roles, space, and time. The first proposal was to create a school position of deputy head teacher (professional development) to lead and coordinate SIP activities. The second was to group the subject departments into five faculties, led by faculty heads for the purposes of SIP curriculum and professional development activity. The five faculties were language (English, Kiswahili), mathematics, science (biology, chemistry, physics), social studies (history, geography, political education), and commerce (accounts, economics, typing). The plan included creation of an in-school Teacher Resource Center (TRC) in the area used as SIP project space. The TRC was to serve as a professional development site, and to house a professional library, materials production resources, and audiovisual resources.

During Phase 3, teachers in all five faculties were to be involved in weekly professional activity workshops led by the deputy head teacher (professional development) and faculty heads. The plan proposed several professional development strategies. Some were continuations from before, such as joint lesson planning and materials development in workshops, whole-staff seminars, and demonstrations of teaching methods. Others were new, including joint study of professional readings, peer observation and feedback on teaching (peer coaching), teacher research on SIP needs and effects (action research), external study tours (e.g., inter-school visits and university courses), and an induction process for teachers new to the school.

The Phase 3 proposal placed less emphasis on outreach. The plan was to continue the FIELOC initiative at existing level, and to continue inviting teachers from neighboring schools to attend workshops and seminars with no structured follow-up. Allsop recommended a moratorium on further outreach while teachers concentrated on improving their own teaching through school-based in-service training.

Phase 3 Implementation Process

From July 1990 to September 1991, the SIP operated under the head teacher and senior teachers, with occasional visits from the former English specialist and SIP coordinator to monitor progress and to assist with Phase 3 plans. Some Phase 3 proposals were put into place prior to formal approval of plans. Teachers were reorganized into five faculties for purposes of professional development. Faculty heads were appointed. (Department heads were retained for administrative and departmental matters.) Weekly half-day faculty workshops led by faculty heads and scheduled during normal working hours began. Steps were undertaken to create a Teacher Resource Center.

An experienced teacher and staff developer from Canada was appointed deputy head (professional development) in September 1991, with the expectation that the consultant would train someone on staff to take over the position by the end of Phase 3. The position was defined as a staff position, not as a special project position. The person occupying the position was expected to teach eight periods a week and to coordinate and facilitate SIP-related professional development and outreach activities.

Under the new deputy head teacher (professional development), implementation of the SIP Phase 3 plan began in earnest. The Teacher Resource Center was completed. Weekly workshops under the joint leadership of faculty heads and the deputy head teacher continued, though with activities focused on Phase 3 teacher development objectives. Monthly meetings of faculty heads, the two deputy head teachers and the head teacher were scheduled, to talk about SIP implementation. End-of-term 'reflection' meetings were introduced in the faculty workshops to enable teachers to participate in planning SIP activities. Plans for teachers to coach one another in the use of selected teaching methods were put into practice. Study tours were arranged for teacher leaders to visit Aga-Khan-supported schools in Kenya and Pakistan and to take courses related to SIP efforts in Britain, Canada, and Pakistan. Teachers began to undertake action research studies on school or classroom practices and concerns. Monthly Saturday seminars organized on a rotating basis by the five faculties were introduced. These were whole-staff development days when selected teachers and guests presented topics to other Mzizima School teachers and visitors from other schools. Pre-service teacher education links were established in order to offer a planned orientation and mentoring process for student teachers at Mzizima School. Use of the FIELOC program continued at Mzizima School, in nearby schools, and in southern Tanzania. Participation of non-Mzizima-School teachers in workshops and seminars proceeded on an invitational basis.

In February 1993, the head of the biology department was designated understudy for the deputy head teacher position. She succeeded the first deputy head (professional development) in September 1993. SIP activities continued under the internally appointed deputy head, the head teacher, and faculty heads. Funding for the position shifted to the school. The final months of Phase 3 (September 1993-June 1994) were a time for consolidation and continuation of SIP structures and processes. Some new ventures began to take shape. These included practice teaching arrangements with local teacher training institutions and a process for inducting newly appointed teachers into the school and into its expectations for teaching, professional development, professionalism and leadership. Mzizima School staff served as trainers in an AKES in-service project to upgrade the professional knowledge and skills of teachers at a new Aga Khan primary school.

Framework for Evaluation of Teacher Development and School Improvement

We used Fullan, Bennett, and Rolheiser-Bennett's (1990) framework for classroom and school improvement as an organizer for our analysis and evaluation of the SIP process and outcomes. One component of this framework identifies four focuses for teacher development in the classroom: content (e.g., knowledge of subject, of learning theory), classroom management, instructional skills (e.g., lesson planning, questioning), and instructional strategies (e.g., lecturing, cooperative learning). A second component targets four organizational focuses for school improvement: consensus on goals, norms of collegiality, norms of continuous improvement, and organizational structures to enable development. At the center of the framework, the concept of teacher-as-learner highlights the centrality of teachers' professional development to both classroom and school improvement. Teacher learning can focus on mastery of a repertoire of teaching methods, on learning to collaborate, on learning to do research on classroom and school-wide concerns, and on acquiring the attitudes and skills of reflective practice. Leadership for improvement is the fourth component of the framework. Any school or teacher development project might be directed toward one or a combination of focuses. In theory, the more components addressed and the closer the ties between them, the greater the chances for positive results in teacher growth, school improvement, and student learning.

Our evaluation begins with the elements of school culture and organization identified by Fullan et al. as essential to mobilizing and sustaining teacher learning, classroom improvement, and improved student outcomes. Then we consider SIP teacher development outcomes in terms of the 'teacher-as-learner' and 'classroom improvement' components of the Fullan et al. framework. Next we discuss the analysis of the SIP's impact on students. We end with our conclusions about the effectiveness of SIP outreach efforts.

Evaluation of School Improvement Structures and Processes

Organizational Structures for Teacher and School Development

Here we describe and analyze the SIP-related organizational strategies for ongoing teacher development. These include the workshops and seminars, peer coaching, action research, the Teacher Resource Center, teacher induction, and leadership development. We evaluate these innovations in light of project goals and research on educational change. Special attention is given to the potential for institutionalizing these structures and processes for classroom and school improvement. Institutionalization is more likely when (a) change is linked to local problems and priorities, (b) administrators are committed and provide support, (c) more and more staff take part in planning and implementation, (d) observable benefits are evident, (e) actions are taken to minimize the effects of staff turnover, (f) the change completes key organizational passages (e.g., shift to regular funding, incorporation into local policies), and (g) the change results in organizational structures that support continuation (Anderson & Stiegelbauer, 1994).

Faculty Workshops

Faculty workshops occurred on a weekly schedule during each school year of Phase 3, as follows: Monday – social studies; Tuesday – science; Wednesday – languages; Thursday – mathematics; and Friday – commerce. Teachers in each faculty shared a 3-hour block of workshop time on their designated day. The workshops were based in the TRC. Analysis of records, site visit interviews, and observations revealed five categories of workshop activity: presentations, subject content development, observation of teaching, SIP evaluation activities, and computer training.

Presentations consisted of individual teachers making oral presentations to their peers accompanied by handouts (e.g., journal or newspaper articles) on topics related to teaching methods, lesson planning, discipline, student evaluation, learning processes, and teacher development (e.g., teacher evaluation tools, action research). The presentations supported professional knowledge growth and teacher leadership, while also providing opportunities for teachers to improve their English language proficiency. Subject content development refers to teacher involvement in workshop activities focused on curriculum analysis, joint planning and critique of subject-specific lessons, the development of student assessment tools, the production of teaching and learning aids, discussion of methods for teaching specific topics, and analysis of student performance in the subjects identified with a particular faculty. Phase 3 workshop activity focused initially on general teaching and learning processes. A renewed emphasis on subject content development emerged in the latter half of Phase 3. During Phases 1 and 2, teachers sometimes observed or gave demonstration lessons highlighting the use of teaching methods and teaching/learning materials. Under the new deputy head teacher (professional development), observation of teaching became a more structured teacher development activity (see Peer Coaching below). In the workshops, teachers determined the purposes of teacher observation, developed observation and feedback tools, and practiced observing and providing feedback on teaching behaviors. Staff evaluation of the progress and results of the SIP teacher development activities was a recurring agenda item in Phase 3 faculty workshops. These sessions included discussion of project activities and plans, and questionnaires given to teachers. One thrust of in-service training during Phase 3 focused on computer literacy development. We interpreted the incorporation of computer training into the workshop agendas as an indicator of institutionalization of the workshop structure as a site for teacher development activities in which priorities might evolve over time.

Workshop implementation was not free of challenges. Identified concerns related to leadership, teacher socialization and commitment, perceived need and focus, practical benefits, balancing curriculum and teaching aims, and classroom follow-up. Faculty workshops provided an important setting for teacher leadership development. Faculty heads led the workshop sessions. The deputy head teacher (professional development) acted as a facilitator, providing leadership coaching behind the scenes, accessing professional resources (e.g., readings), contributing to discussions, mediating conflicts, and arranging amenities (e.g., refreshments). Opportunities for classroom teachers to take the lead in presentations and curriculum development activities contributed to leadership development. By the end of Phase 3, in-school teacher leaders had completely replaced the former dependence on external specialists in workshops.

Ownership of the workshop process was built up by involving teachers in planning workshop activities and by distributing responsibility for presentations and related events. Workshop commitment was developmental, as explained by one of the deputy head teachers: "With new staff it takes about a year or two for them to see how important the workshops are to them. Many of them don't see the point at the beginning … It takes time for people to see the faults in their teaching. Many of them don't even know what they don't know … questions about the meaning of education, reading habits, work orientation. It's entirely new to them." The head teacher's success in reducing teacher turnover and his expectations for participation of both new and experienced teachers (see Teacher Induction below) were critical factors contributing to teacher socialization and commitment to the workshops.

A cluster of concerns about the ongoing purpose, focus, and benefits of teacher workshops surfaced during Phase 3. These concerns were rooted in differences of opinion about the most effective approach to in-service teacher development. The emphasis in Phases 1 and 2 was on teacher growth through involvement in curriculum development activities with subject-matter peers. Talk about and demonstrations of changes in teaching methods tended to focus on subject-specific teaching objectives (i.e., how to teach the subject). In line with the Phase 3 plan, the new deputy head introduced an in-service-training approach to expanding teachers' mastery of generic instructional methods, such as cooperative group learning and effective questioning practices. By the end of Phase 3, a pattern evolved in which workshop sessions on 'generic' topics were followed by more faculty-specific activities. The revival of subject development sessions reflected the teachers' desire for time to address subject concerns, as well as general teaching methods and knowledge.

We concluded that the prospects for institutionalization of the faculty workshop as a structure and process for ongoing improvement were very positive. Workshop implementation was aided by the availability of space, adequate resources (e.g., furniture, furniture, supplies, and equipment), and by the administrative and collegial will to schedule workshops during the regular workday. The high degree of teacher participation was influenced by the support and involvement of the head teacher and senior teachers in the workshops, and by the perceived benefits of workshop participation commonly cited by teachers. These included sharing ideas, acquiring knowledge, improving teaching methods and materials, solving classroom problems, and greater collegiality. Workshop participation was viewed as a professional norm. Leadership had shifted from expatriate specialists to the deputy head teacher (professional development) and faculty heads. The deputy head teacher role was essential to workshop coordination, to leadership support for the faculty heads, and to keeping a school-wide perspective (beyond each faculty). The professional development intent of workshops was protected by reserving administrative matters for staff and subject department meetings. Costs for resources were within the school's normal operating budget. Box 1 depicts a workshop observed in our site visit.

Box 1: Faculty Workshop Vignette (November 1994)

Science Faculty Workshop. 11:15 a.m.-1:55 p.m. Teacher Resource Center seminar area. Eight science teachers and the deputy head teacher (professional development) are seated around the conference table. There are tea and cookies on the table. The science department head distributes a newspaper article, "K-12 Teachers Return to Learning," which describes a group of science teachers in the US learning and practicing new teaching methods. The department head asks each teacher to read one or two paragraphs out loud. At the end of the article, the department head invites them to "discuss what we got from the article. How does it relate to us at Mzizima?" One teacher compares the description of a science activity to experiments by Mzizima School students at a recent Science Olympics. Others note that they, like the teachers in the article, are learning to use new teaching methods by working together through group discussion, team teaching, and peer coaching. This leads to talk about changes between the way they teach now and the way they taught before, and from they way they were taught themselves. The faculty head distributes a second article, about learning outcomes, and the process is repeated. At the end of the first hour, the faculty head brings in four boys and four girls for a demonstration of a physics lesson he plans to teach the next day. The deputy head teacher distributes observation forms to the other faculty. The students sit around the conference table. The teacher stands by the blackboard. The observers sit in the rear of the seminar area making notes and tallies on their observation forms. The lesson takes 40 minutes, after which the students leave and the teachers resume their places around the conference table. The faculty head asks them to critique the session: "I can learn a lot. I expect to hear how else it could be taught." This invites feedback, discussion, and suggestions about the use of practical versus imaginary examples, clarity of lesson delivery and visual aids, teacher use of questions, question waiting time, strategies for enabling student discussion, and teacher-student rapport. The deputy head teacher invites the faculty head to do a self-evaluation of the lesson. The group evaluates the lesson in terms of the learning objectives stated in the lesson plan, based on observation of student responses during the lesson. For the final 10 minutes of the workshop, the group reads and briefly discusses a one-page case study of a slow learner.

The major issue for workshop continuation identified in our evaluation was that workshops appeared to have taken on a routine quality. There were established protocols for different activities (e.g., presentations, peer observation, lesson development), which teachers knew and followed. Teachers had become 'good' at 'doing workshops'; however, they were missing a sense of explicit goals at the school or faculty level that could provide workshop continuity and clear results over a term or year. We recommended that faculties identify specific workshop goals and projects for curriculum development, sustained in-service training on specific teaching methods, or joint action research.

Peer Coaching

Peer coaching is a teacher learning strategy that involves teachers engaging in mutual observation of and feedback about one another's teaching, as well as joint planning and problem-solving to support the transfer of new ideas and methods into practice (Joyce & Showers, 1980). Peer-coached teachers are more likely to transfer new methods into practice than those receiving only information, skills training, and demonstrations. Peer coaching experts stress the need for teachers to be trained in how to coach, as well as in the teaching methods that are the focus of coaching.

Some elements of peer coaching were already in place at Mzizima School, such as joint lesson development and demonstrations of teaching in workshops. The key change in Phase 3 was to introduce tools, expectations, and opportunities for peer observation and feedback. The hope was that this would lead to greater implementation of the ideas, materials, and methods initially introduced and discussed in the faculty workshops. The groundwork for peer observation was laid in the first year of the new deputy head teacher's appointment. This included maintaining the positive climate for professional growth, developing norms of collaboration among teachers new to the workshop process, providing in-service training on peer coaching, and developing an observation tool based on workshop discussion of desirable teaching behaviors. Peer observation began toward the end of 1992.

Peer coaching at Mzizima School took four forms. Three occurred as planned activities for all teachers in faculty workshops: (a) observation and critique of teaching in a regular classroom setting, (b) observation and critique of microteaching episodes with small groups of students in the TRC seminar area (Box 1), and (c) observation and critique of videotapes of teachers in regular classroom situations. Over 50 classroom teaching videos involving 35 teachers were prepared for viewing and discussion in workshops. Teachers were also urged to coach each other regularly outside the workshops.

Peer coaching in faculty workshops served the dual purpose of teaching the value and skills of peer coaching (e.g., observing, giving feedback, self-analysis), while supporting the use of teaching methods associated with SIP goals. Of 76 weekly workshop agendas from 1992 through 1994 on file, observation of teaching appeared 13 times and critique of teaching videos an additional 11 times. The SIP project files contained 113 completed classroom observation checklists associated with 24 observation episodes during two terms of 1994 (a single episode might involve several teachers observing the same peer). From these records and our teacher interviews, we concluded that peer observation and feedback were regular features of professional development activity at Mzizima School. The overall incidence of teacher-teacher observation was remarkable by any standard, though the extent of teacher participation within and across faculties was variable. Analysis of the 1994 observation files revealed that, whereas a large percentage of science, social studies, and commerce teachers participated as observers, only about half the mathematics teachers observed their peers, and peer observation in the language faculty was limited to Kiswahili teachers.

The institutionalization of peer observation and feedback, at least within the context of the workshops, was achieved despite a complex set of implementation issues. Three of these issues: norms of collaboration and improvement, administrative support, and a climate for risk-taking, are preconditions for successful peer coaching in any school. At Mzizima School, norms of teacher collaboration for school improvement were already in place as a result of teacher workshop and seminar participation in the SIP Phases 1 and 2. Administrative support for professional development was also well established. The head teacher and deputy head teachers not only endorsed peer coaching, they participated fully in it, allowing classroom teachers to observe and discuss their teaching, as well as observing and providing feedback to others. A key factor shaping the climate for professional growth was the separation of peer coaching from teacher evaluation.

Peer coaching is more likely to contribute to change in teaching practice when it focuses on specific behaviors, and when teachers have access to in-service training for the desired changes. The first observation checklist developed by Mzizima School teachers was too complex. After a few trials, the deputy head teacher and teachers simplified the checklist to focus explicitly on questioning strategies, with less detailed attention to other areas of practice in line with SIP goals (e.g., use of teaching aids, variety of teaching strategies). This decision was supported by periodic workshop sessions on questioning strategies.

The material and human resources required to enable peer coaching are not great. In developing countries, however, even modest resource requirements cannot be taken for granted. Production of observation forms, for example, requires paper, access to photocopiers, and money for supplies and service. When the photocopier at Mzizima School broke down, peer coaching went on hold. Videotaping was possible because there was money for equipment and materials and to arrange for an audio-visual technician. When the technician left, videotaping stopped. When teacher development strategies are dependent on initial and ongoing funding for resources, those costs, modest as they might be, must figure into plans for sustainability.

Teachers are not naturally skilled in observing and coaching one another. As explained by the deputy head teacher (professional development), "It takes six months to 1 year to feel confident about workshops and observation ... At first they can't gauge what their peers are thinking ... It's not easy to admit mistakes ... They are afraid of what you're going to say. And they don't have the habit of reflective thinking ... It takes time." Workshop-based practice in peer observation, feedback and self-critique provided an ongoing context for Mzizima School teachers to acquire peer coaching attitudes and skills. Efforts to encourage teachers to engage in peer coaching outside the workshop context at Mzizima School, however, were not particularly successful, despite scheduling nonteaching time into teachers' timetables, providing tools and practice, and communicating supervisor expectations for teachers to partner and observe each other on a weekly basis. Peer coaching takes time away from other things that teachers might be expected or want to do, such as planning, marking, extracurricular duties, and giving individual attention to students.

A final issue for peer coaching at Mzizima School relates to the purpose of peer coaching as a teacher learning strategy. Teachers uniformly talked about what they learned from observing others (ideas for practice as well as awareness of their own weaknesses), not about how they benefited from feedback from peers who had observed them. From our field notes of post-observation debriefings in workshops, it was evident that Mzizima School teachers had become skilled at critiquing each other's teaching practice. One can be an expert connoisseur of teaching, however, without being able to practice the methods being critiqued. The power of coaching as a tool to support personal objectives for growth in practice appeared to have slipped from view. One of our recommendations was for faculty heads and teachers to refocus the use of peer coaching to support the acquisition and transfer of specific teaching strategies or the refinement of teaching skills, aligned with evolving individual or group priorities for improvement.

Action Research

Action research involves teachers, individually or in small groups, conducting small-scale research studies on issues of practical importance to them in the classroom and school (Sagor, 1992). Research may be designed initially to increase understanding of issues as a basis for action. Ideally, research is also undertaken to study the implementation and outcomes of actions taken to address targeted issues.

In Phase 3, teachers were expected to engage in action research into the impact of their efforts to improve teaching and learning as a basis for future interventions. Plans were drafted for a school-wide study of change in teachers' questioning strategies with and without peer coaching. The plan was eventually abandoned because SIP priorities and activities were competing for time and attention. During the final year, an attempt was made to require all teachers to select a topic, implement a study, and present a written report of their study to other members of the staff. Assistance was limited to a few readings and presentations on action research in workshops, an outline for research design and reports, and an offer of assistance from the deputy head teacher (professional development).

By the end of Phase 3, only six individual studies and one departmental program review were complete. Teachers explained the delays in terms of lack of research skills, heavy teaching loads and limited time, and low motivation: "Teachers do action research because it's a requirement and not because they like to do it." They viewed the action research requirement as a one-shot expectation, not as the beginning of a cyclical process of professional inquiry and growth. While the studies completed or underway focused on issues of importance to the teachers concerned (e.g., why students arrive late, teacher opinions about field trips, student attitudes toward mathematics), they did not examine the impact of new teaching strategies or other SIP-related innovations, as originally intended.

We concluded that action research as practiced in Phase 3 had little impact on school and teacher development at Mzizima School, and that continuation of action research was unlikely. We recommended that, instead of individual research projects, the school could identify periodic research priorities and form teams of teachers to carry out the research. The projects might involve a mix of research to clarify issues of common concern (e.g., Mzizima School graduate destinations and success) and research into the results of classroom and school improvement initiatives. Individual action research is a legitimate vehicle for personal professional growth. Studies of effective and improving schools, however, suggest that collaborative inquiry holds greater potential for school-wide improvement (Joyce, Calhoun, & Hopkins, 1999; Sagor, 1992). In either case, teachers initially require ready access to technical assistance in school-based inquiry during the action research process (see Stuart, 1991; Kunje & Stuart; 1999; for action research in an African context). Access to research expertise and experience was missing at Mzizima School.

Teacher Resource Center

If school-based teacher development is to be effective, there must be places for teachers to interact professionally to further their own growth and school improvement. The Mzizima Teacher Resource Center (TRC) provided a school-based site and material resources (e.g., supplies, professional library) for the enactment of SIP in-service training activities. Effective use of the TRC, however, was dependent on creating professional time for teachers, on providing material resources for TRC in-service training activities, and on leadership in the use of the time and resources. Our evaluation focused on what was happening in the TRC, not on the TRC as a distinct school and teacher development innovation.

Teacher Induction Program

During the final year of Phase 3, a Teacher Induction Committee chaired by a faculty head, with assistance from the deputy head teacher (professional development), was created to manage the integration of newly hired teachers. As a private school, responsibility for teacher recruitment and induction fell to the head teacher and staff. The head teacher reported that recruitment of good teachers by word-of-mouth was more effective than advertisements, which attracted "job seekers" rather than "teachers who really want to work." Candidates for teaching positions were first observed in demonstration lessons by the head teacher or deputy head teachers and the relevant faculty and subject heads. They then met with the Induction Committee chair for an initial orientation to the school. Teachers offered a position remained on probation for 3 months before being offered a contract of 1 to 3 years. For the first 2 weeks, they observed experienced teachers in the classroom and participated in faculty workshops. During this period the Induction Committee provided a fuller orientation to school routines and expectations for professional duties, lesson planning, workshops, peer observation, and so on. Once new teachers began to teach, they were observed and coached by the deputy head teacher (professional development) and their faculty heads, and were encouraged to consult other teachers about teaching methods and concerns. After 3 months, the Induction Committee reported to the head teacher with a recommendation to hire or not. New teachers were exposed from the start to expectations for professional conduct, teaching, and growth established through the SIP. The deputy head teacher (professional development) described the experience of incoming teachers as follows:

> When teachers first come to the school they often only teach by lecture method … or writing notes on the board for students to copy. The chair of the Induction Committee and I work with them. We ask them to teach a few lessons and we observe. It's almost always the lecture method and maybe two or three questions that are just pure recall … no wait time … nothing. We ask them, "Is that the way you teach?" and they say, "Yeah, that's teaching." Then we tell them at Mzizima they'll have to teach differently. I talk to them about questioning techniques, activity methods, and small group learning. You can see that they don't see the point unless they have observed somebody else's lesson. We show them some of our teaching aids. Many have never even heard of anything like this at university, or that we have a teacher manual. Then they begin to realize how much they have not implemented … things they may have heard of but have not really understood.

Teachers quickly learned that talk about teaching and learning, seeking and giving advice about teaching, peer observation, and joint lesson planning were normal practice at Mzizima School. We concluded that the teacher induction process worked well in recruiting and socializing committed teachers to the norms for professional conduct and practice.

Teacher Leadership Development

An explicit goal of the SIP's Phase 3 was to wean the school from dependence on expatriate leadership for ongoing teacher and school development. Key leadership development strategies built into the SIP initiative included a school commitment to the hiring and promotion of Tanzanian teachers, the creation of new leadership roles (e.g., deputy head teacher (professional development), faculty heads, FIELOC coordinator, coordinator of teacher induction), the appointment of teachers to those roles, external professional development opportunities for current and potential leaders, mentoring of teachers appointed to new leadership positions, and opportunities through workshops and other SIP activities for teachers to demonstrate and develop leadership qualities and skills. The head teacher appointed at the outset of Phase 3 contributed significantly to the development of teacher leadership. He delegated school responsibilities, trusted in teachers to carry out those jobs, and supported the creation of new leadership roles.

One of the leadership development strategies carried out in Phase 3 was sending selected teachers on study tours, including 2-week visits to AKES secondary schools in Kenya, certificate courses in education at universities in Britain and Canada, and scholarships for a couple of teachers to complete master's degree at the Aga Khan University in Karachi and at the Ontario Institute for Studies in Education of the University of Toronto. A total of 19 teachers participated in study tours between 1991 and 1993. Teachers described their study tour experiences positively in terms of broadening professional awareness, sharing ideas with other colleagues, learning about alternative programs and organizational arrangements, and comparing Mzizima School practices to those of teachers and students elsewhere. The opportunity for study tours served as a powerful incentive to keep teachers at Mzizima School. The professional growth and incentive value of study tours notwithstanding, this staff development strategy was too expensive to be continued within the school budget, and was not replicable in most Tanzanian schools.

Our data led us to firmly conclude that the SIP initiative had contributed to the development of in-school teacher leadership for school improvement, and that a successful transition from dependence on external expertise (expatriate consultants) had occurred during the SIP's Phase 3. That said, there remained several issues related to continuation and to the potential for replication of the Mzizima SIP experience. Of these, the contributions of the head teacher appointed at the end of Phase 2 must be highlighted. The school governing authorities under Aga Khan Education Services recruited a highly experienced and effective head teacher to run the school. He successfully addressed a host of school-wide management and climate issues that had hindered the progress of improvement during Phases 1 and 2, particularly with regard to teacher conduct, hiring, student discipline, and parental support. The head teacher was not intimidated by the expatriate consultants, and was able to work in partnership with them to incorporate the SIP structures and processes into the ongoing culture and operations of the school. He was also very supportive of expanding teacher leadership and sharing decision-making. The successes of the Mzizima SIP were due in part to the presence and effectiveness of this head teacher. Unlike Aga-Khan-supported SIP projects in other areas of East Africa (Kisumu and Mombasa, Kenya; Kampala, Uganda), there was no urgent need to include head teacher 'management training' in the school improvement effort. Assumptions about and lack of attention to head teacher effectiveness was one of the major obstacles to the dissemination of Mzizima SIP interventions to other schools. Insomuch as the head teacher at Mzizima was approaching retirement, continuation and further development of the SIP structures and processes were in jeopardy unless explicit attention was given to this dimension of school management and leadership for change.

A second key issue for leadership at Mzizima School concerned the continuing and future relationship with external expertise and knowledge, to support ongoing school-based development. While teachers had confidently taken control of their own and the school's development, this did not mean that they had no further need for access to outside ideas and expertise. The cumulative wisdom of research on effective schools and school development indicates that the most successful schools do not isolate themselves in their own private visions and practices. Rather, they engage in strategic interaction with and contribute to outside sources of knowledge (personnel from other schools, university consultants, professional literature, etc.) in relation to school priorities for development and change (Hargreaves & Fullan, 1998). They also develop partnerships with outside entities (e.g., community, educational organizations) to strengthen in-school resources and leadership for change. This was the least well developed dimension of leadership for school improvement at Mzizima School. Once the expatriate consultants departed and funding for study tours reached its end, contacts and relationships with educational consultants and personnel from other interested schools declined. The persistent failure of local dissemination efforts to engage teachers from other schools in Mzizima SIP activities (besides the FIELOC program) contributed to a sense of isolation. We suggested that periodic input from external consultants with specifically targeted expertise was still valid in the context of school priorities for teacher development. We also recommended that future dissemination efforts be designed as in-country partnerships with schools pursuing similar aims for improvement, rather than unidirectional concentration on the transfer of expertise from Mzizima to other schools.

The final leadership issue identified in our evaluation concerned the status of and potential improvement in teachers' conditions of service: salaries and benefits, opportunities for career advancement, and further education, as illustrated in the following teacher interview excerpts:

> Teachers at Mzizima have gained a lot ... experience, methods ... The big question is how do we keep these teachers here?

> I remember asking about salaries and being told you're getting professional development not money. [That] is not enough. Professional development and working conditions are what is needed. The total welfare of the teacher and his [sic] family.

Teacher salaries and benefits at Mzizima School were generous in comparison with many private and public schools, but competition for teachers in the private school sector presented a threat to the long-term investment and outcomes of teacher development activity at the school. The induction process was able to replenish a modest degree of teacher turnover annually. Were a large number of staff to depart, however, past gains in professionalism and expertise would be seriously challenged. We recommended that the school governing authorities give renewed priority and attention to teacher concerns about salaries and other incentives.

Summary: School Improvement Structures
Phase 3 of the Mzizima SIP involved multiple interventions focused on teachers' professional development: the establishment of an in-school resource center; reorganization of staff into discipline-oriented faculties; the appointment of a deputy head teacher (professional development) and faculty heads; weekly faculty workshops involving all teachers; peer observation and coaching of teachers; action research; teacher leadership development, including external study tours; and an induction process for newly appointed teachers. From a 'process' point of view, the Phase 3 plan was implemented with a high degree of fidelity to the original proposal. Of the teacher development structures and processes carried over or initiated in Phase 3, the evidence for institutionalization was greatest for faculty workshops, peer coaching (in the context of faculty workshops), the new teacher induction process, the Teacher Resource Center, and the new leadership positions. Continuation of study tours was financially unsustainable. We recommended that further attempts at action research be undertaken collaboratively by teams of teachers within and across faculties, and that this research be focused on shared priorities for improvement, with access to research assistance.

Shared Goals for Improvement
The SIP developers, administrators and teachers expressed consensus on the broad aims of the Mzizima SIP, and on the evolution of those aims from Phases 1 and 2 to Phase 3. They characterized the initial phases of school improvement in terms of school-based curriculum development by teachers. The emphasis in Phase 3 shifted toward the school-based development of teachers' instructional competence, in order to support the implementation of the courses, lessons, and materials created through the SIP. The deliberate shift in emphasis was based on internal and external evaluations of SIP progress and outcomes. It reflected not only the consensus on broad goals for school improvement, but the practice of evolutionary planning associated with effective long-term school improvement (Louis & Miles, 1990; Fullan & Stiegelbauer, 1991).

Goal consensus notwithstanding, there was a distinct difference in how SIP goals were represented in earlier and later phases of change. This difference is captured in the distinction between a long-term vision for teaching, learning, and teacher professionalism at the school, and the more specific and time-bound goals or objectives that define school improvement activity within a specific phase of development. At Mzizima School, the long-term vision of creating a school characterized by sustainable organizational structures for continuous school-based development, shared teacher leadership for improvement, and a more student-centered, inquiry, and activity-based pedagogy in the classrooms was consistent and persistent over time. In Phases 1 and 2, this vision was translated into the development and enactment of tangible programs and projects: the FIELOC program, class libraries, schemes of work aligned with national syllabuses, and new lessons and teaching aids. In Phase 3, the outcome orientation of goals for improvement was less evident. Specific goals were set for the development, implementation, and institutionalization of new organizational structures and processes for teacher development: teacher leadership positions, peer coaching, action research, and teacher induction. What was to be accomplished through the enactment of those organizational innovations remained defined in terms of the broad vision for teaching and learning. While teachers were exposed to a wide variety of teaching methods in workshops, consistent with the overall vision for teaching and learning, the only sustained support for implementation was attention to teachers' questioning practices. We recommended that more explicit short-term goals for improvement consistent with the overall vision be defined through the process of collaborative action research, and through intensive in-service training on specific teaching strategies.

Teacher Collegiality

Little (1987) describes four indicators of a high degree of professional collegiality: teachers talk together about teaching practice, teachers jointly plan and solve problems related to their teaching, teachers learn together and teach one another, teachers observe and discuss each other's practice. She argues that, for teacher collegiality to exist and persist, two conditions are necessary: interdependence – teachers must believe that interaction with colleagues is essential to being an effective teacher; and opportunity – the school organization must provide ongoing opportunities and support for collaborative work. These indicators of collegiality and key enabling conditions were clearly present at Mzizima School. While our evaluation confirmed that school norms of collegiality had evolved, joint work occurred mainly among teachers within each faculty cluster. We suggested that collaborative work across faculties was unlikely to develop without a rationale and goals for curriculum integration and teaming, and without organizational arrangements to enable teachers to gather in cross-faculty work groups for specific tasks.

Norms of Continuous Improvement

Our analysis led us to conclude that the Mzizima SIP had succeeded in establishing organizational structures and processes that both foster and sustain professional norms and practices of continuous improvement. At the same time, however, we judged that the school improvement processes had become somewhat routine, without the clearly targeted outcomes for improvement that had characterized earlier phases of the SIP (hence our recommendations for short-term school-wide goals, joint action research, and in-service training initiatives). During Phase 3, the support system for change also became increasingly oriented toward an individual growth model, dependent on the will of each teacher to define and pursue specific goals for personal improvement in teaching, on an incremental basis. We argued for a balance between support for collective efforts focusing on shared priorities for school or faculty-wide improvement, and continued support for the more individual professional development interests and needs of teachers.

Teacher Development Outcomes

Here we examine SIP-related teacher outcomes in terms of the 'classroom improvement' and 'teacher-as-learner' components of the Fullan et al. framework. Our data included the results of a 1992 teacher survey about the impact of SIP activities on teachers' professional knowledge and skills, SIP classroom observation records, our 1994 teacher and student interviews, and our observations of classroom practice.

Teacher-as-Learner Outcomes

We considered SIP outcomes relating to four aspects of teachers-as-learners: collaboration skills, reflective practice skills, research skills, and professional growth orientation.

The 1992 survey showed a high degree of consensus about the positive impact of SIP activities on teacher collaboration. Improvements in teacher collaboration were linked to talk in faculty workshops about teaching strengths and weaknesses, and to peer observation and discussion of teaching. Our 1994 interviews and observations further confirmed the SIP effects on teachers' collaboration skills and attitudes.

> It has improved collaboration in my department. After meeting in workshops, the collegiality grows. Nobody's right or wrong. We're there to help each other. Somebody's weaknesses are not something to look down on.

> Sometimes difficulties with specific students. We come and ask other teachers what would you do in my place. If we have a topic we aren't comfortable with, we ask other teachers to demonstrate it. We learn from them. It's nice to work with other people. You can get ideas, instead of having no one to talk about your problems.

The evidence was consistent with research on teacher collegiality (Little, 1987; Rosenholtz, 1989) as it relates to teachers' professional development and confidence in their effectiveness.

The 1992 teacher survey included items related to the teacher research and reflective practice. About half the respondents ($N = 21$) agreed that SIP activities had helped them learn to do and present research on educational topics for workshops and seminars (action research was not yet underway). The development of reflective practice skills (e.g., from the survey, "ability to critically observe teaching, ... awareness of school strengths and weaknesses") was associated with peer observation and with participation in external study tours. Our 1994 teacher interviews confirmed the development of teacher capacity for reflective practice, especially through peer observation.

> When other teachers are doing microteaching, we learn a lot about our own weaknesses by observing teaching.

> Sometimes we lecture but don't explain. So when I've observed, I see how other teachers give examples or explain, and I take them back and try them in the classroom.

Teachers did not, however, indicate progress in the development of teacher researcher skills. This was consistent with the reported problems in implementing action research during Phase 3. We suggested that teacher capacity for reflective practice could be strengthened by reorienting peer coaching toward evolving individual and shared goals for growth in teaching practice. The focus on teacher questioning had not changed for 2 years, and our classroom observations suggested that many teachers had become reasonably skilled in this domain. We recommended that further action research be done collaboratively by teams of teachers within and across faculties, and that this research be focused on shared priorities for improvement, with adequate assistance for research design, implementation, analysis, and presentation.

Two thirds or more of the teachers responding to the 1992 survey reported positive changes in attitude toward school, personal standards of teaching, willingness to try new teaching methods, and belief in their capacity to become better teachers. Our 1994 teacher interview data highlighted three key dimensions of SIP impact on teacher professionalism at Mzizima School: (a) teachers' commitment to teaching, (b) teachers' confidence in themselves as teachers, and (c) teachers' belief in continuous learning as a defining characteristic of teacher professionalism.

> It has made me more aware. More of a teacher. Initially, I was looking for employment. After working at this school, I've found I'm really a professional teacher.

> Most of us have gained confidence in teaching, lesson notes, schemes of work.

> If we didn't have this project, I don't think I would have ever learned anything new, or learned my own weaknesses as a result of observing others' lessons.

The consensus on growth in teacher confidence in the classroom and in discussions with peers was remarkable. These outcomes are consistent with North American research that reports positive correlations among teacher efficacy, collegiality, commitment, and teacher learning (Rosenholtz, 1989). We concluded that the positive impact on teacher professionalism was not attributable to any one SIP innovation, rather to the high expectations for and evidence of participation in varied teacher development activities.

Classroom Improvement Outcomes

Staff development researchers describe different levels of outcomes as teachers learn new teaching methods: awareness and knowledge, skill acquisition, and transfer of learning into the classroom (Joyce & Showers, 1980). The 1992 survey data indicated that a large majority of Mzizima School teachers believed that they had gained more knowledge about specific teaching skills, including questioning techniques, use of teaching aids, and student evaluation. They reported a general increase in knowledge about different teaching strategies and specific growth in knowledge about alternatives to chalk-and-talk approaches to teaching, with an emphasis on more interaction with and among students. There was widespread agreement on knowledge growth in classroom management practices. In addition to knowledge growth, the 1992 survey included items about improvement in teaching skills. The data indicated improvement in lesson planning and preparation, questioning, and the ability to deal with problem students. About 40% to 50% of the respondents, however, reported no improvement in preparation and use of teaching aids (perhaps these respondents teachers who did not benefit from workshops in SIP Phases 1 and 2). The survey did not ask about teachers' actual use of more interactive teaching strategies. Our 1994 interview data confirmed the 1992 surveys regarding SIP-related improvements in teaching skills, such as preparation of lessons and teaching aids, use of teaching aids, presentation and questioning skills, and indicated explicit growth in teaching strategies from teacher-centered lectures to more interactive and student-centered methods.

> I've learned to prepare lessons before going to class.
>
> Use new methodologies. Use overhead projectors. Sometimes watch videos. I also use graphs and visual aids.
>
> Conducting a lesson in group work. I'm no longer afraid that noise means lack of control. Organizing classes with small group discussion, not just lecture.
>
> Conducting a lesson in groupwork, questioning techniques, wait time … and especially monitoring students' progress. If I see a student is not doing well, I can call the student and talk to him [sic] about what the reasons are.

We conducted two student focus group interviews (students in their fifth year at Mzizima School, students entering Form I). The students confirmed that Mzizima School teachers, on the whole, did not fit stereotyped images of Tanzanian teachers who simply transmit content by lecture, notes, or texts, and whose questioning is limited to checking for rote answers. The students portrayed teachers who asked questions for understanding and who encouraged learning through discussion. They described classrooms characterized by positive teacher-student relations. They reported variation, however, in the degree to which teachers employed more student-centered and interactive teaching methods.

SIP records of teacher peer observations in 1994 supported the reported uses of interactive teaching strategies. Most of the teachers used multiple strategies while observed. 'Discussion' was marked on all observations. About half the teacher observations in each faculty included 'groupwork' or 'demonstrations', or both. While these records were nonspecific and nonjudgmental about the form and quality of use of strategies such as groupwork and discussion, we interpreted these data as providing useful verification of the teacher self-reporting data from the 1992 survey and our interviews.

During our site visit, we observed 15 teachers selected for variation in faculty, experience, Form, years at Mzizima School, sex, and qualifications. Our observations provided a positive image of teaching practice relative to the vision of teaching and learning espoused in SIP goals. The teachers all provided written lesson plans for the lessons observed (teachers knew we were observing but were not told in advance which teachers would be observed and when). The lessons taught fit into schemes of work, building upon previous lessons and connected to future ones. Most teachers made use of teaching aids (e.g., overhead projector, poster, objects, diagrams on the blackboard) to help explain concepts. The predominant approach was one of teachers asking questions and students responding, interspersed with mini-lectures. Some were quite skilled in building a lesson through Socratic questioning. Others relied more on factual questions. Teachers did not take student understanding for granted. At a minimum, they asked if students understood. Many challenged students to explain responses orally or with visual aids (e.g., drawings on the blackboard). Most of the teachers made connections between abstract concepts in the lesson and the local context. A majority involved some or all students through demonstrations, work at the blackboard, role plays, or small-group activities. In-class use of textbooks and other written materials was minimal. There was little use of recitation, note copying, or worksheets. Most of classes were devoted entirely to engaged learning time between teachers and pupils. Teachers generally used positive classroom management, particularly in the way they encouraged and affirmed student participation.

We observed little student-student interaction in the lessons. The use of cooperative learning methods was limited to one regular class and a science lab. A few teachers talked about the difficulties of using small-group work due to short periods (45 minutes), the time needed to rearrange desks and put them back in rows for the next teacher (students stayed put while teachers rotated), and teacher concerns about noise. Discussions were mostly teacher-student exchanges. Whole- or small-group discussion among students was rare. We observed few examples of open-ended inquiry and discovery teaching methods. Student learning was mostly directed or guided toward preordained conclusions. Some teachers, however, demonstrated skill in the use of questioning strategies that enabled students to 'discover' the right answers.

We offered the following conclusions about improvement in teaching. The patterns of classroom practice, in terms of content, teaching skills, and classroom management, were consistent with or in the direction of the goals for teaching practice described in Mzizima SIP proposals. Teacher skill in implementing those practices varied among both veteran teachers and those new to the school. Teachers were more skilled at lecture and discussion than the one-way, chalk-and-talk images of teaching described in earlier project documents. The shift toward more student-centered interactive teaching was progressing incrementally, through the addition and refinement of teaching skills. While teacher awareness of alternative teaching strategies had increased, there was less progress in implementing these strategies (e.g., inquiry learning, cooperative group learning). This suggested a direction for future in-service training.

Our analysis of the SIP's impact on *change* in teaching was limited by the lack of data about prior teaching practices. Project plans, monitoring reports, and previous evaluations had made generalizations about teaching practice with little or no supporting data. Some used descriptions of teaching from other studies of African classrooms to represent typical teaching at Mzizima School. Those limitations notwithstanding, we offer some snapshots of the teaching we observed at Mzizima School in Box 2. The Mzizima School examples provide a more optimistic view of the potential for introducing alternatives to teacher-directed rote learning methods than other accounts of teaching and attempts to alter teaching methods in Africa (e.g., Arthur, 1998; Fuller & Snyder, 1991; Prophet, 1995; Rowell & Prophet, 1990; Kunje & Stuart, 1999).

Box 2: Vignettes of Teaching and Learning at Aga Khan Mzizima Secondary School

Science. Form I. Thirty-nine students (20 girls, 19 boys) seated at individual desks in six rows. The teacher (male) begins, "Last week we were looking at social insects ... Today, we are going to look at how insects have been successful." He writes on the board, "The success and importance of insects." Teacher displays a teacher-made poster that shows the proportion of all animals that are insect species. He uses Socratic questioning to explore the general question, "What makes insects successful?" Student: "By the number of species." Teacher: "What can you say about their success in terms of the number of species?" Student: "They are successful because they outnumber all other animals." The teacher paraphrases the response on the blackboard and continues, "What reasons can you give for insects outnumbering all animals?" Student: "They reproduce a lot." The teacher notes this on the board and invites students to suggest other reasons for insect success. No response. He waits, and provides a clue: "What do you know about insect bodies that might help explain success?" Student: "They have wings." Teacher: "Why does the ability to fly help success?" Student: "They can fly to find food." The class proceeds in the same questioning format, with the teacher inviting student ideas, challenging them to explain their ideas, and outlining key points as they emerge in the discussion. Students struggle with some questions. The teacher walks about the room, smiling and patiently calling on different students until someone offers an explanation. No one is told they are wrong, but only certain points get added to the outline. Students copy the outline into their notebooks. During the last few minutes of the lesson the teacher passes a microscope and slides around the room, which show how insects can camouflage themselves.

The bell rings. The teacher tells the students that next class they will be learning about why insects are important.

Social Studies. Form V. Geography class. Nine students (6 boys, 3 girls) and teacher (male) seated around four tables arranged in a rectangle. Students with textbooks open. The teacher reviews the past lesson on 'coastal landforms,' asking different students to name "erosion landforms" and "methods of erosion." Today's lesson is "to look at positional features of the coast, but first look at erosion transportation agencies on the coast." Student: "Waves." Teacher: "OK, waves are the most important. How does that work?" Student: "Waves move sand from the shore." Student: "By swash and backwash." The teacher asks the boy to draw a picture on the backboard and explain "swash and backwash." Teacher: "Any other agents?" Student: "What about tidal patterns?" Teacher: "How does that help?" One student tries to explain. Teacher: "Anybody else?" Silence. Students leaf through textbooks scanning for answers. The teacher goes to the board and says, "I think what happens is this", draws a diagram and explains how tidal currents can contribute to erosion. The teacher shifts discussion to the topic of coastal landforms. Students name them and he writes on the board, "barrier beach, coastal sand dunes, spits, bars … " Students scan textbooks for other examples. Teacher: "Now let's look at how these features are formed." He calls on a girl to explain how a beach is formed. Other students elaborate on her response. Teacher: "Where is the source of material for a beach?" Student: "Erosion." Teacher: "From where?" Student: "From along the coast." Teacher illustrates on the board. Teacher: "What's another source?" Student: "The sea bed." Teacher: "How is that possible?" Student: "From the wave action." Teacher: "How does this actually take place? How would you draw it?" The discussion proceeds through different coastal features in similar fashion. Two girls and one boy do not participate in the discussion and are not called upon. Discussion shifts in the last half hour to "human and economic aspects of the coast." Teacher: "Use examples of uses and misuses along our coast." Open discussion ensues. It ends up with the students concluding that some 'positive' uses by humans, such as land reclamation, may have 'negative' consequences for nature. The period ends. The teacher announces that they will look at two types of coast next time. Students ask about a field trip to the coast. The teacher promises to confirm the date and transportation with the head teacher.

Student Outcomes

As should now be evident, our evaluations of the organizational and teacher development outcomes of the Mzizima SIP were generally positive in terms of the results obtained and the potential for continuation. The ultimate impact of all those years of effort on student engagement and learning, however, remained naggingly uncertain. Attempts in Phases 1 and 2 to engage part-time research consultants from a local university to help study the quality and impact of SIP development activities on teachers and on Mzizima students did not work. Thoughts of implementing action research studies on the use and impact of new teaching methods were not carried out. No in-house system for monitoring and correlating student characteristics and retention to student performance on national examinations was set up. Thus, in the context of our evaluation, we had little basis from which to assess claims or to arrive at independent judgments of SIP effects on student learning. We examined aggregate data on Form IV student examination results from 1985 through 1993 from school files. The number and percentage of students scoring in the upper three passing quartiles (referred to as divisions) increased from about 30% to slightly more than 50%. Subject-by-subject rankings of Mzizima School students in comparison with students from all secondary schools in the country, from 1989 to 1993, showed improved performance in geography, English, physics, chemistry, biology, mathematics, commerce, and accounts. Relative performance in Kiswahili, history, and political education, however, declined. The major difficulty in correlating student exam results with the curriculum and teacher development accomplishments of the SIP was the lack of data on stability and change in student characteristics. Anecdotally, we heard that gradual improvements in the school's reputation (due in part to the high-profile SIP activities), coupled with a long-term crisis in the state of public education (e.g., teacher strikes, overcrowding, underfunding), had made the school more attractive to academically strong students than in the past. The analysis of student impact was further complicated by a reported increase in the proportion of students entering from English-medium schools. Such changes in the academic quality and linguistic proficiency of the student population could have contributed significantly to the improved examination results. These cautionary notes notwithstanding, it seemed reasonable to conclude that the changes in teaching methods, professionalism, student discipline, and school climate associated with the final years of the Mzizima SIP had contributed to the improved Form-IV examination results in most subjects. As one teacher explained, teacher retention, combined with SIP activity, positively affected student results because "each year we've been correcting and revising." Cumulative improvement is impossible in schools with high turnover and no processes for ongoing refinement of courses and lessons.

Outreach

Outreach in Phases 1 and 2 focused on dissemination of the FIELOC program and on invitations to nearby schools to send teachers to Mzizima SIP workshops and seminars. These activities continued in Phase 3 without much change, as the staff concentrated on their own teacher development at Mzizima School. New outreach initiatives were linked to pre-service teacher education programs and a primary-teacher training project.

FIELOC outreach efforts in Phase 3 were directed toward support for continuation of the program in participating schools, not on expanding to more schools. Major issues associated with FIELOC dissemination included funding, replacement and revision of materials, turnover and training of new teachers, coordination of FIELOC support, and competition from a government program. A Mzizima School English teacher was appointed FIELOC coordinator (in addition to his teaching duties) in Phase 3. His role included an annual, 2-day, FIELOC in-service workshop for teachers from Mzizima School and schools in Dar-es-Salaam, an orientation in-service session and a follow-up visit to support implementation in the southern Tanzanian schools. The SIP project continued to temporarily fund the coordinator's activities and in-service training costs. Another donor picked up the costs of reproducing the course materials. The Ministry of Education provided training sites in the south of Tanzania. Due to funding and timeline constraints, the program materials were not updated to reflect syllabus revisions postdating the previous printing. The possibility of shifting the production of the FIELOC program materials to a commercial publishing house was explored, but to no avail. The key factor affecting FIELOC dissemination in Phase 3 was that the Ministry of Education (aided by the British Council) established a network of secondary school Teacher Development Meeting Centers (TDMCs) and introduced its own Form I English language orientation course and text. Teacher participation in FIELOC in-service training sessions in Dar-es-Salaam and in the south declined, as the government program was introduced and supported through the new TDMCs. While continuation of FIELOC outreach seemed unlikely, the pioneering efforts of the Mzizima SIP to demonstrate the viability of a Form I English language orientation course had a major influence on the government's decision to develop its own program. On a separate front, the FIELOC project contributed to the mobilization of a similar program in Zanzibar (see Chapter 8).

Outside participation in Mzizima SIP workshops and seminars continued to decline in Phase 3. A single mathematics teacher from another school regularly took part in workshops. While 18 visitors attended the first Saturday seminar in 1993, by 1994 outside attendance at Mzizima SIP seminars had dwindled to a few teachers from two private schools. Various reasons were given by Mzizima staff and staff from other schools visited during our on-site visit, including inability of outreach schools to schedule release time for teachers to attend faculty workshops during the workday, lack of head teacher understanding and support for participation, lack of follow-up support in outreach schools, and lack of perceived need among teachers. We concluded that there was a fundamental flaw in the Mzizima SIP workshop-seminar outreach strategy. The Mzizima SIP was a whole-school initiative involving all teachers. The dissemination strategy, however, relied on sporadic contact with individual teachers, with no organized support for utilization in the outreach schools. Few of the in-school conditions that contributed to the beneficial teacher outcomes of Mzizima SIP in-service activities were present in the outreach schools.

A third approach to outreach initiated in Phase 3 involved links to pre-service teacher education programs at the University of Dar-es-Salaam and a local teachers college. Arrangements were made for a two-phase practice teaching experience, beginning with a 1-week orientation to Mzizima for interested student teachers and followed by an 8-week practice teaching block for about half those students. The orientation week included an introduction to the school, observation of mentor teachers, teaching with feedback from mentor teachers, attending faculty workshops, and an end-of-the week meeting to reflect on the experience. The intent was to promote Mzizima School as a demonstration site for progressive teaching and teacher development; to ensure that student teachers coming to Mzizima School understood the expectations for teacher conduct, practice, and learning; and to allow Mzizima School staff to identify promising teachers. After two cycles, the program reportedly worked well, though the number of students admitted for orientation was reduced from 20 to 10 to make it more manageable for Mzizima School staff.

During the final year of Phase 3, AKES initiated a separate in-service teacher development project for teachers from the newly established Aga Khan Primary School (AKPS). While this project was not part of the Mzizima SIP plan, it was designed to take advantage of teacher expertise developed through Mzizima SIP. Teachers from Mzizima School were asked to design and deliver in-service sessions to upper-grade-level teachers from the primary school. Teachers from the Aga Khan Nursery Schools in Tanzania were asked to do the same for the lower primary teachers. Seven Mzizima School teachers presented afternoon in-service sessions to AKPS teachers on child development, teaching styles, teaching methods (e.g., inquiry methods, lecture method), lesson planning, student evaluation, and error correction in English. It was too soon and beyond the scope of our evaluation to evaluate the impact of this primary teacher training venture. From the perspective of Mzizima School teachers, issues related to their involvement in such outreach activities included nonparticipation in decision-making about their role; lack of compensation for work beyond their normal job descriptions; loss of time for other professional and personal activities (e.g., after school-tutoring); and questions about the fit of their expertise with the needs of primary teachers.

Reflections on the Mzizima SIP

We conclude this case study with some reflections on the significance of the Mzizima SIP for school improvement in developing countries and elsewhere. To begin, we acknowledge that we have not emphasized the financial contributions of the Aga Khan Foundation and other donor agencies who contributed funds to the SIP project over its 9-year history. The financial resources provided through project funds cannot be discounted. In our view, however, the key teacher development strategies enacted were more dependent on the will and skill for change of those in leadership positions than on large financial and material investments. A timetable that allows teams of teachers to meet in 'workshops' on a regular basis during the school day, for example, can be created at minimal cost if teachers, school administrators, and relevant governing and supervisory authorities agree on the importance of time for collaborative professional work, and are committed to finding ways to make that time available. Other nontraditional strategies to support school-based teacher learning and change, such as peer coaching and collaborative action research, require access to appropriate training and expertise in the use of those strategies, at least initially; however, the use of those strategies on an ongoing basis is more dependent on user understanding, skill, and commitment to those processes of development than on inputs of money.

As a potential model for school-based improvement, one of the key lessons of the Mzizima SIP experience is the importance of establishing consensus on a multi-dimensional vision for school and teacher development. At Mzizima School the game plan for school improvement addressed curriculum, instructional methods, teacher learning, teacher leadership, school climate, and school organizational structures and policies. Efforts in each of these areas were undertaken in a coordinated and integrated manner over time. As noted in our analysis of teacher leadership development, the project benefited greatly in its final phase from the presence of an effective head teacher who was not recruited and trained as part of the SIP plan. Attention to head teacher quality and to head teacher management training would be an important added focus for leadership development if this SIP model were replicated in other schools. The same argument could be made for school governing authorities, whose long-term support for the Mzizima SIP remained understated, but was no doubt critical to sustaining the vision and process over time.

The long-term commitment to school improvement at Mzizima was remarkable. The story told in this chapter encompasses a decade of work by the project developers and school personnel. The 'news' that it can take 5 to 10 years to accomplish significant and sustainable school-wide change is not new, but is worth reiterating for education reformers at any level who might be seeking quick fixes for schools. In this vein, it seems important to highlight the incremental and evolutionary character of the Mzizima SIP. The intent from the start was that the SIP process should involve all teachers and students. However, the process was piloted initially with selected subunits (the English and mathematics departments) and then phased into other departments as school personnel learned how to effectively use the teacher development structures and practices associated with the SIP process, especially the teacher workshops. This phased-in approach turned out to be quite effective in terms of building capacity within the school to lead and sustain the SIP process. At the same time, we note that at several points in the Mzizima SIP story, the overall initiative was at risk of being redefined as something other than a whole-school improvement process. During Phases 1 and 2, for example, the SIP initiative was treated and seen by many as a temporary 'project' through which participating teachers jointly developed courses, lessons, and teaching and learning materials for use in class. This was particularly the case for mathematics and science teachers. A separate but concurrent thrust of SIP activity fostered a 'program' view of school improvement associated with the English-language-development class library and Form I English Language Orientation Course programs. It was really not until Phase 3 that the vision of SIP as a school-based structure and process for continuous school and teacher development stabilized. Within this structure and process, short-term projects and new programmatic ventures could be undertaken as focuses and products of the SIP in light of evolving needs and school priorities for development and change. In Phase 3, for example, the focus of teacher development shifted from curriculum development activity to more concentrated training and reflection on teaching practices, and to the specific short-term needs of teachers for computer literacy training. The distinctions between school improvement as a project, a program, and a process were particularly evident in the disappointing outcomes of Mzizima SIP outreach. The outreach strategies focused on dissemination of knowledge and products through outside teacher involvement in SIP project activities and SIP-developed programs, but neglected to tackle the more difficult and ultimately more important challenge of assisting in the mobilization and implementation of SIP structures and processes in participating schools.

In closing, we are mindful of the care taken in the Mzizima SIP experience to local capacity building and to the institutionalization of the organizational structures and capacity for continuous improvement. While there were infusions of outside funds and external expertise in the form of expatriate consultants during the developmental phases of the SIP initiative, the ultimate vision of developing the capacity within the school to continue on its own after the end of special project status remained a priority throughout. Institutionalization of the SIP process became a reality as a result of various strands of intervention, such as hiring and committing to the long-term development of local teachers rather than expatriate teachers, as in many Tanzanian private schools; extending SIP participation and benefits to all teachers; minimizing the attritional effects of teacher turnover through an organized process for teacher induction into school norms and practices; investing in school leadership development; incorporating SIP-related organizational structures and leadership positions into school policies and operating procedures; and anticipating and phasing out reliance on external funding over an extended period of time. We identified two key areas of potential concern for the future vitality and sustainability of the Mzizima SIP process. One was the apparent need to establish clear short-term focuses for improvement within the broader vision and process for school and teacher development. The second was a need to anticipate the loss of the head teacher, whose administrative skills, support and leadership created such a positive managerial context for school improvement work to proceed and to flourish over time.

The Mzizima SIP experience presents a positive portrayal of the potential for school improvement in developing countries. The story offers insights and lessons for school reformers that may be applicable to both public- and private-sector schooling worldwide. At the same time, it remains the story of a single school. The difficulties of scaling up the process of improvement from a single school to an entire school system were not very effectively addressed in the design and implementation of outreach through the Mzizima SIP project. The Mzizima SIP experience, however, led project planners and developers within the Aga Khan Foundation and Aga Khan Education Services to approach other SIP projects in East Africa less from a lighthouse or demonstration school basis, and to begin experimenting with SIP designs that involved long-term development in expanding clusters and systems of schools within specific educational jurisdictions.

4

Conflict Between National Curriculum Standards and Efforts to Improve Teaching

Joanne Capper, Shelom Nderitu and Paul Ogula

This evaluation study was commissioned by the Aga Khan Foundation (AKF) and the Institute for Policy Research to evaluate the effectiveness, cost-effectiveness and likelihood of sustainability of the second phase of the Kisumu School Improvement Program (SIP), operating in western Kenya. AKF's goals for this study were (a) to evaluate the extent to which the project was implemented as intended, (b) whether it had an impact on changing teachers' behaviors toward a more child-centered approach, (c) whether the change in behaviors resulted in any improvement in student learning, (d) whether the cost per teacher trained was reduced, and (e) to estimate the likelihood that the project would be sustained when handed over to the local government. To measure the extent to which these goals were achieved, tests were administered to pupils in Standards (grades) III and VI; classrooms were observed; and interviews were conducted with teachers, head teachers, school staff, municipal government staff, parents, community members, pupils and project staff.

Project Description

The Kisumu School Improvement Program - Aga Khan (SIPAK or SIP) began in January 1990 and was designed to improve the quality of teaching and learning in primary schools in the Kisumu Municipality by promoting the adoption of child-centered teaching. Child-centered teaching was defined by the project as increased levels of student participation in activity-based learning and group discussions, including increased use of materials to support learning and encouragement of student questioning. The project's strategy involved training teachers, providing classroom-based coaching and support to teachers as they attempted to integrate what they learned into their day-to-day teaching behaviors, and providing instructional materials. The SIP's emphasis on child-centered methods was consistent with the goals advocated by the Kenya Ministry of Education (MOE) and the pre-service approach of the Kenyan teacher training colleges, but not consistent with the format of end-of-primary-cycle examinations, which are a major influence in Kenya's school system.

The project began in response to declining scores on the national primary school leaving examination, which leads to the Kenya Certificate of Primary Education (KCPE). In 1988, the Municipal Education Office (MEO) in Kisumu commissioned a survey of the primary schools in the municipality to try to identify the reasons for the decline. The results indicated several contributors: (a) instruction was dominated by teachers who placed an emphasis on choral responses and rote memorization and offered students little opportunity to discuss or interact, (b) group work was infrequent and there were no opportunities for pupils to engage in discovery or problem-solving, (c) there was a shortage of textbooks and other materials in the classrooms, (d) the classroom culture was dominated by examinations and completion of the national curriculum, (e) physical facilities such as classrooms and desks were in short supply, (f) teacher morale was low, (g) pupil-teacher ratios were high, (h) school management was inadequate and ineffective, (i) parents and communities had little commitment to raise the funds needed to operate schools, and (j) parents and other local leaders were not involved in schooling nor did they understand 'modern' educational methods.

The SIP was designed to counteract many of these identified shortcomings and focused on (a) involving teachers in developing curriculum materials for their own use in the classroom, and (b) training and coaching teachers in new instructional skills and approaches within the context of their own classrooms, with a particular emphasis on child-centered teaching, activity-based learning and students working in small groups.

Other areas of emphasis emerged during the project, such as staff training for the extended school community, including parents, inspectors and tutors who worked in Teacher Assistance Centers (TACs) established by the Kenya Ministry of Education. At the time of the study (May 1996), the SIP had been in operation for six years. Phase 1 was from 1990 through the end of 1992, and Phase 2 from early 1993 to August 1996.

Phase 1

The SIP worked with 15 schools during Phase 1, beginning with teachers working in Standards I to IV, but dropping Standard IV after 1 year, since teachers began to specialize at that level. They began with three schools, adding three in the second year and nine in the third. One of the first selections was a high-cost school, another had fairly high-quality facilities but a fee structure typical of the low-cost schools, which made it able to accommodate more low-income parents.[6] Selection factors included semi-permanent buildings with mud walls, a positive attitude toward the project goals, proximity and accessibility of the school to the SIP project headquarters, community support and involvement, need for improvement, and school security to protect the materials that would be introduced to the classrooms through the project.

An evaluation was conducted at the end of Phase 1 and concluded that there was evidence of substantial change in classroom practice but that teachers' use of individualization and groupwork was limited. The evaluators also noted that the project was not encouraging sufficient depth in teachers' thinking about the teaching methods encouraged by the SIP, was not providing sufficient attention to the school as a whole, and was not providing adequate training, guidance or support for the role of the head teacher as instructional leader.

Phase 2

Many of the recommendations of the Phase 1 evaluation were incorporated into the second phase of the SIP. During Phase 2, the project focused on teachers in both the lower and upper primary levels, including Standards I through VI. Each academic term, four new schools were selected for focused attention. During that term (4 months), the SIP project staff, called program officers (POs) worked with teachers on a daily basis, but at the end of the term, this attention was reduced substantially, as four new schools became the focus of service. The earlier schools received intermittent assistance and support depending on their needs, requests, and time available.

Phase 2 schools were selected based on several factors, including positive head teacher and teacher attitudes toward the program, secure classrooms to house learning materials, adequate desks, and zonal representation. Schools that did not have lockable doors and windows or sufficient desks were required to obtain them as a requirement for participation.

New leadership was brought in to guide the SIP during the second phase of the project, and, under this leadership, the SIP expanded, extending into the upper grades, out to parents, community, MEO staff (including inspectors and Teacher Assistance Center tutors), head teachers, local universities and teacher training colleges. During the 3 years of Phase 1, SIP staff worked in 15 schools, but during Phase 2, 27 new schools were served, with occasional visits to the original 15, for a total of 42 schools served. In addition, teachers and head teachers from a number of other schools attended open SIP workshops.

Phase 2 introduced the role of school coordinators. Coordinators were teachers in the target schools, selected by each school's head teacher as potential leaders. Coordinators served as resident experts, working alongside their colleagues in their own classrooms. They received additional, specialized training to help them assist teachers in implementing SIP activities once the project team moved on to other schools.

Workshops for teachers included child-centered teaching and questioning strategies, teaching of reading and Kiswahili, small-group work, curriculum interpretation, art and crafts, and use of learning centers. Many of these workshops served two purposes: (a) training the SIP staff, and (b) training the range of other educators within the municipality, including teachers, head teachers, school coordinators, inspectors, and TAC tutors. The workshops were followed by classroom coaching that included teacher observations, discussions and demonstrations by the SIP program officers. Inspectors and TAC tutors were offered special workshops on their role as staff developers and advisors. School coordinators were offered courses in syllabus interpretation, writing schemes of work, leadership, classroom management and organization, and peer coaching. School committees and parent-teacher associations were offered training in school management, bookkeeping, budgeting, and the role of the school committee.

As of May 1996, the SIP reported having offered 47 workshops during Phase 2, with an overall attendance of 1,742. It is important to note that attendance at these workshops was voluntary, and therefore the number and type of workshops attended by teachers (or head teachers) within a school varied considerably. There was no guarantee that all teachers within or across SIP schools received the same amount or type of training.

In early 1994, SIP began offering training and technical assistance to teachers at the upper primary level (Standards IV-VI). At the time of the evaluation, upper primary teachers in 13 SIP schools had received training and coaching assistance. Workshop topics included questioning strategies, pupil assessment, and interpretation and use of examination results.

Training and technical assistance was not typically offered to an entire school staff, although, if staff requested assistance or the PO felt there was a specific school-wide need, school-based sessions were offered. In this regard, the title of the project is a misnomer. The focus of improvement was much more on the individual teacher than on the school as a whole. 'Teacher Development Project' would have been more apt. From late 1994 onward, a greater focus was put on the training of head teachers. Workshops were offered in teacher evaluation, school management and organization, bookkeeping and funds management, leadership, approaches to teaching and learning, qualities of an effective school, and syllabus interpretation. The inspectors and TAC tutors were encouraged to provide follow-up support to head teachers.

Before bringing a school into the SIP, the SIP staff met with each school's PTA/school committee to explain the purpose of the SIP and to encourage financial and academic support. They also facilitated parents' visits to other SIP schools to see the impact of SIP services.

Steps were taken to ensure that SIP efforts would be sustained. Increasingly, responsibilities for SIP activities involved or were handed over to the MEO, including the planning of workshop sessions, conducting of workshops, replacing, upgrading, and training inspectors and TAC tutors, and the creation of a revised reporting format for inspectors that was more consistent with SIP instructional strategies.

In addition, a system was established to place the procurement of instructional materials in the hands of the schools and the MOE office. A central feature of the SIP was to provide SIP teachers with classroom materials (generally raw materials) and work with teachers to show them how to construct and use the materials as learning aids and learning centers. Beginning in 1995, each parent was asked to contribute Ksh 50 (US $1) per child per year for the purchase of consumable materials such as manila sheets, newsprint, glue and other items that the SIP supplied. This money went into a central municipal fund to allow the purchase of materials in bulk at reduced costs. Materials were then distributed to the schools based on the amount of money submitted.

Study Design

Overall Design

A treatment-group/control-group design was used, with three levels of SIP intervention forming three treatment groups, and a control group drawn from a nearby district (since SIP staff reported that all schools within Kisumu had been impacted by the SIP in one way or another). In addition, extensive qualitative data were collected from teachers, head teachers, parents and pupils.

Procedures

During the last week of May 1996, local data collection staff were hired and trained. The instruments were pilot tested and revised. In the month of June, each of the 17 sample schools was visited by a team of two senior researchers from about 7:45 a.m. until mid-afternoon. (Only 15 of these schools were used in the final analysis). Although all SIP schools had been notified that an evaluation team would be collecting data during the month of June, none knew if they were in the sample, nor when the site visit would be conducted, to ensure that teachers did not prepare lessons that were atypical of their day-to-day behavior.

Each day began with a brief meeting with the head teacher to indicate the purpose of our visit and to schedule the day's activities. In each school, one Standard III and one Standard VI teacher was observed typically for two class periods each, about 75 minutes. All of the Standard III (treatment) teachers were SIP trained, with the exception of two schools where the Standard III teacher had been recently transferred. In the latter schools, we observed a Standard II SIP-trained teacher, but tested the pupils of the former Standard III SIP teacher. On the first day of the last week of the data-collection effort, tests were administered to pupils in all sample classrooms.

Description of Treatment and Control Groups

Eleven treatment schools were drawn from SIP schools located in the Kisumu Municipality. However, it was not possible to find control schools within Kisumu since all schools had in some way been impacted by the SIP, especially with the high rate of teacher and head teacher transfers and the fact that SIP workshops were open to all who chose to attend. Although a number of schools had recently been added to the Kisumu district, they all were rural schools and so were determined to be less similar to SIP's urban and peri-urban schools than those in the nearby district of Siaya. Each of these communities is described below.

The Kisumu Municipality is located on the edge of Lake Victoria on Kenya's border with Uganda. It is the largest town in Kenya and scheduled to become a city. It is the headquarters of Nyanza Province, the commercial center of Western Kenya. In 1994, the population of Kisumu Municipality was about 600,000, the majority belonging to the Luo ethnic group, with about 3,000 Southeast Asians. Several industries are located in Kisumu, including baking, fishing, brewing, pharmaceuticals, and agricultural processing. A large number of the population are self-employed through fishing, agriculture, transportation, trading, bicycle and motor vehicle repair, and tailoring.

Siaya District is a 1-hour drive from Kisumu. It has a population of about 2,000, primarily from the Luo ethnic group. It is primarily a peri-urban and rural district, with 18 schools. The area is very dry, with little rain, and so does not support farming. Businesses include trading, bicycle and motor vehicle repair, tailoring, pottery, and blacksmithing. In the Siaya sample, all four of the control schools selected were urban or peri-urban. Two were selected to be compared with the top schools in Kisumu (among the top hundred in the country) on the primary school leaving examination leading to the Kenya Certificate of Primary Education (KCPE). The other two were taken to be average and therefore comparable with the average schools in Kisumu. Table 1 shows that, while the control district had higher repetition rates than did the treatment district, it had lower dropout rates.

Table 1: Primary-Level Repetition and Dropout Rates, 1994-1995

Location	Repetition rate	Dropout rate
Kisumu	10.2%	5.8%
Siaya	16.8	4.5
National	15.4	5.4

Note. From Kenya Ministry of Education, Statistics Section.

Selection and Sample

Treatment schools selected for study were chosen using several factors, but primarily length of time in the SIP. Other factors included whether the teachers and head teachers who were originally trained at a school were still working at that school, head teacher longevity at the school, and whether the treatment had been offered to both lower and upper primary teachers. In addition, the socioeconomic and geographic status of the schools was considered.

The first selection cut was based on length of time a school could have been exposed to SIP influence. Since SIP had been operating for six years, schools that had participated in the program during the early years had had more opportunity to engage in SIP-sponsored workshops and to integrate the intended teaching.

T1: Group T1 represents schools that had the potential for the greatest amount of influence by SIP, their teachers having initially received training and coaching between 1990 and 1992 during Phase 1, with possible participation in subsequent SIP workshops throughout Phase 2. This category provided the opportunity to assess the durability of the SIP treatment.

T2: Schools in Group T2 began their SIP involvement in the early stages of Phase 2 (September 1993), thereby having a 3-year period to attend additional workshops and for the training and coaching to be internalized.

T3: Group T3 schools had only participated in SIP for 4½ months as of early May 1996, although some of their teachers or head teachers may have attended SIP-sponsored workshops in the preceding years.

To summarize, all treatment schools had 4 months of intensive classroom-based coaching, but varying workshop attendance and opportunities for intermittent follow-up depending upon when they began the program. The sample categories then are:

T1 Phase 1 – initiated in 1990 (4 schools)
T2 Phase 2 – initiated in 1993 (4 schools)
T3 Phase 2 – initiated in early 1996 – 4½ months of treatment (3 schools)
C Control group – Siaya (4 schools)

Within Standards III and VI, selection of classrooms was based on teachers' participation in SIP and, if there were more than two SIP teachers at a school, teachers who were teaching English or mathematics at times compatible with the site visits.

Location: Most of the schools that were involved with the SIP were urban or peri-urban. To maintain consistency across the three treatment groups, only urban or peri-urban schools were selected for study. This is one of the reasons that Siaya was selected to serve as a control site, since it is a peri-urban area. All four of the control schools were peri-urban schools and, therefore, more similar to the treatment schools than rural schools would be in Kisumu.

One school in each of Groups T1 and T2 was a high-cost school, and analyses reported later in this chapter were conducted both with and without these schools to assess the impact of a more well-endowed school facility and parent population. The SIP trained teachers at the upper primary level in two of the T1 schools, all of the T2 schools, and two of the T3 schools. All schools had head teachers who had been working in those schools for at least one year, most for several years.

Instruments

The following data were collected in all 15 schools described above (11 treatment and 4 Siaya control schools). All instruments were developed by the three senior members of the evaluation team and were pilot tested and revised.

Classroom observations: This instrument focused primarily on teacher and pupil behaviors associated with classroom management and child-centered teaching and learning. Although there is considerable variability in definitions and views of child-centered learning, the instrument was developed to document teacher use of the behaviors conveyed in the SIP training. Therefore, the SIP training manual, which consisted primarily of a collection of handouts provided at the various workshops, was the major source used to determine the behaviors to be observed. In addition, the observers documented the language of instruction, the language of pupils' responses, the types of questions teachers asked and their responses to students' answers.

The observers took notes on the lessons and, at the end of the observation, provided an overall rating and description of the lessons, some of which are described later in this chapter.

Household survey: A survey of households that had been used in several other recent studies was modified for use in the Kisumu context. This instrument was intended to capture estimates of family income, family education levels, familiarity with and attitudes toward the SIP program, and whether parents may have sent their children to SIP schools because they were considered better.

Pupil interviews and questionnaire: The pupil questionnaire consisted of 33 items about the nursery school the child attended, language(s) spoken in various settings, school climate, locus of control, homework, reading, and TV watching. Fifteen pupils from each of the Standard III classrooms were interviewed in their local language using the questionnaire, while all of the pupils in the Standard VI classrooms observed were given the questionnaire to complete in a group setting after the classroom observations had been completed. One of the observers was available to answer questions in the local language.

School staff focus group: A focus-group interview protocol was developed for use with school staff. Our interest was in trying to gauge the impact that the SIP had on schools as a whole and on staffs' views of SIP. We also had an interest in the role that head teachers played as instructional leaders, so staff were asked about frequency of staff meetings, topics discussed at those meetings, frequency of head teachers' visits to classrooms, and changes in the frequency and topics of formal and informal staff discussion about issues of teaching and learning.

Head teacher questionnaire, and teacher and head teacher interview protocols: The teacher and head teacher instruments were intended to capture views and perceived impact of the SIP. Instruments also included questions about parent involvement, views of the curriculum, language of instruction, teacher and head teacher academic and professional qualifications, perceived locus of control, and perceptions and definitions of child-centered teaching and learning.

School climate checklist: Indicators of an effective school not captured through the instruments described above were included on the checklist, to capture indicators such as running water, cleanliness and attractiveness of school grounds and classrooms, and number of teachers absent.

Unstructured interviews: Unstructured interviews were held with the SIP staff, with the Municipal Education Officer (MEO), the Deputy MEO, and several TAC tutors.

Tests: Tests were administered to Standard III pupils in English and mathematics and to Standard VI pupils in English, mathematics, and Kiswahili and were designed to be consistent with the Kenyan national curriculum. One open-ended test item was added for each of the subjects and grade levels, primarily designed to assess whether students were able to read simple passages and write coherent responses, and whether they were able to apply their computational skills to simple but real-life problems.

The Standard III English and mathematics tests consisted of two parts: (a) short-answer and matching items, and (b) one or two open-ended passages or questions. The English short-answer test measured picture recognition; knowledge of plurals, word endings, antonyms, and the past tense; use of 'a' or 'an'; and selection of the correct word to complete a sentence. The mathematics test measured basic addition, subtraction, multiplication, and division skills; use of fractions; and ability to tell time in 5-minute intervals.

The open-ended passages had two purposes: to show how tests can be designed to support child-centered teaching and to measure the more central, integrated skills within each of the two subjects. The English-language open-ended passage was intended primarily to test pupils' ability to read and write English. It used simple words to account for the fact that English was not the pupils' mother tongue and was intended to show how practice tests could promote the teaching of writing, encourage creativity and imagination, and validate children's views and feelings. It was also intended to convey how a writing exercise could be fun for children – all important aspects of child-centered teaching. Below is the open-ended English passage:

> *Read the story and answer the question. Write as much as you can.*
> *Birds can fly. If you could fly, where would you go? What would you do when you got there? Draw a picture of you flying.*

The two open-ended mathematics questions measured whether pupils could communicate and represent mathematical concepts, whether they understood basic mathematical operations when represented in a word problem, and whether they understood Kenyan currency. Questions were written to 'connect' to children's lives. The open-ended questions were:

> *Draw a picture to show your friend how many 10-cent coins are in a shilling.*
> *Draw a picture that shows that you have three more sweets than your brother. Your brother has six sweets.*

Standard VI - English. The Standard VI English and mathematics tests used in this study had been developed and used in the evaluation of a primary school improvement intervention in Busia, a town about 2 hours' drive from Kisumu. Both were based on the national curriculum and each consisted of two parts: (a) a multiple-choice test of 50 items, and (b) an open-ended question or passage.

The open-ended mathematics question was added by the SIP evaluators and had three parts, although we later discovered some problems with the question:

> *Pretend that you are going to plant a garden of maize. The garden is 3 meters long by 3 meters wide. You have 15 maize seeds and each maize plant will need 1 meter of space to grow.*
> *Draw a picture to show the garden plot and how you would arrange the maize seeds so that each one has 1 meter of space.*
> *Will you be able to fit all 15 maize seeds into the garden plot?*
> *If not, how many will be left out?*

Standard VI - Kiswahili. No multiple-choice or short-answer questions were given in Kiswahili, only an open-ended question designed to measure whether the pupils could read a simple passage in Kiswahili and write a response in Kiswahili.

The Kiswahili markers were trained for the Kiswahili portion of the KCPE examination. Initially, they proposed to use the scheme for marking English and Kiswahili compositions in the KCPE examination, which consisted of the following scales and point allocations: reading - 2 points; comprehension - 2 points; writing, spelling, punctuation relevancy, vocabulary, and grammar - 6 points; total points possible - 10.

The marking scheme finally used was quite different and much more detailed. It deliberately illustrated its possible instructional value in communicating the strengths and weaknesses pupils have in writing and in suggesting to teachers ways to improve the teaching of writing.

Table 2 is a summary of the instruments used in this study, the target groups interviewed, observed or tested, and the sample size for each group.

Table 2: Summary of Instruments, Target Groups and Sample Sizes

Instrument	Target group	Sample size
Classroom observation	Standard III and VI teachers	30[a]
Teacher interviews	Standard III and VI teachers	30
Head teacher interviews	Head teachers	15
Head teacher questionnaires	Head teachers	15
Parent interviews	Parents of pupils tested	510
Pupil interviews (Std. III)	Pupils in Standard III sample	255
Pupil questionnaire (Std. VI)	Pupils in Standard VI sample	634
School committee interviews	School committee representatives	15
Focus groups	School staff	15
English and math tests (Std. III)	Pupils in Standard III sample	617
English, Kiswahili and math tests (Std. VI)	Pupils in Standard VI sample	610

[a] Data were collected for two non-SIP but high-workshop-attendance schools. Ultimately, these schools were not included in the analysis.

Findings

The SIP's Impact on Student Learning

Short-Answer Items

An analysis of variance (ANOVA) was conducted to detect simple differences in test scores across groups. At the time the tests were given in mid-June 1996, students had received about 5 months of instruction, so one would expect that they had not mastered all of the content for that year. Two of the treatment groups (T1 and T2) contained high-cost schools, but no high-cost schools were in the T3 or control groups. It is likely that the parents of pupils at the high-cost schools have higher incomes and educational levels, may be more likely to read to their children, help them with homework, have books and magazines in the home, and provide textbooks and other instructional materials for their children. It is also more likely that classes in these schools are smaller and have other instructional inputs that may enhance learning. All of these factors related to attending a high-cost school may serve to increase test scores beyond those of low-cost schools and beyond what the SIP may have contributed, so the reporting of test results compares the treatment and control schools with and without the two high-cost schools.

Across subjects, pupils in T1 schools performed significantly better than all other groups or better than T3 and control groups, followed by pupils in T2 or control schools, or both. The T3 schools (those that have been in the SIP for only 4-5 months) performed less well than the other treatment schools and almost always less well than the control schools (Table 3). However, one of the T3 schools achieved a mean of only 8 points on the English short-answer test, bringing the average score down substantially for T3. The other two schools in T3 scored means of 41 and 71 points on this test. If the low-scoring school's score was removed from the analysis, the mean for T3 would be 55, higher than the control group mean but still lower than the means for T1 and T2 on this test.

One explanation for the superior performance of the T1 (Phase 1) schools is that the teachers continued attending SIP workshops during Phase 2, and so received more training and reinforcement than did Phase 2 teachers. SIP records show that teachers in three of the four T1 schools attended 115 workshops during Phase 2.

Standard III students performed reasonably well on the short-answer tests (means of 60 percent in English and 52 percent in mathematics), suggesting that they are learning curriculum content in the way it is presented in the curriculum and textbooks. Scores on all subjects and at both grade levels were somewhat lower in T1 schools when the high-cost schools were not in the analysis (2-5 points), and substantially lower in T2 schools (6-11 points), with the exception of Kiswahili in T1 schools. The range of mean scores across groups was much narrower for the mathematics tests (15 points) than for the English tests (34 points).

Table 3: Summary of Test Scores by Class, Subject and Question Type With and Without High-Cost Schools, Mean Percentage Correct and Standard Deviations [a]

Test	T1	T2	T3	Control	Total mean
		Standard III (Number of pupils)			
English SA (617)	79 (17) [c]	64 (24) [b]	45 (29)	51 (26)	60 (27)
w/o HCS (588)	76 (16) [c]	53 (21) [a]			52 (29)
English OE (601)	28 (24) [c]	9 (15)	4 (8)	12 (13) [a]	13 (18)
w/o HCS (588)	19 (16) [c]	4 (8)			10 (13)
Math SA (610)	59 (19) [c]	51 (18) [a]	44 (17)	52 (15) [a]	52 (18)
w/o HCS (604)	57 (18) [b]	45 (16) [a]			48 (17)
Math OE (615)	18 (26) [c]	9 (16) [a]	3 (8)	9 (13) [a]	10 (18)
w/o HCS (607)	12 (15)	5 (11)			7 (12)
		Standard VI (Number of pupils)			
English MC (706)	60 (22) [b]	54 (23) [a]	43 (18)	51 (19) [a]	52 (21)
w/o HCS (630)	55 (20) [b]	43 (17) [a]			46 (19)
English OE (276)	18 (24) [b]	14 (20) [b]	4 (10)	3 (7)	11 (19)
w/o HCS (238)	14 (22) [b]	13 (23) [b]			9 (18)
Math MC (621)	43 (15) [b]	40 (17) [a]	33 (12)	37 (11) [a]	38 (14)
w/o HCS (604)	39 (14) [a]	32 (11)			35 (12)
Math OE (624)	9 (10) [a]	9 (10) [a]	5 (6)	7 (8)	8 (9)
w/o HCS (546)	7 (7)	6 (6)			6 (7)
Kiswahili OE (337)	36 (25) [b]	28 (17) [a]	26 (22)	18 (15)	27 (21)
w/o HCS (299)	45 (21) [a]	25 (17)			26 (21)

Note. All reported tests of significance are at the $p < .10$ level unless otherwise noted. SA = short-answer test; OE = open-ended test; MC = multiple-choice test; HCS = high-cost school; w/o = without.

[a] Significantly better than one group. [b] Significantly better than two other groups. [c] Significantly better than all other groups.

However, at the Standard III level, pupils in the control schools, which did not receive any special project staff development during that 6-year period of the SIP treatment, performed better than the T3 pupils on all tests, the same as T2 pupils on one test, and better than T2 pupils on two tests. The T2 pupils only outperformed the control-school pupils on the English short-answer test, and their edge disappeared when the high-cost school was removed from the analysis. Although T1 schools performed significantly better than all other schools on all tests, when the high-cost school was removed, there are only two tests where they significantly outperformed control-group schools: the English short-answer and open-ended tests. In short, then, there appears to be little achievement advantage to being in a SIP school until a school has been in the project for at least 5 or 6 years, and then only on performance in English. And, this advantage may be due in part to higher numbers of workshops attended by T1 teachers.

Scores were lower overall at the Standard VI level (means of 52 percent in English, 38 percent in mathematics and 27 percent in Kiswahili), although the tests for this level were more difficult than those for Standard III. Across all tests except the mathematics open-ended T1 schools performed significantly better than T3 and control schools but never significantly better than T2 schools. T2 schools only performed significantly better than control schools on the English open-ended test, although their scores on this test were only 14 percent. The T3 schools consistently performed significantly worse than the T1 and T2 schools and usually worse than the control schools.

Open-Ended Items

The most noticeable finding is the stark difference between students' scores on the short-answer test items compared with the open-ended items, where scores averaged only around 10 percent across groups, although they reached 28 percent for the T1 Standard III students when high-cost schools were included, and an average of 28 percent in Kiswahili across all groups. Since almost one third of the 14 possible points were allocated to drawing a picture, it is likely that the scores for reading and writing were even lower than the scores recorded in Table 3. This means that, for students at both grade levels, there is clear evidence that they are not learning how to read, write, or communicate in English or Kiswahili, or both, based on their performance on open-ended test items. Nor are they learning to apply simple mathematical concepts and skills to real-life problems, as measured in the open-ended items.

During the classroom observations, we noted that, when teachers were teaching English, it was usually a grammar lesson. Only a couple of the teachers actually had the children read, and that simply involved children taking turns standing up and reading a paragraph. They were not asked to predict what the passage might be about, to connect it to their lives or past knowledge, to summarize a story, or to create their own stories. We saw no Standard III teacher teaching children to write, other than to do the exercises in their textbooks. The exercises were generally fill-in-the-blank or matching and did not involve pupils in communicating; in generating sentences, ideas, or thoughts; or in expressing their own views and feelings. Below is a sample of the range of responses to the English reading and writing prompts.

Prompt (Standard III)
Read the story and answer the question. Write as much as you can. *Birds can fly. If you could fly, where would you go? What would you do when you got there? Draw a picture of you flying.*

Pupil 1 (Score = 1 for written passage; 3 for picture)
I cant fly. ane kiat canfly I cana fly above aboat can flia I can fay if yianf lay I cane slay
This pupil answered 29 of the 36 short-answer questions correctly and was able to draw a picture of himself flying.

Pupil 2 (Score = 8; 4 for picture)
I'd go to england I'd go to calofonia and also america then to atlanta[a] I'd go to atlanta for skhing playing with ice and having a lot more fun
[a] This test was given just before the Olympic games in Atlanta. This high-cost school pupil answered 35 of the 36 short-answer questions correctly.

Pupil 3 (Score = 1; 0 for picture)
birds hen I sing Iypiay batafigns duke ositrh
This pupil must have recognized the words "birds" and "fly." She answered 23 of the 36 short-answer questions correctly but drew a bird rather than a picture of herself flying.

Pupil 4 (Score = 9; 4 for picture)

If I could fly, I could go
high up in the sky and see the
earth below me. I could see the
heaven and I could worship
the lord my God. I could get
blessings from God and see
how he could handle the
worle world by his power,
and how he destroyed the evil singl
handedly. And there is no more that the end.

This pupil answered all of the short-answer questions correctly, drew a colorful picture of himself flying in a superman costume and wrote, "I am superman!, flying like a rocket!" He also was from a high-cost school.

Pupil 5 (Score = 6; 4 for picture)

I can't fly
if I could
fly I could go
where I want to go.
Birds can fly.

This pupil answered 35 of the 36 short-answer questions correctly and drew a colorful, charming picture of himself flying with the words "this is me I am flying" next to his drawing. He was also from a high-cost school.

Pupil 6 (Score = 2; 1 for picture)

I am fyirng to goin the god
and bird is fiyiny to go in the god
I am going in your mother in home
I am fly to gothomonog
I an gathr it griant gthora
Ianiohatgtritognhat
Iaragiriohnoagtfiahi

This T1 pupil answered only 20 of the 36 questions correctly but some of her passage was coherent. (It was difficult to tell whether there were spaces between the words.)

Almost all instruction and pupils' responses were in English, even at the Standard III level. SIP staff indicated that parents expect teachers to teach in English beginning in Standard I, despite the fact that national policy is to teach Standard I in the pupils' mother tongue, gradually transit into English and teach primarily in English in Standard IV. English and Kiswahili are to be taught as second and third languages beginning in Standard I. It was not clear how much of the pupils' failure to respond to the open-ended questions in the tests given for this study was due to their lack of understanding in English, although most pupils performed fairly well on the short-answer and multiple-choice tests that were administered in English.

The Standard III pupils also performed very poorly on the two open-ended questions for mathematics. Out of a possible 10 points, most received less than 1 point. Pupils in Standard VI averaged less than 1 point out of a possible 12 for their responses to the open-ended questions. When compared with their short-answer and multiple-choice test scores, these results suggest that pupils can perform some basic mathematical calculations but do not know how to apply them to real-life problems, nor represent basic mathematical ideas.

Controlling for External Factors

The test results reported above do not take into account a variety of factors that can influence student learning, such as parents' income and education, teachers' experience, pupil-teacher ratios, or access to textbooks. Extensive data were collected on a range of family, teacher and school variables and regression analyses run to account for these factors in student test performance. However, missing data on individual cases reduced the regression sample from 620 per grade to around 100 or less, thereby weakening the validity of the findings that could be generated from such analyses.

In general though, the findings showed that, for Standard III pupils, when family income and mother's education are controlled for, being in a SIP T1 school is no longer a significant contributor to learning achievement. However being in a T2 school does appear to have a positive influence on test scores in both English and mathematics, although not significantly in English. Although being in a T2 school is not statistically significant, the coefficient for test scores in English is 12.4. This means that a pupil's score is likely to increase by 12.4 percentage points because he or she is in a SIP T2 school. However, the coefficient for the T1 schools drops from 21.6 to only 3.1, indicating that workshop attendance and classroom management account for a much larger effect on T1 schools than on T2 or T3 schools. A similar effect is evident in the mathematics test scores. The T3 coefficient increases considerably when the number of years a teacher has been teaching is added, suggesting that these schools have teachers with less experience than do T2 schools. This means that progress made in a T3 school is more likely due to SIP than to an experienced teacher.

Workshop attendance had a significant and positive influence on pupil performance in both the English and mathematics tests. Test scores were estimated to increase by over two percentage points for each workshop a teacher attended, and the T1 teachers attended significantly more workshops than did other teachers. Child-centered teaching was not a significant contributor to test scores. Classroom management behaviors were a significant influence on English scores but not on mathematics. The variable most consistently influencing test scores is the number of years a teacher has been teaching.

At the Standard VI level, when family variables are considered, the SIP had no influence on student learning as measured by these tests. The number of years a teacher has been teaching was not significant, as it was at the Standard III level, and the statistics for the other variables were unreliable for all Standard VI tests and all open-ended tests at both levels.

Based on the limited sample available for regression analyses, it appears that the child-centered teaching behaviors promoted by the SIP did not have a positive influence on test scores at either grade level, although further analyses of these behaviors follows.

The SIP's Impact on Teaching

The classroom observation instrument was designed to assess the child-centered-teaching and classroom-management strategies contained in the SIP workshop materials. One Standard III and one Standard VI classroom were observed by two evaluators in each of the treatment and control schools. Two lessons were observed for each class, and, in most cases, the lessons observed were English and mathematics. In the upper grades, several science lessons were observed. In the following discussion, teachers' behaviors are reported in three ways: a brief overview of the typical teacher behavior observed in treatment classrooms; index scores of child-centered teaching and classroom-management behaviors, which consisted of a collection of related classroom behaviors; and detailed descriptions of individual teaching behaviors considered to be aspects of child-centered teaching.

The classroom observation instrument was designed to assess two main features of a teacher's behavior: behaviors that could be characterized as child-centered teaching, and how the teacher managed the classroom. For both constructs, specific behaviors were generally drawn from the SIP training materials. An index was created for each of these sets of behaviors, calculated for each teacher and summed across schools within each treatment group based on a weighted sum of selected behaviors. Child-centered teaching/learning behaviors included (a) active learning in small groups, (b) encouraging pupils to interact or help each other and discuss, or both, (c) conveying high expectations, (d) using learning aids, encouraging pupils to pose their own questions and seek answers to those questions, (e) encouraging pupils to think, (f) attending to individual students' needs, and (g) pupils intellectually engaging with tasks, creating, shaping, and integrating what they are learning with what they already know, and so on. Classroom-management behaviors included (a) clearly communicating expectations for student behavior, (b) starting and stopping the class on time, (c) using time efficiently for learning, (d) making sure that all pupils are paying attention and are engaged, (e) having control over the class, (f) having all materials ready and available for each lesson, (g) ensuring that lessons flow smoothly, and (h) moving smoothly between lessons or other activities.

Overview of the Typical Teacher Behavior Observed

Overall, the teaching in the SIP classrooms was not particularly child-centered. There was remarkable similarity among teachers. Most lessons began with a 10-minute explanation of the topic, with teachers asking a rapid series of closed, factual questions that did little to engage pupils in higher levels of thinking or reasoning. An assignment was then given, pupils collected their exercise books from the teacher, and spent the remainder of the lesson completing exercises in their textbooks or ones that the teacher had written on the board. Usually, the teacher went around the room correcting pupils' work and occasionally offered an explanation. Teachers seldom attempted to find out whether individual pupils really understood what the teacher was teaching. Seldom did we see pupils working together in small groups or talking with each other, although most SIP classrooms had pupils seated in groups. Quite a few of the teachers did use learning aids (manipulatives) to help make lessons more concrete, particularly in mathematics. We almost never saw pupils using the learning centers that SIP had helped teachers to develop.

Index Scores: Child-Centered Teaching and Classroom-Management Behaviors

At the Standard III level, teachers in T2 schools exhibited substantially more child-centered teaching (61%) and classroom management behaviors (88%) than any other group, followed by the more recently involved T3 teachers, suggesting that the SIP's work with teachers during Phase 2 did contribute to those teachers' improved classroom practices. The regression analyses, however, indicated that child-centered teaching did not contribute significantly to increased test scores, but that classroom management behaviors and workshop attendance did. The index scores for child-centered teaching and classroom management behaviors for the T1 teachers were lower than those for the other two treatment groups addressed during Phase 2 and, in fact, the control group teachers' classroom management scores were higher than those of the T1 teachers.

All Standard VI teachers exhibited few child-centered teaching behaviors and there was little difference across the groups in either main category of teaching behavior. Interestingly, these teaching behaviors are not consistent with the pattern found in the unadjusted test results for Standard III teachers, where the pupils in T1 schools obtained the highest scores, followed by those in T2 and the control schools, with pupils in T3 lagging behind. However, it is unlikely that the T3 teachers' new teaching strategies would have had sufficient time to have an impact on student learning since they had only been introduced to these strategies within the preceding 4 to 5 months.

So, while the SIP was able to get teachers to use the desired child-centered and classroom management behaviors on a short-term basis, the new behaviors seemed to fade over time and appear to have had less influence on improvements in short-answer and multiple-choice test performance than did teaching experience and workshop attendance.

Table 4: Indices of Child-Centered Teaching and Classroom Management Behavior, Mean Percentage and Standard Deviation, by Group and Class

Group	Child-centered teaching		Classroom management	
	Std. III (*SD*)	Std. VI (*SD*)	Std. III (*SD*)	Std. VI (*SD*)
T1 (4 per class)[a]	42 (4)	44 (9)	44 (1)	60 (5)
T2 (4 per class)	61 (15)	34 (7)	88 (4)	53 (4)
T3 (3 per class)	52 (15)	39 (9)	65 (4)	56 (5)
Control (4 per class)	31 (5)	35 (12)	58 (4)	65 (4)
Total mean (30)	46 (12)	38(9)	63 (4)	59 (4)

Note. Std. = Standard; *SD* = standard deviation.
[a] Number of teachers observed per class.

Components of Child-Centered Teaching

In the next section, selected observations of behaviors that comprise both good teaching generally and child-centered teaching specifically are described. These include (a) teacher questioning, (b) how teachers respond to students' answers, (c) teacher feedback, (d) group learning, (e) availability and use of textbooks, (f) use of learning aids (manipulatives), and (g) use of classroom materials.

Teacher Questioning

Observers documented the types of questions a teacher asked, the type of responses a teacher gave to correct or incorrect answers, and whether the question was asked of a boy or a girl. This was recorded for a 10-minute segment of each class about 5 minutes after the class began.

Questions were defined as closed or open in keeping with the definitions in the SIP training materials. A closed question was defined as one that has a very limited number of acceptable answers (usually just one) and typically elicits a one-word or short-phrase response. An example of a closed question is, "What is the body of a fish covered with?" Open questions were defined in the SIP training materials as questions that "… anticipate a wider range of acceptable responses. They draw on the students' past experience but they also cause students to give opinions and their reasons for their opinions."

Open questions were classified as 'open simple' or 'open complex.' An open-simple question may have more than one right answer, but does not particularly encourage children to think and analyze. An example might be, "What are sources of energy?" Open complex questions are intended to engage students in complex thought. For example, the following types of questions may require that students draw upon their own views or cause them to reason and analyze, have multiple possible answers, and debate: "Why do you think that? How could that have been prevented? What other options are possible?"

Table 5: Types of Questions Asked by Teachers, by Group and Grade Level[a]

Question type[b]	T1		T2		T3		Control		Totals
	S3[d]	S6[d]	S3	S6	S3	S6	S3	S6	Total
C-B[c]	25	11	20	17	14	17	27	17	148
C-G[c]	25	5	16	17	13	11	28	21	136
CCN-B	2	0	0	0	0	1	0	0	3
CCN-G	1	0	0	0	0	0	0	0	1
OSC-B	5	5	11	7	1	14	2	2	47
OSC-G	1	4	12	6	1	1	4	4	33
OSNC-B	0	0	25	2	0	1	4	1	33
OSNC-G	0	0	5	2	0	0	1	0	8
OC-B	0	3	0	0	0	1	0	0	4
OC-G	1	0	0	0	0	0	0	0	1

[a] Four teachers per class in T1, T2 and control schools. Three teachers per class in T3 schools. [b] C = closed not connect; CCN = closed connect; OSC = open simple connect; OSNC = open simple not connect; OC = open complex. [c] B = boy; G = girl. [d] S3 = Standard III; S6 = Standard VI.

Most of the questions asked by teachers were closed questions requiring short, predictable answers and little or no analytical thinking on the part of the pupils (Table 5). Teachers asked very few questions that attempted to connect to pupils' lives or prior knowledge: only about one per class, most of which were open-simple questions. Only about one percent of the 414 questions asked during the ten-minute observation segments were open-complex questions. Boys were asked 14% more questions than girls. A high number of questions were posed to the whole class, requiring choral responses and closed, one-word, predictable responses. Teachers' responses were generally to acknowledge an answer as correct and move on to the next question. This was often done with a "Yes" or "Good" and seldom with enthusiasm. Only six harsh (critical or dismissive) responses were documented.

The quality of feedback provided by most teachers could be described as minimalist. Teachers seldom attempted to understand what children were thinking, to identify their misconceptions or even accurate conceptions, or to provide explanations that might clarify children's understanding. Most of the dialogue with pupils was perfunctory, narrow, intellectually unengaging, and lacked a personal connection. The only classrooms where there was any substantive communication among pupils was in the two high-cost schools. In these schools, the pupils appeared to be much more comfortable and relaxed about interacting with each other.

Box 1: Classroom Profile – Standard III SIP Teacher

This newly trained SIP teacher was teaching English vocabulary. She used a knife as a learning aid to teach the words 'knife', 'blade', 'handle', 'blunt', and 'sharp.' She had individual pupils come up to feel the sharp and blunt ends, but often waited too long after asking a question before selecting a pupil to respond. While research has found that waiting 3 seconds instead of 1 second after asking a question results in several cognitive benefits for pupils, these benefits derive only from questions that may have several responses or require more thinking on the part of pupils. In this instance, the teacher was looking for only one correct response and her delay in calling on a pupil only served to waste valuable learning time. We observed several teachers waiting too long after asking a question and before calling on a pupil to respond.

In this class, as in many other SIP classrooms, the pupils were seated in groups but did not work in them. The teacher's written lesson plan was typical of most teachers we observed: vague and general. It consisted of statements such as "introduce by asking question," "write the words on the board," "give them work," "check their work."

When the teacher began the mathematics class, many pupils were not paying attention because they were searching through a stack of books to find their individual exercise books.

Group Learning

Although children were seated in groups in many of the SIP classrooms, only 4 of the 22 SIP teachers had pupils working together in their groups. This does not represent a significant change from the situation found in the evaluation of Phase 1, which reported that teachers' use of individual and group work was limited. This may be an example of how teachers can adopt the 'trappings' of an idea without incorporating the essence into their teaching repertoire. Organizing pupils into groups is much easier than understanding how to engage them in effective learning activities, although the SIP did provide teachers with a number of materials and ideas for small group activities. Small-group work is often dismissed as unworkable in developing countries because of large class sizes. However, many of the SIP classrooms had reasonable enrollments. Teachers who engaged pupils in small groups had classes ranging in size from 28 to 38 pupils.

Box 2: Classroom Profile – Standard III SIP Teacher

This teacher taught two excellent lessons that exemplified child-centered teaching. The mathematics and English lessons were highly integrated and the pupils were actively involved. The teacher was friendly and the children seemed quite happy. In the mathematics lesson, she was teaching the topic of shapes. She began by asking pupils to recap the previous lesson on shapes by picking out a particular object, discussing its shape and the properties of the shape, which she wrote on the board, for example, rectangle - two opposite sides equal: square; all sides equal.

She then had pupils work in groups to cut out shapes. During the cutting, the pupils discussed what they were doing, for example, one pupil said, "No, all sides equal." When a group finished one shape, they moved on to another without being told. During this time, the teacher moved around the room, expressing interest in each group's work. She summarized the lesson by having the pupils come to the front of the room to fit cutout shapes into a large sheet of paper from which the shapes had been cut.

Even more impressive was her connection between the mathematics and English lesson. The English lesson was on the concept of a circle and she began by having the pupils join hands to make a circle, then break into small groups to make several small circles. She then had them do other things with or about circles, such as making a circle from a rope and then making a sentence to describe what they had done. She attributed her approach to SIP training and indicated that, while she had acquired some of these methods at college, she had stopped using them until SIP reminded and retrained her.

Classroom Materials

SIP classrooms were far richer in materials than were control classrooms. This was the most distinguishing characteristic of SIP classrooms. The materials provided a much warmer, more interesting environment for the pupils – and for the observers. We received substantial feedback about the value of the materials from teachers, head teachers, and parents; a major concern was finding the funds to provide the non-SIP classrooms with materials. The control classrooms and the Standard VI classrooms that had not participated in SIP were bare. Usually there were no materials of any kind other than teachers' and pupils' books, desks, and a chalkboard. However, in a few of the SIP classrooms, the materials were faded, dirty and bedraggled and sometimes hanging off the wall. We never observed a teacher in these classrooms referring to or using these materials in any way, other than the learning aids described earlier.

Box 3: Classroom Profile – Standard III SIP Teacher

This teacher was also the SIP coordinator at her school. The first lesson was Kiswahili and she was teaching word use using several objects. She held up two objects, such as a large and a small knife and then asked, "Ni up mkubwa? (Which is the big one?)" and "Ni up mdogo? (Which is the small one?)." She then asked for the reasons for pupils' choices: "Kwa nini utachagua kisu hiki? (Why do you choose this knife?)." Children conversed using the words, comparing their choices and making informed preferences.

In the mathematics lesson, children sat in groups and used counters to make sets as indicated by the teacher, for example, "Three groups of two." She then moved to columns and had the children arrange bottle caps in columns as she called out, "Show me four rows with two bottle caps each." Although each child was working with his or her own counters, they were discussing with the others in their groups and cross-checking with their peers' arrangements, suggesting that a culture of communication and consultation had already been established.

This teacher also indicated that she had learned some of these methods previously but had stopped using them until the SIP came along and reminded her of them: "Knowing something is one thing, but practicing it is another.

Textbooks

In 29 of the 34 classrooms, teachers used a textbook during instruction. Pupils used textbooks in only about half of the 34 classes, with about 60% of treatment classes using textbooks, compared with only 37% of control classes. Accurate figures are not available for Standard VI classes since, in many cases, the pupils had with them only the books for the class they were currently attending.

Table 6 shows that pupils in Group T1 were more likely than any of other group to have textbooks. Only about half as many pupils in Group T3 had English and mathematics books as in Groups T1 and T2, and even fewer pupils had Kiswahili books (21%). The pupils in the control group schools were more likely to have fewer books than pupils in Groups T1 and T2 but slightly more books than pupils in T3, with the exception of Kiswahili books (21% of pupils vs. 12%). The regression analysis described earlier in this chapter revealed little influence of textbooks on test scores, although the correlation of textbooks with mothers' educational level may have counteracted the influence of textbooks when the two variables were entered into the regression equation.

Table 6: Percentage of Standard III Pupils Having Textbooks, by Group

Group	English class	Mathematics class	Kiswahili class
T1 (4)	67 % (21)	71 % (14)	80 % (62)
T2 (4)	59 (38)	65 (33)	40 (29)
T3 (3)	34 (18)	35 (15)	21 (23)
Control (4)	42 (8)	36 (17)	12 (7)

In many classes, pupils did not have their own books and shared with other pupils. Generally, each pupil shared a book with one other pupil, but, in eight classrooms, there were nine instances each in which a pupil shared a book with more than two other pupils, and four classrooms where about half of the pupils shared with more than two other pupils. In a number of classrooms there were so many pupils sharing a book that several could not see it.

Learning Aids

Two thirds of the SIP teachers used learning aids during their lessons, compared with only one quarter of the control group teachers. Learning aids were more often used in mathematics classes at the Standard III level. In half of the SIP classes, pupils were encouraged to use learning aids. These often included bottle caps and sticks for multiplication exercises and small pieces of cardboard with words on them for making sentences in English classes. There was remarkable similarity in the use of the learning aids in the SIP classrooms, likely the result of SIP influence. In no instances were boys more likely than girls to have access to learning aids.

Box 4: Classroom Profile – Standard VI SIP Teacher

This teacher's lesson illustrates several activities or behaviors encouraged by SIP. She began by asking the pupils a series of questions aimed at teaching them the use of the conditional tense ("I would" or "I'd"). Almost all of her questions were connected to the pupils' lives and to their feelings. She began by asking, "Who has seen a rat?" She then asked, "Where did you see a rat? In which room of your house did you see a rat?" After several pupils gave their responses, she asked, "What would you do if you found a rat in your bedroom? … in the kitchen?" and so on and wrote each pupil's response on the board. Before class began, she drew several scenes on the board, such as withering crops or bountiful crops. After a series of questions related to pictures eliciting responses that began with "I would," she then had pupils pair up and ask each other questions using the conditional "would", such as "What would happen if there was no rain?" or "What would happen to the farmers' crops?"

This teacher actively engaged her pupils, connected the questions to their experiences, and had them work together in small groups generating and responding to each other's questions. However, the questions they were assigned to ask each other were all scripted according to the drawings she had put on the board. Perhaps if the pupils had been able to generate their own questions and answers they would have been more likely to learn a larger variety of contexts for using the conditional tense. In addition, the teacher did not place the lesson within the larger context of tenses and grammar, nor did she use the opportunity to develop pupils' writing skills by, perhaps, having them write short passages or stories using the conditional tense.

Box 5: Classroom Profile – Standards III and VI Control Group Teachers

In this control group school, the Standard III teacher was teaching about the local administrative system. The lesson seemed too advanced for these young children. It included how the township administration was organized, the heads of departments, the county council, and the role of each of these offices.

The Standard VI classroom was the only one observed where several of the pupils did not have desks and were sitting on the floor. None of the control classrooms had materials displayed on the walls, except perhaps a lone photograph from a magazine or an old, dusty chart of some sort. However, this Standard VI teacher used several desirable instructional strategies that she had learned in her coursework at teacher training college, from which she had recently graduated. She was teaching a lesson on the volume of cubes and cuboids and used a cardboard box as a learning aid. She drew a cube on the board with the length of each side labeled. She then measured the sides of the box and showed that all sides were equal. She distinguished between a cube and a cuboid, put the formula for each on the board and asked pupils to put the formulas in their exercise books. She gave them several exercises to do. After the exercises were completed, she put pupils' answers on the board and reviewed them.

Her lesson was coherent and well organized, although she did not have boxes available for the pupils to use, nor did the pupils work in small groups or discuss what they were doing. When she called on one pupil who refused to respond, she said, "You should be sitting in the front [on the floor]." She gave him a second chance and, when there was still no response, asked, "You don't know at all? Can't speak? Sit down," clearly embarrassed and annoyed by the boy's lack of response.

Summary and Discussion

The classroom observations revealed that, although SIP teachers practiced more child-centered teaching behaviors than did control teachers, their use of them was limited and often shallow. The more important, yet difficult, behaviors such as asking questions that engage pupils in thinking analytically, having pupils be active learners working in small groups, encouraging children to express themselves and explore ideas, or using language to communicate and understand, were very rare in all classrooms. Why is this? There are hints that can be derived from previous research, interviews with teachers, and focus-group sessions with school staff, each of which is discussed below.

Inadequate Conceptions of Child-Centered Teaching

Most teachers' descriptions of child-centered teaching touched on an element of it, but most also were incomplete. They suggest that teachers did not fully understand the concept. For example, several teachers responded that it is a method in which children find things out by themselves: "Children do most of the discoveries by themselves with the supervision of the teacher. … Children can learn on their own. … Children work when the teacher is absent." Did these teachers mean that children develop a deeper understanding by engaging in relevant activities that allow them opportunities to discover knowledge? Or, do they literally mean that a teacher's role is diminished and minimal, that he or she simply sets out materials and the children do most of the learning on their own? If the second interpretation is close to what they believe child-centered teaching to be, it may help to explain why they were reluctant to use it. It is not likely that they could accept that schooling involves leaving children to learn on their own without planning, organizing, and structuring by the teacher. If teachers believe that child-centered teaching means a 'hands-off' approach, then this could be incompatible with their sense of responsibility as teachers and cause them to abandon or substantially adapt the child-centered strategies promoted by the SIP.

Influence of National Curriculum and Examinations

Cohen's (1988) research found that teachers must believe that the changes they are expected to make will result in a better situation for their pupils. Some innovations that may be intellectually appealing to teachers may be difficult to implement or inconsistent with their instructional repertoire (Schwille et al., 1982). This may cause teachers to perceive such innovations as a threat to their students' success and so to be resisted. In most African countries, students' and teachers' success is judged by students' scores on national examinations. End-of-primary examinations determine which students are allowed to continue to secondary school, and the pressure of the exam exerts substantial influence on teachers' practice. Comments made during the teacher interviews and the focus groups with school staff suggest that the curriculum and examination exert a more powerful influence on teachers' behavior than does the pedagogic appeal of child-centered teaching:

> We don't do what SIP wants because we need to cover the syllabus and we only have 35-minute periods. … If they cut back on the amount of material covered in the curriculum, we would have more time to engage the children in thinking, but now we avoid asking questions that might require any extra time. Some of the topics are too detailed for primary level and many topics are repeated in several subjects. … SIP methods require too much time.

The success of pupils is schools' highest priority, and success is judged by performance on the KCPE and on interim regional and local examinations.

The SIP worked with teachers and other educators to produce practice examinations similar to the national school-leaving examinations. However, these practice examinations were not at all child centered and actually undermined the child-centered teaching strategies the SIP was trying to promote. Moreover, there were very few items on the exams that measured higher order thinking skills, and there were also serious problems with the examinations, despite the fact that they were reviewed by the Kenya National Examinations Council. It is quite likely that the examination questions were reviewed solely for their statistical properties and not for their pedagogical validity. For example, the Standard I science questions were in English, despite the fact that Kenyan children have not learned English by the end of Standard I. One of the questions even had the word 'germinate' – far too difficult a word even for native English speakers at that age. A Standard IV reading comprehension passage seemed entirely inappropriate for young children. It was a short story about a "very black man" who had fierce eyes and chased children asking for food until the men in the village chased after him with spears and told him not to come to the village again. In the last sentence of the story, the women who went for water could not find their children, suggesting that the hungry man had kidnapped them. Not only does this story lack compassion for the poor and hungry and instill fear in the children, it also suggests that being very black inclines one to be dangerous and someone to be afraid of. Another question required four words to be arranged to form a sentence. Despite two possible correct answers, only one was allowed.

SIP staff's explanations confirmed what we were hearing from school staff:

> Examinations work against child-centered methods. Examinations are a high priority in Kenya – perhaps the highest priority when it comes to schooling, and teachers do what they think will cause their children to get higher scores on the exams. I think it's difficult for them to see the direct link between child-centered methods and increased test scores. … Another reason is that child-centered approaches require more work for the teacher – at least until they have established a repertoire of materials and behaviors and a level of comfort with the new strategies. Often, teachers complain that these strategies are just too much work. Teacher and head teacher transfers are another problem. There are a tremendous number of transfers in the Municipality. The curriculum is also a problem. Teachers have only 35 minutes per core subject per day. This

can make it difficult to do activity-based lessons. We have tried to show them how to do block schedules and some schools are using it.

The Kenyan curriculum is very full and the end-of-cycle examinations are comprehensive, factually oriented, and not particularly child-centered. This may be one of the most powerful explanations for teachers' behaviors. It has been an issue of debate throughout Kenya for quite some time, and the debate is not likely to be resolved by the SIP. Moreover, although there is a lot in the curriculum, we could not find a coherent approach in either the textbooks or the curriculum that would induce teachers to teach reading and writing. There is no subject called 'reading' or even 'language arts.' The English textbooks almost exclusively teach grammar. We observed that children did have readers, but the stories were dull and uninspired, and the teachers did not seem to know how to use them to help children actually understand and make meaning out of what they were reading.

Insufficient Mastery of Desired Changes

The fact that most of the teachers seldom exhibited the most central child-centered teaching behaviors during our observations (e.g., having students work together in small groups and discuss with one another), suggests that the SIP may have moved too fast with teachers in their race to meet the project's goal of working in 27 schools. The project staff confirmed this view: "The schedule we have set up to work with schools may be unrealistic. We typically work with 3 to 4 schools per academic term [4 months]. We would prefer to work with a school for a year, in teams of two POs working with two schools per year. In 4 months, teachers are not able to fully integrate the new behaviors and, as soon as we are not there to support and coach them, they fall back into their familiar ways."

The SIP's Work with Head Teachers

The Phase 1 evaluation report expressed a concern that the SIP had not provided sufficient attention to the school as a whole or to the role of the head teacher, in its school improvement efforts. Despite this concern, the SIP seldom worked with schools as a whole and did not begin training head teachers until late 1994, 18 months into Phase 2. In interviews with teachers, head teachers and school staff, all agreed that head teachers did not serve as instructional leaders, seldom visited teachers' classrooms, and, in fact, did not even see these activities as part of their role.

SIP Staff Training

SIP program officers (POs) were selected from the ranks of classroom teachers within the Municipality; most had been considered exemplary teachers. They were the front-line contact with teachers and head teachers in SIP schools. The evaluation team met several times with the POs to obtain their views regarding the project and training they received to do their jobs. They reported that they received no training for their jobs during Phase 1, but, in Phase 2, they received training in classroom management, discipline, tracking of pupils' progress, child-centered methods, learning centers, reading, and imagined literacy. The project director (a North American teacher educator previously involved with AKF's SIP activities in Tanzania) conducted much of the training for the POs, including training in the whole-language approach to literacy. The POs reported that most of the training was conducted in a lecture format with numerous activities and group discussions, but that they were not offered opportunities to read or discuss original source material in conjunction with any of their training sessions, such as research articles or summaries of research. Instead, they were given handouts that summarized the points covered in the workshop sessions. Moreover, they reported that they were not provided with opportunities to reflect on what they learned as a group, nor did they receive guidelines or training in what to do when they visited a school, how to do a demonstration lesson, or how to give feedback to teachers.

When asked to rate the training they received to do their jobs, the evaluators created a rating scale with endpoints of 1 and 10. Point 1 represented training where only the theory was explained, and Point 10 represented opportunities to master the content, skills and procedures that they were expected to impart to teachers, including PO practice and feedback as they worked with teachers. The mid-point of 5 represented training that provided theory, opportunities to discuss, and group activities, but not observation, coaching or feedback as POs worked with teachers. The POs all rated their training between 4 and 6, and indicated that they would have preferred training and coaching that gave them time to reflect in groups, to read and discuss original source documents, and to have the expert(s) observe them in the classroom as they worked with teachers.

The project officers did report that they had weekly meetings, before which each PO had prepared a written report of what had happened in the field during that week. They discussed their reports each week and the project director offered suggestions for how to deal with various situations. POs also reported that they received feedback and support from the project director on the workshops they conducted for teachers, and that the director reviewed the workshop materials with them prior to the workshops. In addition, most of the workshops were videotaped, and the videotapes were reviewed and critiqued by SIP staff.

Summary

In summing up, the SIP was quite successful in engendering broad-based support for what it was trying to accomplish. Attitudes toward the project were positive and widespread. The SIP was exceptionally successful in getting all facets of the Kisumu education system to rally around the project and to want to try to make it work. Everyone we spoke with knew about the SIP and, almost to a person, was enthusiastic about it. The enthusiasm seemed most directly aimed at the materials and secondarily at the cooperative and helpful spirit of the SIP staff. However, there were several main constraints to increasing the project's success in changing teachers' behaviors, in improving schools' effectiveness, and in increasing student learning. Some of these were due to project management decisions and some were beyond the project's control but within their realm of influence: (a) SIP staff waited until very late to work with head teachers, (b) SIP staff did not focus on the whole school as the unit of improvement, (c) SIP program officers were not adequately trained, and (d) there was a high rate of staff turnover. Other constraints included (a) a high incidence of school and MEO staff transfer, (b) a high incidence of school closing for multiple examinations, athletic events and music festivals, and (c) a high incidence of teacher absence, in part due to the high incidence of HIV/AIDS-related illness and death in Kisumu, but also due to lax supervision on the part of the head teachers and the MEO. Non-academic activities take place during the school day and are frequent and extended. For example, when we visited one of the newer SIP schools, the teacher was sitting at her desk relaxing while the children worked independently. When we told her that we would like to observe her teaching a lesson, she told us that she had not prepared because she was exhausted from the activities of the music festivals and that this was her first day at school in 2 weeks! In fact, one senior MEO official was unavailable for several days because he was attending athletic events in another district. There were about 3 days out of the 3-week data-collection period when some or all schools were closed due to events. Although national policy limits school closures for these purposes and requires that festivals and athletic events be held after school hours, this policy is ignored and not enforced by the MEO.

An Unstable Target Audience

Western Kenya has a particularly high incidence of HIV/AIDS and Kisumu is no exception. In many of our visits to schools, we encountered staff taking collections to help pay for funerals, teachers on long-term leave due to illness, or absent so they could attend a funeral. The former project director reported a high number of HIV/AIDS-related teacher deaths in project schools. In addition, teacher and head teacher transfers were numerous and frequent. These two factors combined suggest that SIP-trained teachers and head teachers may not remain in the Kisumu schools for one reason or another and that large numbers of new staff will need to be trained. If replacements are not trained to a high level of expertise, then whatever impact that the SIP is able to generate will dissipate rapidly.

Although SIP teachers exhibited substantially more child-centered teaching behaviors than did the control teachers, in most cases their use of these behaviors was limited and often shallow. The more important, yet difficult, behaviors such as asking questions that engage pupils in thinking analytically, having pupils be active learners working in small groups, encouraging children to express themselves and explore ideas or use language to communicate and understand, were rare in most classrooms observed. However, a few teachers did exhibit exemplary teaching behaviors. Although SIP teachers did engage in a more child-centered approach to teaching, their new behaviors did not appear to have a positive influence on test scores at either grade level tested. Teachers seemed to believe in the value of child-centered teaching strategies but were reluctant to fully adopt these strategies because they felt pressure to cover the curriculum and ensure that pupils were prepared to take and succeed in the national primary school leaving examinations. They were not able to reconcile these goals. In addition, teachers may have misconceived the meaning of child-centered teaching and learning, which further inhibited their use of these strategies.

Moreover, based on performance on open-ended tests of all groups, there is clear evidence that children are not learning how to read, write, or communicate in English or Kiswahili. Nor are they learning to apply simple mathematical concepts and skills to real-life problems. Limited evidence suggests that the SIP's efforts to encourage teachers to use child-centered teaching strategies had little impact on student achievement.

Recommendations

The evaluators met several times with the SIP program officers to obtain their views of what they would do differently if they were designing a new school improvement program, or what recommendations they would offer to other communities designing a SIP-type project. Their suggestions were remarkably similar to the informal thoughts and discussions that the evaluation team had as we visited schools, observed classrooms, and met with various representatives of the community. Although we have no structured data to support this contention, we all felt that the head teacher made the difference in the school. Where head teachers were enthusiastic and highly engaged, school activities seemed to be more organized, efficient, and engaged. Where the head teacher was either uninterested or lax, the ethos of the school was also lax: Teachers were absent, there was more milling about, facilities were in disrepair, and so on. POs' views were consistent with our views and with the extensive research documenting the importance of the head teacher as instructional leader.

Program Officers' Recommendations

Program officers' recommendations included the following:

1. Make courses practical, based in schools, with follow-up provided immediately. Workshops were offered in central locations and there was often considerable time between workshops and follow-up.
2. Train all school staff together, including head teachers. POs felt that school-based training would result in a more sustainable intervention and felt that it was a serious mistake to not train head teachers early on. They saw that the SIP had little impact in schools where the head teacher resisted it.
3. Offer a 1-month intensive training session for POs before they begin working with teachers. POs suggested that the training be sequenced so they receive instructions in one or two topics, practice them in the schools or as they conduct workshops, receive feedback and coaching, and then move on to the next topics.
4. Service schools with the most dire needs first and vary the level of intervention, based on schools' needs.
5. Evaluate schools every 18 months and make changes in the SIP schools based on evaluations.
6. Reduce the number of schools to be serviced. Phase 2 specified that the project work with 42 schools in a 3-year period. Both the project staff and the

project director reported feeling stressed by having to meet the 'numbers,' when they often felt they were moving out of schools before the schools had fully integrated the new philosophies and practices.

Evaluators' Recommendations

Study Quality Dissipation and Improve Training of Project Staff

We believe that a major reason for the SIP's limited impact is that project staff were not trained to a level of proficiency sufficient to do their jobs well. This is not intended as an indictment of the program officers. They were dedicated, very smart, hard working and eager to learn. However, this is a problem that has plagued many similar projects, and we believe it is a problem that has not been faced head on in the field of development. Very little is known about how much of the original training intent and quality are maintained as training proceeds down the line from the 'expert' to other trainers and end users, but we suspect that quality dissipation is quite high.

A related issue is the length of time it takes to achieve a level of proficiency sufficient for something to take hold. POs reported that the teachers they worked with often had not solidified their understanding of the SIP strategies before the POs had to move on to another school. If teachers do not quite understand how something is done, it is unlikely that they will incorporate the new skills into their teaching repertoires. An ethnographic study, conducted across several sites to assess the dissipation of training quality as it moves down through the levels, and to determine what it takes to retain quality, could be valuable in informing subsequent project designs.

Promote Craft Study

Recent research on staff development reveals that it is important for educators to be actively engaged in studying their craft, rather than to be passive recipients of training (Joyce, Calhoun & Hopkins, 1999). In order to study one's craft, it is necessary to read relevant professional literature. Although SIP staff had some access to research literature through the SIP's resource center, the materials were limited and often out of date. The research materials that the evaluators brought with them were repeatedly borrowed and copied by the POs. They seemed hungry for more. It is difficult to visualize new teaching or training strategies when one has seldom seen or read about them. These concepts are vague. Reading research in an area and seeing how others have addressed it in the form of exemplary materials provides mental images that trainers and teachers can draw upon to guide their professional development. The training of staff at all levels (project staff, inspectors, head teachers and teachers) should include (a) reading either original research articles, popularized research summaries and descriptions, or copies of exemplary materials, (b) discussing these in various group settings, (c) talking about what the findings mean for their settings, and how they might try out new strategies, (d) trying them out, (e) observing and coaching each other, and (f) meeting again several times to refine their thinking and use of what they have read.

Facilitate Teacher Networking
Teacher networking has been found to have a much greater impact on teacher practice than traditional staff development (Lieberman, 1996). Teacher networks involve teachers meeting with other teachers on a regular basis to develop curriculum materials, share experiences, and further their understanding of an innovation. On-line, Internet-based networks may facilitate such sharing and may provide access to research, materials and other relevant resources.

Address the School Rather than the Teacher
The focus of SIP efforts should be on schools rather than on individual teachers within schools. School staff should be nurtured to work as a team, to identify and solve problems and to renew the school. Rather than focusing primarily on child-centered teaching strategies, child-centered teaching should be promoted within a broader framework of continuous school improvement and the development of teachers' subject expertise.

Develop the Head Teacher as Instructional Leader
In order for a school to pursue incremental improvement, the head teacher must be able to create and sustain an academic focus and an ethos of improvement, and be able to motivate the community to provide resources and other forms of support to the school.

Establish an Effective School Peer-Review Process
Research-based indicators of effective schools have been identified and could be used to establish an evaluation and improvement review process within the Municipality. Based on the school review findings, the school staff and peer-evaluation team could jointly establish specific goals for improvement during the next evaluation period. Ideally, staff from other schools, parents and community members, as well as faculty from universities and teacher training colleges would all serve on review teams.

Develop, Administer, and Report on Child-Centered Examinations
Since examinations have a powerful hold on teaching and learning, the SIP could harness that power by developing municipal-level examinations that are consistent with child-centered teaching practices and that measure higher levels of thinking and problem solving, reading and writing, and the application of knowledge and skills to real-life problems and situations. Central to using examinations as levers is the timely reporting of examination results in a way that points out strengths and weaknesses and suggests teaching strategies that will both lead to better examination performance in the future and be child centered.

Attempt to Influence the National Primary School Leaving Examination and the National Curriculum
Although the primary school leaving examination is a quality test in many regards, especially in that the National Examinations Council prepares and disseminates de-

tailed reports of examination results with suggestions for instructional strategies, the examination is not child centered. This is particularly true with regard to the extensiveness of the topics and subjects covered by the examination, which is, in part, a reflection of the controversially overburdened curriculum.

5

Supporting Child-Centered Teaching under Universal Primary Education in Kampala, Uganda

Iram Siraj-Blatchford, Matthew Odada and Martin Omagor

In the 1960s, the educational system of Uganda was considered to be one of the best in East and Central Africa. But from the second half of the 1960s to the mid-1980s, the once-admired system drastically declined due to civil wars, civil unrest and general instability, which disrupted all aspects of Uganda's infrastructure, including education. Intellectual suffocation was perhaps one of the most devastating aspects of these difficult times. Books, professional journals, other educational reading materials and office or lab equipment became grossly inadequate. Education was devalued and suffered greatly (USAID, 1996).

Primary schools did not escape the traumatic experiences. But, as part of the nationwide reforms embedded in the current government's *Ten Point Programme*, the Ministry of Education and Sports (MES) commissioned the *Education Policy Review Commission Report* in 1989 and a *Government White Paper* in 1992. These documents now provide the backbone of a number of reforms that aim at improving students' mastery of literacy, numeracy and other basic life skills; improving school administration, management and accountability; and reducing inequities in children's participation in primary schooling (Carasco, Munene, Kasente & Odada, 1996).

According to Carasco et al. (1996), primary education in Uganda is still in need of considerable improvement. The reasons for this are manifold and include decentralization, which has marginalized education in some districts; the lack of libraries in most schools (more than 80%); the lack of latrines for girls; a shortage of textbooks, where one might be shared between 40 to 55 pupils depending on the subject and grade level; and lack of physically suitable classrooms. Carasco et al.'s study also revealed that the majority of primary head teachers in Uganda behaved in autocratic ways and held little regard for the feelings or wellbeing of staff or pupils, and that about 71% of the schools never followed timetables. Many teachers were perceived to be demoralized because of the aforementioned conditions, the salary delays, and inadequate remuneration of their services.

Teacher questioning, lecturing and class recitation dominate the teaching methods that are applied in primary schools. This is a reflection of teacher preparation, in-service training, resource availability and use, and the emphasis on national/public examinations (Carasco et al., 1996; UNCC, 1994). Primary education is better in urban areas than in rural areas and is generally believed to be better in Kampala than in any other area of the country. The total literacy rate in Kampala is 88%, compared with 54% for Uganda (UNCC, 1994).

At the time the evaluation was conducted, education in Uganda was at a critical point in its history and NGO/donor support was considered vital. The provision of Universal Primary Education (UPE) had in some cases doubled school class sizes. UPE, coupled with the very poor building and material infrastructure for schools, exacerbated problems of class management for teachers. In addition, teachers' salaries had been affected by the constraints levied on PTA contributions. At the time the evaluation was carried out, many teachers had not been paid for some time.

The Kampala School Improvement Project

The evaluation concerned the Kampala School Improvement Project (SIP), which began in November 1994 and ran for a period of 3 years. It was funded jointly by the Commission of the European Communities and the Aga Khan Foundation (AKF). The purpose of the project was to improve the quality of teaching and learning in selected nursery and primary schools in the capital Kampala, by first promoting and then institutionalizing the adoption of child-centered teaching methods and resources in project schools. The project strategy involved providing introductory sessions for head teachers, the training of teachers in workshop settings and providing instructional materials to support learning. Teachers were also provided with classroom-based technical assistance (on-the-job training), and they were supported as they attempted to integrate what they learned into their day-to-day teaching. The aim was to increase the extent of activity-based learning through greater pupil participation during lessons and group discussions. Children were encouraged to be more active learners; teachers were encouraged to have pupils work in small groups and to make and use materials to support learning.

The SIP's emphasis on child-centered methods was consistent with the goals advocated by the Ugandan Ministry of Education and Sports and the pre-service approach of the Ugandan teacher training colleges and departments. Similar projects have been conducted in other parts of East Africa: Mombasa and Kisumu, Kenya; and Dar-es-Salaam in Tanzania. Many lessons have been learned from other SIPs in East Africa, to the advantage of the Kampala SIP and AKF.

The 3-year Phase 1 of the project had three stages: the first stage entailed setting up the project, employing project staff, and working with three schools neighboring the Makerere Road Complex (MRC). This complex included a Teacher Resource Center (TRC), which initially had to be rehabilitated and subsequently became the administrative and training base for the SIP. Stage 2 involved four more schools. Finally, in Stage 3 work began in January 1997 (the final year of the project) in the last cohort of eight schools. It should be noted therefore, that eight of the schools had been in the project for less than 6 months when the evaluation took place. Each of the schools was asked to appoint a coordinator to liaise with the SIP team, although some schools were unable to respond to this request.

Evaluation Procedures

The terms of reference addressed by the evaluation team included an overview of the process of project implementation and the cost-effectiveness of the Kampala SIP. The key questions addressed by the evaluation therefore included

- the significance of the SIP in relation to the wider education context in Uganda;
- the appropriateness of the project strategy and implications for expansion;
- how far the SIP activities had led to the achievement of both the original and revised objectives;
- whether the SIP infrastructure was adequately developed to (a) meet the needs of the project and (b) sustain the impact of the project in the longer term;
- the evaluation of the major expert inputs and outputs of the SIP to teachers, schools and the TRC; and
- the making of recommendations regarding further development towards a second phase of the project.

In the process of conducting the evaluation, other questions were addressed and primacy was given to the views of those most affected by the SIP: children, parents, school teachers and school managers. Due regard was also given to the views of policy-makers and those responsible for implementing change in schools and teacher education.

The evaluation took place over an intensive period of 2 weeks. Prior to this, the terms of reference for the evaluation were developed by correspondence and during one afternoon meeting in London, England. The 2-week period in Kampala included

- the writing, piloting and implementation of six research instruments;
- half-day visits to eight project schools to undertake data collection;
- preliminary visits to two nonproject schools and three project schools (in addition to the evaluation sample);
- analysis of the SIP evaluation documents and records, fieldwork and report writing;
- observation of two project assistants (PAs) conducting workshops in SIP schools for teachers;
- observation of one workshop run by a PA at the TRC for early childhood educators; and
- meetings with key personnel from support agencies, NGOs and government departments, which were particularly helpful in the assessment of the sustainability and future prospects of the Kampala SIP.

The evaluation team received all relevant documentation, including interim reports and the detailed records of teaching materials; records of SIP internal and external meetings and evaluations of workshops; and SIP schools' timelines of achievement. These were used alongside the data gathered during fieldwork.

The SIP in Kampala is a whole-school 'improvement' project that aims to encourage qualitative change using measures of SIP success. Quantitative pre- and post-test results (associated with some 'school effectiveness' research) were therefore considered inappropriate to the evaluation of Phase 1. The project was not developed as a scientific inquiry with the stringent requirements of baseline measures but more as a practice-development project using discursive approaches and qualitative assessments of achievement. It aimed to influence teacher practice and classroom processes by example. The SIP in Kampala is a largely whole-school-based, in-service intervention program aimed at improving teacher practice. We (the evaluators) found little evidence of quantitative baseline measures, and the project schools had not been matched with control schools. No assessments had been conducted on the children or teachers that could be used to measure the 'value-added' effects of the SIP. We therefore ruled out the possibility of using control schools in our evaluation, but did visit nonproject schools that have characteristics similar to some of the project schools before the SIP involvement.

Robertson and Sammons (1997) cite Fullan (1991):

> The investigation of school and classroom process undertaken by school effectiveness research has been largely correlational in nature and causal inferences have rarely been directly drawn. The large-scale nature of such studies (necessary for the statistical analysis of effectiveness) are not suited to the study of

how change can be best supported and instituted, nor of the detailed processes of classroom interaction and support for this at whole school level.

Even where we do find quantitative evidence of value being added, that evidence needs to be supported by "a rich description of practice" (Reynolds, 1992). The evaluation team was keen to investigate how those who are affected by the SIP actually 'experienced' it.

The evaluation applied both qualitative and quantitative methods of data collection. Several research instruments were devised to provide data that would shed light on the successes and limitations of the SIP in schools. These included a semi-structured group-focused questionnaire for parents. We wanted to include parent perceptions of the SIP because so much of the Ugandan education system depends on the participation of communities. This did mean, however, that we had to inform the SIP schools of our arrival time in advance, and that we had to inform them of our intention to interview the parents. The schools only had 3 days to arrange this. It was only in this way that schools could arrange for parents to be present, but we were well aware that we would not receive a large or random sample in these circumstances. It was clearly difficult for parents to attend on short notice, so our interview sample was consequently small and made up largely of parent-teachers. It is therefore unrepresentative. However, we have analyzed and included parents' views having taken this into account, because we are concerned that parent (and community) involvement in the SIP has been one of the less successful areas of Stage 1, due to the constraints of time, and will require significant development in any future phase of the project.

The evaluation team also developed a semi-structured group-focused questionnaire for children, to get the views of children across the primary age group. Children were randomly selected from classrooms across the age groups and interviewed in a separate room to ensure confidentiality and a friendly atmosphere.

In each school, questionnaires were administered to teachers in the SIP schools, to teachers assigned the role of SIP coordinator, and to the head teachers in SIP schools. In addition, the head teachers and SIP coordinators were interviewed separately.

At the same time as the interviews and questionnaires were administered, a SIP Classroom Observation Checklist was completed in the eight schools by two observers. Seventy-two classes across the age groups were observed for approximately 20-35 minutes each, which included a short discussion with the class teacher. On average, approximately eight classes were observed per school, except, for instance, in the case of nurseries, where there were only five classes and all of these were observed.

The schools sampled for the evaluation study included a mix of those from Stages 1 to 3 of the SIP. We sampled one of the two nursery schools, two of three private fee-paying schools and 5 of the 10 government schools. The schools were also sampled according to the project team's assessment of how successfully the schools had adopted the SIP; those that had limited or very limited success with the SIP were included alongside those that were considered to be more successful in their adoption of the SIP.

The ultimate beneficiaries of the SIP in Kampala will be the children who attend the SIP schools. Assessing the true impact on children's learning and their social and behavioral development was not practical in this particular evaluation for two reasons:

- there were no baseline measures of children's academic, social or behavioral development; and
- there were no control schools.

However, our study was designed to evaluate Phase 1 of the SIP and to seek the *perceptions* of children, parents and school staff of the impact of the SIP on children's behavior, their active interest in learning, and how the school had changed with the introduction of more child-centered pedagogies.

The Findings

Prior to the SIP, the teachers in project schools adopted a didactic model of teaching, which we witnessed on our visits to some non-SIP schools. In addition, head teachers of non-SIP schools appeared to see their role as administrators rather than as managers who support their staff through curriculum leadership and innovation.

We found, generally, genuine enthusiasm from all of those involved in the Kampala School Improvement Project. The key factor generating this enthusiasm has been the way in which change has occurred in the SIP schools within a relatively short period of time. Change in teacher quality has traditionally been sought through pre-service and in-service education. What is distinctive about the SIP is that the whole-school approach targets change through a triad of support: workshop training, on-the-job training and the provision and development of educational resources (providing a professional 'meeting place') in Teacher Resource Centers. The Kampala SIP deals directly with teachers' immediate concerns.

Provision of Training

Unlike other teacher development models, in the SIP model training is not only delivered outside the school, leaving the implementation of change to the teachers, it is also followed up proactively by the SIP team of project assistants (PAs) in the schools and classrooms themselves. Traditional teacher development can take many years to show results. The SIP experience suggests that the whole-school approach brings visible changes quickly and sustains the interest of parents, pupils and staff. The model incorporates an emphasis on a child-centered philosophy as well as due regard to the managerial structures that are essential to maintain change.

Our analysis of SIP records revealed that every school received some basic training. The plan was to ensure that all the teachers would ultimately be provided with training. Our observation of workshops run by PAs at the Makerere Road Teacher

Resource Center (TRC) and the TRCs in some schools illustrated how well attended these workshops were. The usual pattern was for members of the SIP team to work collaboratively with school staff and demonstrate effective learner-centered teaching techniques themselves. All of the SIP-associated training at the Makerere Road complex and in the schools was therefore provided by members of the team, and the SIP's strategy involved no specific use of a 'cascade' model of training. The SIP team gave out informative handouts and allowed sufficient time for teachers to work in groups doing the same tasks that children would undertake. At the end of every session, we observed formal evaluations of the workshops being conducted by the SIP PAs. Our analysis of SIP schools' evaluations of the SIP revealed mostly positive comments.

As well as holding workshops in schools, the SIP team members conducted demonstration classroom lessons. For example, the evaluation team observed a demonstration lesson on storytelling by one PA: a story was read to the whole class followed by a discussion and singing, then a written workcard activity about the story was undertaken with groups of children. The classroom teachers observed the lesson throughout. This was followed by an evaluation of the lesson by the teachers and a critique of the PA's approach, bringing together the planning-action-review cycle of teaching. Although we cannot know how typical an example this was of demonstration lessons in general, it was clear that in this particular school the teachers were used to the practice and they responded well. The teachers were appropriately critical. For instance, they pointed out that with the class size of over 80, many of the children could not see the pictures, and that it was hard to keep all the children's interest. The PA accepted criticism and, working collaboratively, the teachers and PA sought solutions to each of the problems.

It seemed to us that the PA was taking on the role of master teacher, working alongside the teachers. According to the SIP records and teachers' evaluation sheets, this is a regular practice in most of the SIP schools. This practice demonstrated the flexibility of the SIP team concerning the needs of schools, because, initially, the aim of the SIP was to develop master teachers from within the schools. However, when this was perceived to be too great a step for many schools, the SIP team cast themselves in this role. From our interviews and discussions with staff in schools, we know that this flexibility and 'practice-what-SIP-preaches' approach was generally admired. We think that this approach has allowed teachers to begin to understand the collective and complex nature of the relationships among lesson planning, classroom organization, the selection of appropriate resources and teaching styles, as these relationships affect pupil motivation and involvement in learning. Teachers' comments on this training were very positive, with the evaluations typically including comments such as "... very useful in guiding us to teach the right way", "... more areas in the curriculum get covered through practical activities", and "... on-the-job training was helpful in correcting, advising and praising (of children)."

Teacher Resource Centers

The Teacher Resource Centers (TRCs) provide a variety of contexts for teacher development. At the Makerere Road TRC, staff provide a book-and-reference-material loan scheme and a place where teachers can meet to study or choose materials for their schools. In our questionnaire to 85 teachers in our study schools, we asked about the benefits of the TRCs, and, although not all the teachers commented, the following were typical responses. These responses match the SIP team's intentions for the role TRCs should play in teacher development. The benefits reported included

- learning and updating skills/knowledge;
- widening teaching approaches;
- making teaching more 'interesting';
- providing materials;
- attending workshops;
- preparing teaching aids;
- accessing cheap materials;
- making teaching easier; and
- demonstrating.

Staff in schools were well aware of the difficulties and resource implications of starting up a local in-school TRC and in all cases they felt strongly that an SIP was needed to initiate the TRCs and provide the starting materials. Another difficulty we noted was the scarcity of space in some schools and the need for security when TRCs were full of resources. This added to the cost burden for the school: some schools simply could not afford their own TRC. It is difficult to estimate how important TRCs are to the success of SIPs, but it is clear that they have a strong support and development role and that the teachers like them.

The rationale behind the development of school-based TRCs, as expressed by the SIP team, is to "develop the culture of material making, develop school and class libraries, encourage an exchange of ideas among teachers, develop school-based staff development, and create easy access to books and materials." Our findings suggest that all of these intentions are beginning to be realized and that TRCs are instrumental in the development and maintenance of the SIP approach. From our discussions with SIP staff, we are confident that they value the role of TRCs in their own work as much as schools do in their work.

Stakeholder Perspectives

Pupils

A total of 144 pupils were interviewed: six pupils from each of the P2, P4 and P6 classes in each school (except for one school, where more pupils insisted they should attend). Boys and girls were equally represented. The children demonstrated a clear interest in working in groups and in helping their friends learn. We found that all of the pupils in Lower, Middle and Upper Primary sections preferred to work in groups. Pupils in one school, who were once placed in groups and were now sitting in rows in a crowded and poorly resourced room, were bitter about having been deprived of groupwork. The pupils revealed that they were happy to support their friends when they were experiencing learning difficulties. They preferred to consult each other first before approaching the teacher. They told us that their teachers visited the groups regularly to check on the groups' progress, to correct, or to mark.

The questions pupils asked us during the interviews were generally related to the conditions in their schools. Pupils in poorly resourced schools wanted to know why they lacked resources and what the government was planning for them. Children in more privileged schools were excited, seemed to know what they wanted out of school, asked content questions and were generally more lively. Pupils who studied in classrooms that were well resourced and where child-friendly groupwork methods were employed, appeared happier and constantly ready for learning.

The pupil interviews revealed a number of issues to be considered by the SIP. Children were aware of a number of personal and administrative problems that had to be tackled if learning was to be enhanced. Pupils in poorly resourced schools thought that they had a number of alternative suggestions to meet some of the challenges. Greater communication with pupils would also help in identifying some of the communication problems with parents. It is likely that such problems often contribute toward pupils dropping out of school. For example, some pupils informed us that they "were beaten at school for reaching school late and yet our parents want us to do work before we come to school."

Parents

Thirty-one parents were interviewed, but 21 (68%) of these were also teachers. The potential for bias in this sample has already been referred to, and clearly the sample was far from representative of the overall parent population. Thirteen parents reported that they held a Diploma in Education, another 13 held Grade III teaching certificates, one had an A-level certificate and six held degrees. The parents responded to several questions, some of which were personal (concerning their families) and others connected to the schools and the programs.

All the parents interviewed told us that they had visited the children's classes. Their expectations varied, but a few priorities were identified. They felt that the schools needed to provide

- "quality schooling" (81% of parents);
- "discipline/good citizenship" (55%); and
- "moral education" (23%).

They felt that the schools should also prepare the children

- "for high school" (34%); and
- "for exams" (23%).

The majority of parents were aware of the SIP, but most of these parents (as observed earlier) were practicing teachers. Five of the parents (non-teachers) revealed that they had learned of the SIP through a parents' meeting, and six learned of it through newsletters. The practicing teachers claimed that it had been introduced to their school without discussion, and two of the parents (both teachers) were skeptical about the whole exercise and its objectives.

However, the parents who had experienced some advantage of the SIP program claimed that it was involving parents, which they felt was good (52% of parents reported this). Many felt it was supportive (45%), educative (42%), and supported pupil responsibility (34%). The majority of parents were clearly pleased that the schools employed trained teachers who were constantly refreshed (apparently by the SIP). They were pleased that the schools used SIP child-centered methods and that they had the necessary facilities to provide 'balanced' schooling.

Generally speaking, the parents whose children had stayed longest with the SIP tended to be the most positive, while other parents were more critical and hopeful that a SIP would improve the situation. Parents were generally positive about the teaching methods being used, although the majority still indicated a need for improvement in team work, curriculum issues, morals and teaching methods (58%). Most of the parents were convinced that the large number of pupils caused by the introduction of Universal Primary Education was having detrimental effects.

The parents reported a number of positive behavior changes in their children that they considered to be specifically due to the increased freedom and group work provided in school. All of the parents reported an improvement in their children's acceptance of responsibility. A large majority also reported their children to be more investigative (87%), more sociable (81%), to have improved language use (61%), and to be more organized (and time conscious) (61%). Others commented upon their children's increased self-control and confidence.

Head Teachers

Each of the eight head teachers was interviewed, and their responses to our questionnaire were analyzed. The head teachers identified the benefits they considered most significant in the implementation of the SIP. They felt that pupil participation was better, that class organization had improved, that library provision was beneficial, and that there was better planning and provision generally. The head teachers also referred positively to improved standards, the use of practical approaches, subject integration and the provision of low-cost resources.

The head teachers were concerned about the effect of Universal Primary Education on the SIP, but felt that the SIP was helpful in alleviating some of the difficulties. The increased provision of materials and the grouping methods helped them deal with the difficulties they were facing due to increasing pupil numbers, higher pupil-teacher ratios, and the constraints on teaching brought about by the increased demand for resources.

The SIP Coordinators

The SIP coordinators made up a young, enthusiastic and energetic team. The majority were volunteers, although a few were nominated by their administrations (their head teachers or other teachers deciding by voting). The coordinators cited a wide range of benefits that had been gained from the SIP. They felt that the workshops had led to an increase in pupil participation in many classrooms, and that this would lead to improvements in the pupils' English-language, literacy and mathematics skills. They felt, however, that at this stage it would be 'too early' to see these learning improvements. They also felt that activity learning and teacher collaboration had been enhanced. In achieving these improvements, the influence of the workshops and the TRCs was especially emphasized.

The SIP coordinators also reported a number of problems and constraints that they encountered:

- insufficient resources;
- administrators who were slow to accept SIP methods;
- the difficulties of mobilizing teachers;
- change of teachers; and
- the need for sufficient teacher numbers.

They felt that the SIP and their role within it could be developed in a number of ways. In particular they could

- make more teaching materials, especially models;
- conduct more workshops;
- monitor teaching;
- find new ways of motivating teachers;
- encourage visits between and among schools; and
- increase parent involvement.

The coordinators also suggested that the SIP should divide schools into infant, middle and upper stages for better management, and arrange more staff meetings and workshops. The coordinators saw the benefits of the SIP to be relevant first to pupils and teachers, then to the school and community. This seems to be in line with SIP managers' expectations.

We asked the coordinators how schools would manage if the SIP support was withdrawn or stopped. They had clearly not given this issue a great deal of thought and the implications were not at all clear to them. In our opinion, greater attention needs to be given to this issue, as planning for withdrawal of SIP support will be vital for sustainability.

Teachers

A total of 85 teachers completed our questionnaire and the majority felt that the SIP methods made their work easier (66%). Only 8% felt SIP methods made their work harder, although 26% seemed undecided on this issue. We explored these issues further: We feel that the teachers' answers are significant, despite the fact that the open-ended questioning made quantification inappropriate. Teachers who found their work easier with SIP methods cited the following reasons:

- SIP "encourages pupils to work alone";
- SIP "simplifies teacher's work"; and
- "the grouping."

Those who found SIP methods more difficult referred to them

- being "difficult to manage";
- being "time wasting/demanding";
- meaning "handling many tasks"; and
- "taking long periods."

The teachers also felt that the SIP had helped them in dealing with large (Universal Primary Education) classes. They told us that some classes now have three teachers, that the grouping had been helpful and that the teaching aids and learning resources were valued. SIP helped them keep children 'busy' and "encouraged brighter pupils to assist weaker ones".

We asked teachers what they considered to be the main benefits of the SIP. The following aspects were cited, in order of importance (i.e., frequency of citation).

1. The improvement of discipline.
2. Making cheap visual aids.
3. Grouping children.
4. Attending demonstration lessons.
5. Working with other teachers.
6. Children learning better.
7. Understanding SIP philosophy.
8. Re-organizing classrooms.
9. Clinical feedback.

The vast majority of teachers who responded to our question on training were satisfied with the standard of their preparation for the SIP, rating it good (57%) or adequate (37%). We also obtained more detailed feedback on the 'on-the-job' training provided. The responses showed that teachers were generally satisfied with SIP training and preparation in terms of the choice of content, the SIP personnel, and the teaching methods employed.

Classroom Practices

Classroom Observations

Using our Classroom Observation Checklist, we observed 72 classes in the eight SIP evaluation schools. Two testers, who had been tested for reliability, undertook all of the observations, each of which lasted 20-35 minutes and was followed by a short question-answer session with the class teacher. The results suggested that the whole-school SIP approach, and the utilization of child-centered resources and pedagogy were more sporadically implemented than the teacher questionnaires suggested. This could be partly due to the furnishings and equipment in classrooms being inadequate, but there was evidence that some teachers lacked the awareness to make the best of their practice according to the SIP approach. For instance, although there were carpeted floor areas, the children were rarely observed actually using these areas, even where there was a shortage of seats.

While many of the more subtle SIP objectives have yet to be realized in the majority of classrooms, given that some schools have only been involved with SIP for a very short period of time, we observed fairly dramatic changes in many of the classroom environments. Teachers were keen to gain further support and to help children learn. SIP has clearly met its Phase 1 objectives to foster more child-oriented practices in project schools, and it will be the reinforcement of these practices that needs to be addressed in the future.

Displays

Central to the SIP approach has been the development, display and use of low-cost educational games and materials that are to be used in a child-centered-approach emphasizing activity. The majority of teachers (86%) had produced at least a few educational displays and most had also made an attempt to place these, along with the children's work, at child height. However, more sophisticated displays such as 3-D constructions were found in only 25% of classrooms observed, and fewer than half the teachers displayed educational wall charts.

It was suggested that schools had recently started term and this could account for the lack of displays, but discussions with the teachers concerned revealed that they believed their classrooms looked normal. In general, the teachers who showed greatest evidence that they had adopted the SIP approach of making resources were found in the lower grades of the school, where the same teachers remain in their own classrooms, unlike the subject-specialist teachers of older children.

Grouping was also something the teachers appeared to be enthusiastic about. But more than half of those observed seating the children in groups did not expect the children to work together. Some of the teachers did say that they used a mixed method of teaching, and that the advantage of groupwork was mainly that children could engage in peer support and tutoring. We believe there is a good deal of research evidence to support this claim. Another problem we observed with grouping in large classes was that some of the teachers had little idea of the optimal size of a workable group. We observed classes where children were sometimes put in groups as large as 18 children, or where the tables formed a rectangle which had a space within it, thus restricting the movement of objects on the table.

Teachers' styles of teaching were observed to lean more towards the didactic approaches that were more familiar to the teachers. Most teachers followed the school timetable and many planned their lessons; however, very few teachers were observed to have given children tasks differentiated by group ability (although grouping by ability for particular subjects was favored by most teachers). The worst aspect in this area was that fewer than one in five teachers took the opportunity to actually teach small groups of children; most preferred to 'patrol' the classroom or 'trouble shoot' from group to group. The SIP team consider this a step toward more child-centered approaches as teachers gain in confidence in the SIP approach. This demonstrates to the evaluation team that schools need a longer period of support than was envisaged in Phase 1. In Phase 2, the AKF/SIP team should consider a follow-up stage of, say, one year for schools who need support, possibly through a 'light-touch' approach, that is working in the schools on a per-month basis, rather than a per-week basis.

The SIP teachers encouraged child-to-child discussion. We saw some evidence of this in 76% of the classrooms, but there was still a good deal of teacher-dominated interaction. Sometimes this was appropriate; for instance, we observed a class of older children who had just received a lesson on an aspect of science that was new and required the teacher to impart factual knowledge. In other cases, teacher-dominated interaction was the teacher's preferred style. The area where more work is required is in encouraging teachers to let children initiate discussion with each other and with teachers. Only a third of classrooms observed showed evidence of child-initiated talk. We were sympathetic to teachers who wanted to make the change gradually because they feared losing control of their large classes.

Most children in SIP classes showed positive self-esteem and behavior: Only in 14% of classes did we observe children to be subdued, and only a similar percentage to be off task. Children were observed to ask for help with their work in about 49% of classrooms.

Most teachers had some lesson objectives, although some clearly did not understand the difference between an educational *objective* (e.g., learning shapes) and a *task* (e.g., completing a worksheet). Most teachers were alert in their classrooms and were able to identify children who were off task and deal with them. We were interested to observe curricular activity and evidence for the presence of materials and games on which the SIP team had given workshops and training. For instance, SIP focused heavily on early literacy approaches and low-cost educational games.

Most teachers had made language, mathematics and science games and displays. While the teaching and learning of language was emphasized, we were pleased to find that a good proportion of teachers were also taking other core subjects seriously. The impact of SIP was clearly evident; for example, child-made resources in classrooms, where these were evident, were almost exclusively zigzag books. However, a lower percentage of classrooms had wordbanks in use. Some teachers had extended the ideas imparted by the SIP team; for instance, wordbanks were developed for particular subjects or topics. In this area, SIP work was in evidence in most classrooms but especially in the lower primary classrooms.

Table 1: Teacher Preparation and Style

Elements	Yes		No	
	Percentage	Frequency	Percentage	Frequency
Pupil tasks differentiated	14	10	86	62
Evidence of lesson planning	79	57	21	15
Learning objectives clear	69	50	31	22
Timetable followed	92	66	8	6
Teacher made workcards	76	55	24	17
Only 'chalk & talk'	45	32	55	40
Teaches small group	17	12	83	60
Teacher 'patrols' the class	53	38	47	34
Teacher 'troubleshoots' from group to group	60	43	40	29
Teacher notices children not on task	88	63	12	9

Unit Costs
The analysis shows that the unit cost per child has dropped dramatically and that the project is beginning to be very cost effective.

Table 2: Breakdown of Costs in Phase 1 of the SIP

Unit cost per child	US$
Stage 1	149.60
Stage 2	53.90
Stage 3	12.60

Note. Exchange rate: 1000 Ugandan Sh = 1 US$.

When the full impact of the project is taken into account and the number of teachers, student teachers, inspectors and teachers outside SIP schools that have benefited from the project are accounted for, it is clear that the project should be continued. At this juncture the SIP is a lean, effective and efficient force for change and one that is ready to be launched on a larger scale.

Conclusions and Recommendations

The SIP has come to reintroduce and reinforce, within schools, what Ugandan pre-service teacher educators had promoted: the behavior, actions, attitudes, skills, knowledge, and values that were the bedrock of primary teacher education in Uganda prior to political instability. Primary teachers trained in Christian- and Muslim-funded and run colleges were encouraged to accept a practical concern for people, information, environment, and resources. The teacher education focus in the pre-civil-war Uganda was to develop humanity and society, and to manage change (Odada, 1997), and the SIP initiative has been seen as an active intervention to reestablish these values. This could go some way toward explaining why all parties within the SIP process have tended to be unreservedly positive about the project.

This whole-school-improvement approach, with some outreach to parents and community, has the advantage of impacting on the schools' ethos as well as effecting change in classrooms. One expected outcome has been change in the management styles of head teachers and the teaching styles of teachers, to optimize learning. A closer look at primary schools in Uganda reveals two broad types of school and classroom management. The first is a form of management by hierarchical authority, where the head teacher is autocratic, and the second is management through operational objectives. The former was common in Kampala SIP schools before the SIP intervention. With SIP, the change is toward the latter style and the majority of the SIP schools demonstrate change for the better as they work toward more collegial relations between head teachers and teachers.

The SIP project team members were clear about their determination to change teacher practice to accommodate more 'child-friendly' teaching and learning strategies, and to develop a professional base for training. They applied a multi-faceted approach to training that was highly successful in most of the schools. One of the most successful aspects of the design was the flexibility that was incorporated into the project through a regular cycle of self-reflection and evaluation. For instance, after six months of the project, new objectives were added in order to respond to head teachers' and schools' conditions and needs. One particular example is the way in which the notion of in-school 'master teachers' was dropped in the face of teacher resistance. Instead, the project assistants (PAs) become teacher-demonstrators of good practice.

The SIP has influenced 15 schools. In the best schools, there is evidence that the SIP has entirely transformed the environment of classrooms and instilled an ethos of improvement, by mobilizing most teachers' expectations of themselves and the children. Many teachers have developed new skills in making low-cost educational materials. New skills have also been gained by many teachers in more child-centered or friendly methods of classroom organization.

The Makerere Road TRC has been successfully established and school-site TRCs have also proven to be popular and effective innovations. The 'on-the-job' training has bridged the gap between theory and practice for teachers. It has provided the vehicle by which teachers have been convinced of the need for change and the process by which teachers have adapted their classroom practices to develop a more child-centered school ethos. It has provided for staff development and the development of school infrastructure. The project director has provided ongoing training for the staff and has developed what we perceive to be a highly trained team of trainers, who are knowledgeable, committed, flexible and sensitive to the needs of SIP schools. The team members have a particular strength in early years education. Similar expertise could be developed for the later primary years in any future phase of the project.

The development of greater participation by parents and community groups is vital to the future of the SIP. More work is needed with key personnel in the local establishment to promote the integration of SIP into the education system in Kampala, and to ensure sustainability during the next phase.

The SIP has clearly met its Phase 1 objectives to foster more 'child-oriented practices' in the project schools. The evaluation team members were therefore able to unreservedly support the ongoing development of this project to Phase 2.

The constitution of an advisory board, in the form of a consultative or steering committee, should be considered for Phase 2. This would provide

- the benefit of representation, and support, from local educationalists and researchers;
- improved local communication; and
- the local community with a stake in the SIP.

Now that the 'word-of-mouth' news about SIP has spread to other Kampala schools, the project methods are being more easily accepted and many more schools are keen to be recruited. The annual unit cost per child has dropped dramatically and the project is now particularly cost effective. When the full impact of the project is taken into account, beyond the SIP schools to teacher development through TRC and conference in-service education, it is clear that the project has a wide ranging impact.

Many teachers do, however, remain unclear about the educational objectives of the *child-centered* approaches being adopted. This should be studied further and efforts made to develop theory and practice within the Ugandan context. In Phase 2, the provision of 'light-touch' follow-up support for schools in their second year should therefore be considered.

We consider it appropriate that Phase 2 should be set within a 5-year development period, and that specific measurement of teacher development, child development, or both, should be included to aid future evaluation.

The role of the TRC has been fundamental in training and in motivating educationalists. It is also a physical embodiment of the SIP approach. It should therefore remain a central part of SIP. The location of the main TRC should be reconsidered, and its role vis-à-vis the school-based TRCs considered further. Prior to Phase 2 of the project, AKF and the SIP team should calculate the costs for the maintenance and development of the main and school-based TRCs.

The SIP should continue to encourage teachers to assist pupils working in groups. There is clearly a need for teachers to be more aware of when their assistance is required by pupils and how to deliver assistance without offending pupils. Pupils revealed that they were most comfortable if teachers presented themselves as members of the group when they gave this form of support.

We consider parental partnership to be crucial in providing effective primary education. The comments reported are encouraging but should be treated with extreme caution, given the potential bias of our sample.

It has been suggested that the development of community links has proven difficult in the first phase and therefore needs to be a top priority in Phase 2. Education in Uganda is heavily reliant on community support, with 476 private primary schools compared with only 78 government primary schools. Mobilizing parents and community groups is therefore vital to the future of SIP. The pattern so far has been to emphasize the role of parents in fundraising. The challenge will be to develop common understandings and a genuine educational partnership.

The suggestion by SIP coordinators to reorganize the project to take account of the differences between infant-, middle- and upper school curriculum management should be carefully considered. The SIP should monitor the effects of its intervention on these different levels of schooling, and we suggest that a more subject-centered (rather than whole-curriculum) approach to the development of group work may be beneficial in the upper years.

In following up teachers' comments about the need for greater official recognition of their SIP work, we asked how and why they wanted SIP experience recognized by the Ministry of Education. Most of the teachers wanted certification by examination (24%) or by continuous assessment (47%). These issues need to be pursued in the next phase and discussed with Ministry officials and local postsecondary institutions, such as Makerere University and the Institute of Teacher Education-Kyambogo (ITEK).

Further Recommendations for Phase 2

The following additional recommendations need to be considered for Phase 2:

1. More teachers need to be convinced that their fear that the implementation of SIP methodology will result in their syllabus not being completed is unfounded.
2. Further effort needs to be made in developing teacher understandings of educational objectives and the benefits to be gained from small group work and displays.

3. SIP implementation seems to be strongest in infant classes, then middle, and lastly upper classes. The reasons for this should be investigated further. We know that the SIP team has particular strengths in early years.
4. Some teachers' views that SIP makes their work 'easier' need to be studied further. We think some teachers have misunderstood child-centered teaching in this respect. The SIP formats for scheming and lesson planning need to be addressed again, as some teachers are still unclear about the educational objectives of their lessons.
5. The SIP is popular and widely accepted as beneficial. As it expands to schools, SIP staff should have a strategy to penetrate teacher education institutions (as they have done successfully with Kibuli Primary Teachers' College and ITEK) so that teacher education institutions use the SIP approach for training in-service teachers. This should have a greater sustainability and multiplier effect.
6. Sustainability is a big concern for the future of not only SIP but other similar projects. Sustainability here means a willingness and an ability to learn and spread all or part of the SIP principles and practices. For instance,
   (a) Schools should be encouraged to take on a gradual but practical commitment to resource allocation for SIP.
   (b) Schools that are successful should be encouraged to strengthen their outreach activities.
   (c) Parental and community participation should be encouraged through persuasion.
   (d) SIP and AKF staff should enter into dialog with the Ministry of Education, district education officer (DEO), relevant NGOs, teachers' colleges, universities and other institutions, with a view to requesting them to begin trialing SIP philosophy and eventually adopt it in pre-service education.
   (e) Both SIP and AKF staff should open up more targeted dialog with possible partners, focusing especially on short courses and working for contacts with creditable institutions for accreditation, certification and examination. The latter seem motivating aspects for teachers in the SIP.
7. The AKF/ SIP team need to revisit fundamental, taken-for-granted concepts such as 'child-centered', 'activity-based', 'teacher resource centers' and evaluate their appropriateness for Phase 2, in the light of Universal Primary Education and the developing-world context, in which basic education is still a priority. Activity-based education with a mixture of whole-class teaching would be a more convincing and rational approach for a 5-year program. AKF should look at the evidence in East Africa on the relative effectiveness of child-centered approaches in the context of large classes with poorly trained and poorly remunerated teachers. Other studies are doubtful.

8. A greater emphasis should be given to management training. Head teachers are instrumental to the change process. Management training is also a way of recruiting schools, after their head teachers' training.
9. There needs to be further analysis of why on-the-job-training is so popular with teachers, and their progress needs to be monitored once the SIP team has left the school. The Kampala SIP is a good model for this in East Africa.
10. Developing community links has been difficult in the first phase due to time constraints and needs to be a top priority in Phase 2.
11. For Phase 2, the role of Aga Khan Education Services Uganda should be considered and defined.
12. AKF should consider developing the SIP with the specific purpose of evaluation in mind. The need for baseline measures for Phase 2 should be built into the project.
13. Large classes undermine the SIP principles and practices. Class size should be carefully monitored, and strategies such as team teaching should be explored where there is more than one teacher in the class, as is often the case.
14. Accreditation and certification provide ways of recognizing SIP contributions and motivating teachers. This should be taken up with the relevant government agencies.
15. In order to streamline and guide the SIP staff in the expansion and consolidation, we suggest that an advisory committee of not more than nine people be introduced. This committee should be charged with the task of doing public relations, encouraging, motivating and collecting partners for joint ventures.
16. SIP staff should be encouraged to enroll in further professional development; for example, masters degree courses at Makerere University or elsewhere, if time and funding permit and the professional development does not interfere with the SIP.

6

Decentralized Partnerships for School-Based Teacher Development in Mombasa, Kenya

Stephen E. Anderson and Shelom Nderitu

The goal of the Mombasa School Improvement Program (MSIP) was to improve the quality of teaching and learning in Mombasa primary schools, with an emphasis on supporting the use of child-centered activity-oriented teaching and learning methods. The plan targeted Standards I to VI in all primary schools (government and private) within the district. The in-service training game plan included workshops based in teacher centers for teacher representatives from all schools, and in-class assistance by project consultants (referred to as project officers) to successive cohorts of schools (designated 'project schools') over the life of the project. Drawing upon insights from earlier SIP projects in Western Kenya and Uganda, the project developers identified several related objectives as important to the sustainability of improvements associated with the in-service training activities, including head teacher management training, mobilization of parental involvement, and capacity building to strengthen the professional support capabilities of the local education system (reactivation of Teacher Advisory Centers (TACs), change facilitator training for district supervisors and teacher developers). Initial funding for the SIP extended from 1994 to 1999.[7] The findings and conclusions in this chapter focus on our evaluation of MSIP capacity building and teacher development outcomes at the end of the fourth year of the project (Anderson & Nderitu, 1999).

Local Education Context

Mombasa District presented a challenging context for school improvement in 1994. There were 112 primary schools. The student population in Standards I to VI was approximately 40,560. The teacher population stood at about 1,014.[8] School settings ranged from urban to rural. Many of the rural schools were located in very poor communities. The district had a history of poor educational performance and a reputation for low community support for education. Nationally, it ranked 60th of 63 districts on Standard VIII examination results. From 1994 to 1999, the number of primary schools increased to 162, a phenomenon attributed to the proliferation of private schools as government funding for and parental confidence in the public system waned.

While subject to national curriculum policies, accountability (i.e., student testing), and organizational requirements, the management of primary education (Standards I to VIII) in urban areas of Kenya is handled by local municipalities under the authority of a municipal education officer (MEO), with the assistance of inspectors and school advisors employed by a municipal council. The MEO and local education officers are mainly responsible for the implementation of education policy, the deployment and in-service development of teachers and head teachers, the supervision of curriculum and its implementation in schools, and the coordination of testing. Their involvement in school facilities development, education funding, education materials, and supplies has been reduced as a result of government cost-sharing policies that have progressively shifted financing of primary school education to the local school level, from 1984 to the present. Currently, central government funding is essentially limited to teachers' salaries. Local education authorities have the power to impose levies on parents to generate additional funding for education materials, school facilities, and the like. These funds are pooled and centrally administered by district authorities. Schools may engage in their own fundraising activities under the auspices of parent-teacher associations (PTAs) and school committees. Government teachers are employed through a national Teachers Service Commission (TSC). Some schools hire extra teachers through the fundraising efforts of PTAs.

The Mombasa Municipal School District is subdivided into five divisions, with two zones per division. The municipal education officer is assisted by a deputy MEO, a district inspector, and a chief school advisor. The latter are responsible for coordinating and supervising the work of 10 zonal school inspectors and five divisional school advisors. During the MSIP project, the municipality did not operate with a full slate of divisional and zonal officers. District funds were used for the MSIP project officer positions, with the expectation that individuals seconded to those positions would rejoin the system as advisors or inspectors at the project's end, ostensibly with a renewed capacity and commitment for teacher development work at the school level.

Primary school Teacher Advisory Centers are a unique feature of the Kenyan education context. Policies to establish and regulate Teacher Advisory Centers date from the early 1970s (Ayot, 1983). By 1998, about 1,300 district and zonal level TACs had been established by Kenyan education authorities (Welford & Khatete, 1998). TACs are a responsibility of municipal (district in rural areas) education officers. Most are located in school compounds but are meant to serve clusters of schools within the jurisdiction of a district or zone. They are staffed by TAC tutors, who must be qualified teachers with several years of teaching experience. The government provides no funding for TACs, other than the tutors' salaries through the TSC. District authorities and schools served by each center are expected to raise funds to equip and maintain the TACs. The roles of inspectors and TAC tutors are supposed to be complementary. According to a study of Kenyan TAC tutors, many tutors see themselves as assistants to inspectors, and view their jobs as stepping stones to the inspectorate (Osler, 1997).

The public education system in Kenya is structured on an 8-4-4 model. Public schools are subject to national curriculum policy (subjects, time allocation, content). Access to secondary education is largely determined by performance on national primary school leaving exams at the end of a pupil's eighth year of primary school. Exam results are publicly disseminated, and the government publishes rankings of school 'positions' based on overall and subject-specific 'mean scores' for students from each school. Educators and parents are strongly oriented toward exam results as an indicator of pupil success and school quality.

Historical Overview of the Mombasa School Improvement Project

The MSIP was initiated in November 1994 as a partnership among the Aga Khan Foundation (AKF), Aga Khan Education Services (AKES), and the Municipality of Mombasa. AKF agreed to seek and provide external funding for the 5-year project. As grantee, AKES agreed to take responsibility for project management: hiring a project director and support staff, financial management, and setting up the Project Implementation Committee to oversee implementation of the partnership. The municipality agreed to provide office and training facilities, two used vehicles, cooperation from the municipal education officer and education personnel, and to appoint and pay the salaries of a team of project officers to serve as in-service trainers as specified in the project design. From a sustainability viewpoint, a key feature in the design was the employment of project officers through the local education system, rather than through external contracts. The project director and community development officer (CDO) positions were funded by the project.

MSIP participants described the project in distinct phases, corresponding to the extension of school-based assistance to successive cohorts of project schools. The number of project schools increased as more project officers were appointed. Phase 1 (1994-1996) included the project mobilization period and the delivery of school-based training in four project schools. During Phase 2 (1996-1997), school-based training extended to another 16 schools. In Phase 3 (1997-1998), 30 project schools were added.

Phase 1

The first year (1995) was largely spent in negotiations and logistical arrangements to get the project mobilized, including establishing a joint committee to oversee and coordinate implementation of the partnership, allocating and refurbishing facilities for an office and training site, hiring a project director, and initiating the process of appointing project officers. The appointment of project officers (POs) was more complicated than anticipated as a result of bureaucratic matters, with only four on board by December 1995.

During 1995, the major areas of project activity included the delivery of project awareness sessions to education officers and head teachers across the district, the preparation of baseline surveys to aid in the selection of project schools and project evaluation, and preparations for school-based in-service work with the first four project schools. Teacher workshops and in-class assistance began in four schools in one zone near the project offices during the first half of 1996. Management training sessions for head teachers, TAC tutors and education officers were also held during this time. Two more project officers were appointed, and professional development for POs took place with help from staff of the AKF-and-AKES-sponsored Kisumu SIP. (See Chapter 4 for a description of the Kisumu SIP.)

The project experienced a major transition midway through 1996. The implementation model was significantly revised in light of the perceived need to develop a more sustainable infrastructure for in-service assistance to teachers. The new model called for decentralization of MSIP activity to the zonal Teacher Advisory Centers. Each PO would be teamed with a TAC tutor and zonal inspector to organize and deliver in-service workshops and school-based in-service assistance to schools in their respective zones. Workshops would be open to all primary schools. POs were to provide in-school assistance to three schools at a time for a full school year (3 terms), rather than 1 term as initially projected. In order for this model to work, 10 zonal TACs and the district TAC had to be operationalized, TAC tutors appointed, and TAC management committees established in all zones. A portion of project funds was reallocated toward refurbishing and equipment for TACs (e.g., typewriters). Responsibility for assisting with reactivation of the TACs was assigned to the MSIP community development officer appointed in October 1996. Workshop activity was suspended while efforts turned toward the mobilization of the zonal TACs.

Phase 2

During Phase 2 (September 1996-August 1997), six POs worked with 20 project schools (three in each of five zones, plus four Phase 1 schools and a new school in the sixth zone). Six zonal TACs and the district TAC were inaugurated between November 1996 and March 1997. Two more TACs were opened during the second term of 1997. With guidance from the MSIP community development officer, TAC management committees representing all primary schools in each zone were established. A system of financing the renovation and activities of TACs through annual contributions from school committees served by each TAC was introduced. The CDO held meetings with parents at project schools to explain the project and to help organize parents to take control of school funding and commit to support for the TACs. Reading workshops for teachers were held in six zones during the first term of 1997. Mathematics workshops were delivered in eight zones in the second term. By Phase 2, two vehicles were available for project use. The project officers established a routine of weekly visits to project schools to do classroom observation and coaching of individual teachers, and to hold in-school meetings and workshops with teachers and head teachers. Toward the end of Phase 2, two more POs were appointed, bringing the total to eight.

Phase 3

Phase 3 (September 1997-August 1998) of the project began with the identification of 30 new project schools (3 schools for each of the 10 zones). Two POs were assigned to work with schools in two zones. TACs were officially opened in the remaining two zones during this period. Project activities were interrupted for several months toward the end of 1997 as a result of a series of unfortunate events, including a nationwide teacher strike, civil disturbances associated with interethnic tensions and national elections, and torrential rains and flooding throughout the coastal region. Workshop activity and Phase 3 project school visits did not return to normal until early in 1998. Reading workshops were delivered in four zones not previously served. Science workshops and oral communications workshops were provided in all zones during the first 2 terms of 1998. Two POs were added during the second term, bringing the total to 10 (one per zone). Midway through this phase, as a result of feedback and follow-up visits to Phase 1 and 2 schools, a decision was made for POs to spend the final term of 1998 revisiting those schools, while maintaining a diminished presence in Phase 3 schools. We gathered evaluation data during that period. Plans were in place to serve 40 more schools from January 1999 until the end of project funding in September.

The Evaluation Study

The MSIP case study evaluation was commissioned in September 1998 as the project entered the final phase of its 5-year mandate. The study was designed to describe the MSIP project and to evaluate its major components and impact on classroom teaching methods, student learning, head teacher management, Teacher Advisory Center services, and parental involvement. Evaluation data included project proposals and implementation reports, and information gathered from interviews, observations, surveys, and project files during a 2½-week site visit in October 1998.

Specific data sources for different focuses of evaluation are noted where appropriate throughout this chapter. The evaluation was framed in relation to project goals identified in project documents and stakeholder interviews, and in relation to relevant educational research literature. We considered the quality of teacher in-service activities, for example, with reference to the known characteristics of effective in-service training (e.g., Joyce & Showers, 1980; Loucks-Horsley et al., 1987; Craig et al., 1998). We employed a developmental model and procedures for assessing teacher implementation of teaching methods advocated in MSIP in-service activities (Hall & Hord, 1987; Anderson, 1997).[9] Information about MSIP inputs to increase parental involvement in schools was interpreted in light of Epstein's framework for school and community partnerships (Epstein, 1995).

The remainder of this chapter is devoted to presentation and discussion of three major focuses of the evaluation study: MSIP efforts to develop local capacity for continuous school improvement; the project in-service training inputs for teachers; and classroom outcomes (i.e., teacher use of desired teaching methods, student outcomes). The chapter ends with a discussion of insights and issues for school improvement based on the MSIP experience and evaluation that might inform similar projects elsewhere.

Evaluation of MSIP Capacity-Building Activities

The MSIP game plan for improving the quality of education in Mombasa primary schools included interventions to strengthen the school system's capacity to sustain ongoing teacher and school improvement. Here we consider project goals, implementation, and outcomes for three key focuses of capacity-building interventions: head teacher management, parental involvement, and the Teacher Advisory Centers.

School Management by Head Teachers

The original project design called for intensive training for head teachers in basic management skills and in management practices associated with school and teacher development. This design was modified to avoid duplication of a national professional development program for head teachers called PRISM (Primary School Management). Head teacher training by PRISM trainers in Mombasa School District began in early 1997, and included a 3-week course, 6 months of practical assignments, and a 1-week follow-up session. The course covered a range of management topics and skills (e.g., financial and personnel management), and gave special emphasis to establishing school development plans that focus on curriculum, staff development, resources, and facilities. As the MSIP evolved, project support for head teachers relied on PRISM for basic management training, and focused on head teacher efforts to carry out their school development and leadership roles in relation to MSIP teacher and parental involvement aims. This occurred through management seminars and field-based consultation with MSIP personnel.

Project Awareness Meetings and Management Training

Zonal meetings were held at the outset to familiarize primary school head teachers with the MSIP goals and plans. Cooperation from the municipal education officer ensured that virtually all head teachers gained an initial awareness of the project's intentions. These meetings were followed up at the start of each phase by MSIP orientation and change management seminars for head teachers, education officers, and school committees associated with each cohort of project schools. These sessions covered topics ranging from the role of the head teacher to specific areas of management activity, such as school development planning, accounting, resources, teacher collaboration, community involvement, lesson planning, teacher evaluation, child-centered teaching, and strategies for improving education quality. Further seminars were held at the end of each phase on issues and strategies for sustainability beyond the project school period of assistance.

Consultation with MSIP Project Officers

MSIP support for head teachers included consultation with POs during their project school visits to coach teachers, as illustrated in excerpts from PO interviews and diaries.

> Discussing with the head teacher after I have observed lessons in their schools ... Advising head teachers on areas that they also could observe, advise, and monitor for progress. *(PO interview)*
>
> Meeting with H/M [headmaster]. The following was discussed: (1) shortage of desks; (2) shortage of materials for teaching aids; (3) large populated classes; (4) sustainability of the activities. Resolutions: (1) to bring more desks, has now bought 120 desks; (2) materials to be bought immediately; (3) he plans to admit two Class 1s and spread other surpluses by repeating the pupils [advised to seek MEO permission]; (4) to meet the school committee for fund allocation [reported that the school has enough money]. *(PO work diary)*
>
> During visits to project schools I normally make a follow-up to head teacher's implementation of knowledge acquired during PRISM seminars ... which addressed school development plans. I usually ask the head what he or she is doing with the SDP [School Development Plan]. *(PO interview)*

The MSIP community development officer provided added assistance to project and nonproject school head teachers with parental mobilization (see Parental Involvement).

Evaluation of Head Teacher Training and Assistance

We investigated head teacher understanding of MSIP instructional goals, their engagement in change-facilitating actions, and improvement in school management.

In our interviews, we found a high degree of consistency in head teacher understanding of the core elements of activity-based child-centered learning advocated through the MSIP. Key pedagogical concepts highlighted by most of the head teachers interviewed included pupil participation in learning activities, pupil exploration and discovery, teachers as learning guides, pupil questioning, pupil-teacher interaction, students learning at their own pace, and reduced emphasis on knowledge transmission by lectures and notes.

We used Hall and Hord's Intervention Taxonomy (Hall & Hord, 1987) as a framework for classifying head teacher change management activities into six categories of action, as follows: developing supportive organizational arrangements, training, providing consultation and reinforcement, monitoring and evaluation, external communication, and dissemination. Our interview data suggested that head teachers were directly involved in developing organizational arrangements conducive to teacher change (e.g., school planning, working with school committees to secure funding and parental support for school development, collaborating in the management of zonal TACs, allocating school funds for materials and professional development, structuring subject panels and other teacher work groups, and organizing teaching assignments to facilitate interaction about new practices). While head teachers did not take a lead role in training teachers in child-centered methods, they supported the training provided in MSIP- and TAC-sponsored workshops by strategically selecting teacher participants, encouraging teachers to cascade their learning to colleagues at the school, supplementing costs of attending in-service activities (e.g., transportation fares), facilitating PO and TAC tutor classroom visits, and providing their own in-class observations and feedback. Head teachers also encouraged and enabled consultation and problem-solving among teachers implementing the changes with POs, TAC tutors, and peers through classroom visits, staff meetings, class meetings, subject panel meetings, and in-school workshops. At the school and zonal levels, head teachers were actively involved in system-wide efforts to coordinate standardized testing of students, and to analyze and use student test results as a basis for classroom and school improvement. Finally, under the direction of the MEO and district education officers, head teachers from neighboring schools in some zones were meeting regularly in clusters to exchange ideas and coordinate activities such as examinations and interschool competitions. While we did not determine how reliably this image reflected the actions of head teachers across the district, it was evident that knowledge, if not skill, in change management was available to head teachers in the district.

To what extent were the reported knowledge and practices of head teachers shaped by MSIP training and support? We asked head teachers and teachers from our school sample about changes in head teacher practices associated with MSIP and PRISM training. The head teachers emphasized improvements in general management skills, such as reflective practice, school development planning, participatory decision-making, and financial and personnel management (e.g., leadership development, conflict resolution).

> I've seen what I was doing wrongly. I've reflected on my management practices. I'm able to focus on certain things I want to put right ... Before we didn't have school mission, aims, school plan ... Before we had plans to build this ... etcetera ... but not systematic ... no priorities ... Now things are more systematic ... I introduced the school development plan in which both the teachers and the PTA participated ... There's a lot we have learned ... for example, financial management ... We weren't very conversant ... Now manage funds better ... It has been easier for me to identify leadership qualities in teachers, for example, who is best suited to do guidance, clubs like scouting, subject allocation. *(Composite quote, head teacher interviews)*

Teachers emphasized improvements in head teacher support for teacher implementation of new teaching methods (e.g., providing materials, aiding in-service participation, in-class observation and feedback, encouraging collegial work among teachers).

We concluded that the reported changes in head teacher management practices resulted from the combined effects of MSIP and PRISM.

> PRISM taught us to become better managers of school. School development plan training was mainly from PRISM. SIP is teaching how to improve the quality of teaching and learning ... to sustain materials, and to develop child-centered approaches.

> Sometimes you attend training and if there's no follow-up you forget all you learned there. SIP has been good for us because it has been kind of facilitating the implementation of the knowledge that we gained in PRISM.

Ultimately, we judged that the greatest impact of MSIP on head teachers had less to do with formal training, than with the creation of a support system for teacher and school development that enabled them to put into practice many of the ideas and methods introduced through management training from different sources. The major components of this support system included the revitalized Teacher Advisory Centers, increased parental contributions to school governance and funding, and ongoing access to external and in-school professional development for teachers and head teachers. We also concluded that the coordination of MSIP support for head teachers with other sources of professional development (e.g., PRISM) was strategically significant for local capacity building and efficient use of available resources for school improvement.

Parental Involvement

This section examines MSIP actions to mobilize parental participation as partners in school improvement. We concentrate here on MSIP efforts at the *school level* to sensitize parents about their role in education, to develop effective school committees, and to secure parental contributions to school funding. Parental involvement in financing and managing Teacher Advisory Centers is addressed in the next section.

To aid our analysis of parental participation we used Epstein's Framework of Types of School-Family-Community Involvement (Epstein, 1995). Epstein identifies six types of involvement: parenting, communicating, volunteering, learning at home, decision-making, and collaborating with community. We added a seventh type: providing financial and material support, to incorporate the ways in which parents and school patrons pay for schooling (e.g., taxes, fees, donations, and fundraising). Epstein argues that effective parent-school-community partnerships not only improve parental understanding and support for school programs and teachers' work, but also facilitate the learning and growth of students. While the MSIP did not set out to affect all types of parental and community involvement in Mombasa primary schools, the framework proved useful to clarify what was done, as well as potential areas for further development.

MSIP Goals and Inputs for School-Level Parental Involvement

Parental participation was addressed as a focus for development and as a strategy for school improvement. The overall objective for parental participation was "to get communities to appreciate, support, and contribute financially and intellectually to the school improvement efforts." Parental involvement was identified in project plans as a key enabling condition for achieving and sustaining effective classroom activity.

The key MSIP strategy for developing parental and community participation was to employ a community development officer (CDO). The CDO was expected to mobilize parental and community involvement through "training, awareness and sensitization of parents." This was enacted in meetings with school parents, and meetings to establish and train school committees.[10] The CDO's functions included assisting the POs, TAC tutors, head teachers, and district officers in their work with parents and community. The person hired as the CDO was an experienced and respected teacher, education officer, community development worker, and community activist. As a result, the project did not need to invest heavily in job-related skills training for the community development officer.

The Work of the Community Development Officer

The CDO worked mainly through meetings with various groups. His records from October 1996 to September 1998 listed a total of 46 school meetings with parents, 13 meetings with school committees, and 42 meetings with teachers. During this same period, he recorded an additional 64 meetings with TAC management committees (see Teacher Advisory Centers below). The CDO's school meetings with parents fell into four types: (a) sensitization meetings with all parents, or parents of pupils in particular classes, (b) organizational meetings with school committees, (c) consultative meetings in response to particular problems, and (d) participation in school events (e.g., sports days, awards presentations). By examining project records we were able to form a picture of the intensity of the CDO's interactions with parents across the 50 project schools and the overall district (162 schools as of 1998). From January 1997 to April 1998, for example, he attended one or more meetings at 13 of the 20 Phase 1 and 2 project schools, and at 13 of the 30 Phase 3 project schools. Of these 26 schools, the CDO met with parents at 16 schools once, and 2 to 5 times at the other 10 schools. In the same period, he participated in parent meetings in seven nonproject schools. After his initial visits to the 20 Phase 1 and 2 project schools, the CDO's school visits occurred on an 'as needed' basis upon request from POs, TAC tutors, head teachers, or parental groups themselves. The statistics bring into perspective the strategic limits to what one person in this role can do. The data also suggested that the CDO's initial open meetings with parents were a key intervention, often the only one if he concluded that further intervention was not required.

The identification of schools in need was a key step in the CDO intervention process. After his initial visits to project schools, weekly project staff meetings served as a key context for communication between POs and the CDO about project school concerns, as reported in these excerpts from PO interviews:

> When the heads complain of an area I feel it's the CDO's, then I discuss with him, particularly during the Friday afternoon meetings ... when I identify a problem in the project schools. For example, no desks, no windows ... which will hinder the ability to develop teaching aids. After talking to the head teacher, I discuss it with the CDO so that, when he meets with the parents, he sensitizes them on the need to attend to such problems ... I went to classes and found that children were not in class. They had been sent home for nonpayment. I went to the CDO. He talked to the parents. They paid and the children came back. *(Composite quote, project officer interviews)*

These data depict teamwork between POs and the CDO related to parental involvement, as well as the demarcation in the POs' minds regarding the teacher development focus and boundaries of their work. The CDO's presence at zonal and district head teacher meetings, and his involvement with TAC management committees were also key opportunities for school personnel to communicate concerns and ask for help.

The CDO's efforts to develop parental participation and support for school improvement can be summarized using Epstein's framework of types of parent-school partnership and Hall and Hord's categories of change facilitator interventions. The major focuses of direct intervention by the CDO were oriented to parental education, parental participation in school governance, and financial and material contributions. The CDO's parental education efforts occurred largely through his presentations at school-level parental awareness and sensitization meetings (a 'training' intervention in change agent terms). Sample minutes from parental meetings revealed that his parental education talks covered most of the types of parental involvement defined in Epstein's framework. Beyond exhorting parents to fulfill their responsibilities with regard to parenting, communications, volunteering, and learning at home, however, the CDO and other project staff were not involved in providing direct assistance in those areas. The CDO did contribute directly to the revitalization of PTA/school committees in many project and some nonproject schools (Hall and Hord's "developing supportive organizational arrangements"), and provided initial operational guidelines and assistance to get those committees up and running ('training'). He helped parent committees set up bank accounts, collection and accounting procedures for school-based funding, and gave advice about community donations. Finally, the CDO made himself available to school personnel and parent committees as an advisor on any implementation issues that pertained to school-parent-community relations (e.g., school complaints about parents, parent complaints about teachers and head teachers, management of school funds, parent-teacher communication, children's learning and welfare).

Evaluation of MSIP Impact on Parental Involvement

We evaluated three areas of school-level parent-community involvement: parental awareness and support for children's learning and schools; parental participation in school management; and sustainability of parental participation in school improvement.

Project reports proclaimed widespread parental awareness and appreciation of the MSIP efforts, a positive shift in parental attitudes toward contributing financially to schools and TACs, and that parents were playing more active roles in their children's education as a result of the CDO's parent education interventions (e.g., communicating with teachers, supervising homework, checking report cards, buying school books). We sought further data in interviews with head teachers and teachers regarding these claimed effects on parental participation. Interviewees from three of four project schools sampled reported positive outcomes on parental awareness and responsibility for children's learning, support for schools, contact and collaboration with teachers, and financial contributions:

> CDO helped parents become more aware. When convinced, their response is quite good in terms of getting involved, such as ensuring that children are well fed before coming to school ... If you send children to bring materials, such as flour for sticking things, before MSIP very few would cooperate. These days they are cooperative, even in small monies ... MSIP has made parents come to school frequently. These days they come freely in connection with progress of their children. If you call a parent through a letter they'll come. They used not to come ... We've been helped in soliciting money from parents or other sources. The money is used to build, to repair, to replace equipment, to buy books and other material, to pay subordinate staff and water, electricity and telephone bills ... Participation in meetings is better. *(Composite quote, project school head teachers and teachers)*

Staff from one project school with a history of low parental involvement reported little change in parental attitudes and participation. In two nonproject schools where the CDO had few prior contacts, faculty were uncertain about MSIP impact on parental involvement.

We were limited in our ability to judge the project's impact on school finances beyond the testimonials provided in interviews and scant information in project records. In sample minutes of the CDO's initial meetings with parents in project schools we found evidence confirming the low level of parental contributions at the point of intervention. One head teacher, for example, "complained that, out of 300 parents, only 42 paid the PTA fund of Ksh 450 (Kenyan shillings) per parent per year." At another school, the head teacher is quoted as reporting a 15% school dues payment rate in 1997. Contrary to statements from school personnel about increased parental contributions, participants in our school committee focus group reported that parents who did not pay school dues before still did not pay them. Several factors were identified as contributing to the low payment rate: financial hardships from a recent drought, misinformation from municipal authorities about school funding, and head teacher misappropriation of funds. Despite these caveats, there was no doubt that MSIP assistance had facilitated a shift in control of parental contributions to schools from the municipality to school committees. While it was unclear whether the level of parental contributions had actually increased, the shift in control resulted in net gains for schools, since there was less danger of funds disappearing into administrative or unknown costs at the municipal government level.

Project records reported "40 PTAs revitalized and reorganized" as a result of MSIP interventions by September 1998. Most were from the 50 designated project schools. This typically occurred in the CDO's parental awareness and sensitization meetings at schools, where, in addition to parental involvement talks, the CDO facilitated elections of school committees as needed. The CDO commonly returned at least once to provide guidance and management training to newly constituted school committees.

We inspected parental meeting minutes to gain a perspective on the focus and scope of parental participation in school management. These data supported claims of collaboration between parents and educational staff in project school management decisions and financial affairs touching on a wide range of concerns (e.g., fundraising, school fee payments, school needs and expenditures, academic standards, pupil performance, employment of extra teachers). We were impressed by the frequent references to parent-teacher discussions about pupil learning. At one school, for example, parents expressed concerns about the shortness of instructional time, and suggested the adoption of one-hour-lesson periods per subject. At another they recommended that remedial teaching be based on assessed needs of children in specific subject areas.

We identified several key issues related to the continued development and sustainability of high levels of parental involvement in school governance, financing, and support for school and teacher development. While approximately a third of the district's primary schools had benefited from direct contact with the project CDO, another 100 schools remained, and many of those originally assisted faced the prospect of new school committee elections and need for management training. The capacity of a single resource person to satisfy the continuing need for initial parent mobilization, while simultaneously monitoring and attending to implementation and renewal needs of working committees (as well as the TAC management committees), had already been exceeded. We argued for incorporation of a team of CDO positions into the municipally funded support structure for schools. We further recommended that steps be undertaken through the MSIP project to develop a resource manual and training program for people expected to carry on this vital work in school-community partnerships. Two quasi-legal issues affecting the capacity of school committees to carry out a greater role in managing school finances were the lack of formal power to require parents to pay school dues and the absence of clear authority beyond the school level for supervising the management of school accounts. These issues required political-legal solutions beyond the power of the MSIP project. Finally, from our interviews with head teachers, we noted that school development plans focused largely on facilities, materials, and teacher development priorities. We argued for the inclusion of parental involvement in school development plans, using Epstein's framework for school-parent-community partnerships as a guide for assessing the need for change within this domain.

Teacher Advisory Centers

This section examines the impact of the MSIP on services provided by primary school Teacher Advisory Centers across the Mombasa District. We review the goals for TAC development and consider findings about TAC facilities, TAC management committees, TAC funding, and TAC tutors in terms of MSIP inputs and outcomes.

MSIP Goals for TACs

Although zonal TACs existed in Mombasa District before MSIP, project reports and interviews conducted in the evaluation revealed that many were inactive, lacking tutors, effective TAC management committees, adequate facilitates, resources, and financial and political support. ("Downstairs was just a storeroom full of dust ... teaching and learning aids in boxes under tables. It was not a place for someone to sit. ... The center used to be empty. ... It used to be an empty room with nothing.") At the time the project began, there was no district Teacher Advisory Center nor district TAC tutor in place.

MSIP support for TAC development focused on four elements of the TAC system: (a) facilities and resources, (b) TAC management, (c) funding, and (d) TAC tutor selection and training. The vision was for TACs funded mainly by schools in each zone to evolve into 'nerve centers' for teacher training and advice, professional resources, preparation of teaching aids, research, and professional meetings.

TAC Mobilization (Facilities, Management, Funding)

The TAC development strategy involved inputs from the external funders, the Municipal Education Office, and schools. The initial aim was to 'refurbish' the TACs in each zone. MSIP promised to provide each TAC with basic equipment (photocopy machine, typewriter, overhead projector, paper cutter) and materials (teacher reference books, textbooks, stationery, materials for teaching aids such as manila paper and felt markers) to get started. In order to receive these goods, however, the center had to have an MEO-deployed TAC tutor, a functioning TAC management committee (TMC), and secure and adequate office, classroom and storage facilities. The TMCs were required to set up bank accounts and to obtain funds for renovations and furnishings through school fees and community donations. Once those conditions were met, TACs were officially opened and the equipment and resources from the MSIP were turned over to the Center.

The most significant project inputs to TAC mobilization occurred through the interventions of the MSIP community development officer. He was instrumental in establishing the TMCs, in organizing elections of TMC executives, and in sensitizing the committees to their TAC management roles. Zonal TMCs consisted of parent representatives from each school committee, the TAC tutor, all head teachers, the zonal inspector, the divisional school advisor, and the neighborhood chief. Project reports documented the CDO's participation in 17 TMC meetings from January to March of 1997 (one to three per committee). Minutes from these meetings depicted the CDO's input to the committees. One objective was to convince them of the need for school communities to take over responsibility for funding of schools and of professional support for teachers. A second was to inform TMCs about the project and the conditions that had to be met to officially 'open' the TAC in each zone. The CDO provided TMCs with guidelines for membership, meeting schedules, and duties (e.g., identifying needs, budgeting, fundraising, maintaining a bank account, accounting procedures, supporting teacher in-service activities, sustaining resources, sensitizing school communities, and collaborating with other TMCs and with education authorities). The mobilization of TACs depended on funding from school committees in each zone for initial renovations and to sustain the TAC's operating costs during the first year. Each TMC set its own fee structure, ranging from Ksh 3,000 to Ksh 7,000 per school. Openings were staged as public ceremonies to build community awareness of TACs and to solicit donations from business and community groups. Impressive donations were made to a few TCs for specific needs, such as furniture, library books, and even building expansion.

In addition to the CDO's work with individual TAC committees, the project staged one district-wide community awareness workshop for TMC executives, TAC tutors, and inspectors in May 1997. Topics addressed included academic standards, the roles of head teachers and school committees, the TMCs' role, and basic planning, management, accounting, and record-keeping skills. After this initial mobilization period, assistance to TMCs proceeded in four ways: (a) CDO participation in TMC meetings, (b) monitoring implementation, (c) helping resolve management problems, and (d) assisting with fundraising. The CDO's school-level work with parents dovetailed with his efforts to support the development of TACs.

MSIP Impact on TAC Facilities and Management

The CDO's initial mobilization efforts were quite successful. TMCs were formed, accounts opened, and meetings underway in all 10 zones and the district TAC within 6 months of the CDO's appointment. Eight zonal TACs and the district TAC were refurbished and inaugurated by September 1997, and the remaining two by mid-1998.

The impact of MSIP interventions on TAC accessibility, facilities and equipment, professional resources for teachers, and community involvement in management were uniformly acknowledged by district education officers, school personnel, and parents.

> We didn't have a TAC before ... The plant had always been there but I knew nothing about it ... SIP has played a vital role ... provision of materials ... machines such as duplicating machine and typing machines ... Never had before ... Before there were no books ... so no borrowing facility ... and no other materials like felt pens, manila for making of learning aids ... Because of the change in management ... community now more responsible ... It's their TAC ... They are contributing ... a big improvement. *(Composite quote, teachers, head teachers, and district education officer interviews)*

We examined our documentary, interview, and TMC focus group data for additional evidence regarding the quality of TAC management – participation, frequency of meetings, and meeting content. Our analysis of TMC records from the CDO's files led to the following conclusions about the participation and frequency of meetings. The public school sector was well represented at TMC meetings; the fast-growing private school sector was not. Head teachers commonly attended, often accompanied by at least one school committee member. It was rare to find parents in attendance without the head teacher from their school. TAC tutors and MSIP staff (i.e., CDO, PO) were routinely present. Municipal education officers were infrequent participants. Comments from head teachers from different zones showed variation in the frequency of TMC meetings. "We meet often, almost every month. … I attend term meetings. … It's been a bit inactive. I don't know why." Overall, it appeared that TMCs were meeting once a term, if not more, and that the frequency of meetings was subject to the initiative of the TAC tutor. We scanned minutes of TMC meetings to get an idea of what TMCs actually do. These records revealed that much TMC time was consumed by matters of funding and resource acquisition to support the TACs, TAC tutors, and TAC activities (e.g., budgets, school fees, purchase of furniture and supplies, soliciting donations, payment of repairs, TAC library resources, funding for clerks and cleaners). TMC minutes also showed awareness and support for TAC and MSIP teacher development activities and needs. We concluded that TMCs had been successfully established and that most were playing an active role in developing and sustaining effective Teacher Advisory Centers.

MSIP Impact on Teacher Advisory Center Funding

Sustainability of funding was the single most important challenge facing the TACs. TMCs were expected to raise two thirds of the necessary TAC funding from schools served by each center, and the remaining third through donations and fundraising activities. Unfortunately, once the TACs were officially opened, school contributions plummeted. Financial records compiled by the CDO showed a decline of 40% to 100% in school contributions in all 10 zones, from 1996 to 1998. The decline in school committee payments to TACs was attributed to multiple factors, with some more relevant than others, depending on each TAC. These included low rates of payment of parent fees to schools, mismanagement of school funds, competing priorities for school funds, ineffective TAC tutor and TMC leadership, poor management of funds at the TAC level, low parental awareness, lapsed commitment to TACs, and ineffective means to enforce fee payment. Multiple solutions were suggested in stakeholder group interviews: closer supervision of TAC funds use, donations, income-generating activities, charging membership and user fees, parent education, replacing uncommitted TAC tutors, recruiting more private-school participation, penalties for schools in arrears, more support from municipal politicians, and joint problem-solving by TMCs and local stakeholders (e.g., head teachers, education officers, community members). We concluded that the most promising proposal for sustainable TAC funding was to shift from voluntary payment to a membership system, with different user fees for members and nonmembers. We also suggested the creation of a district TMC task force to help identify needs and strategies sensitive to zonal differences on a case-by-case basis.

TAC Tutor Development

The MSIP project influenced the appointment of TAC tutors and provided formal training and on-the-job assistance to TAC tutors for their teacher development role. The incorporation of the PO position within the municipal system of education officers led to clarification of the roles and working relationships of TAC tutors, zone inspectors, school advisors, project officers, and TMCs. Once the refurbished TACs were 'opened,' the expectations, skills and attitudes required of an effective TAC tutor became more transparent. TAC tutors were more accountable under the scrutiny of TMCs and, as a result of ongoing contact with MSIP project personnel and of pressure from education officers and head teachers, eager to take advantage of the refurbished TACs. Ineffective tutors were replaced by others whose experience and career motives were more in line with the desired teacher support roles of centers and tutors.

The MSIP provided both formal and on-the-job training for Mombasa TAC tutors. Formal training events included an orientation to MSIP; a change facilitator skills training workshop for POs, head teachers, education officers, and TAC tutors put on by SIP officers from the AKF Kisumu project; a 'community awareness' workshop for TMCs, TAC tutors, and inspectors; and an 'induction' workshop offered to four new TAC tutors in late 1997. Beyond MSIP-sponsored training, TAC tutors received a 1-week course on the TAC tutor role and TAC management through a national primary school quality initiative called Strengthening Primary Education (SPRED). These training events covered such topics as child-centered teaching, developing teaching aids, clinical supervision, record-keeping, accounting, library management, and stakeholder roles in the new support system for education improvement.

Project leaders worked closely with the municipal education officers to clarify roles and came up with a vision of a 'four-person team,' consisting of a divisional school advisor, zonal inspector, TAC tutor, and project officer in each zone. The four-person team concept did not work out, since the municipality was unable to fund education officer positions in at least half the zones. POs and TAC tutors did, however, collaborate closely in most zones, as illustrated in these excerpts from PO interviews.

> At the beginning of every term we sit together with the TAC tutor and write a program of activities ... We go to schools together Tuesday and Thursday. The other days he follows the MEO officer's program ... Once a week we meet at the TAC and discuss what we have observed ... There are days we hold workshops and seminars together ... We meet at the TAC to prepare materials we may be needing to use in school for demonstrations and workshop and to receive teachers who are coming ... Whenever they call a TMC meeting [the CDO] plays the major role, but we attend ... When the TAC is having other work, for example, setting and moderation of exams, we go in to assist them. *(Composite quote, PO interviews)*

Reports of TAC tutors teaming with POs were confirmed in our interviews with TAC tutors.

We concluded that the most significant MSIP assistance to TAC tutors occurred less through the training events than through their teamwork with project officers. Inspection of PO work diaries and monthly reports and TAC tutor monthly reports, however, suggested that the intensity of that teamwork varied for pairs in different zones.

MSIP Impact on TAC Tutors and Utilization of TAC Services

Interview comments from TAC tutors, project officers, and district education officers were uniformly positive about the impact of MSIP interventions on TAC tutors.

> Thank God to SIP for encouragement and motivation to bring up standards. Now you tell me you want a science teaching aid, I will go to the cupboard and get it for you. There were a lot of apparatuses before, but just in dusty

> boxes. More than materials, it's the courses we are conducting ... Creates awareness for teachers ... motivation ... new ideas. We never used to have common exams in zone and division. Now we analyze our own exams. Now there is competition ... prize-giving day last year. Even subject panels are now functioning well. Before they were just there, but no work done. Recently we found Class 5 means not doing well. Now we are moving school-to-school discussing. *(TAC tutor interview)*

In this quote a TAC tutor speaks about the changes in her work: increased professional motivation, increased capacity to access and use materials, greater in-service training involvement with teachers, stronger sense of efficacy about TAC tutors' impact on teachers, and expanded range of professional activities across the zone. These comments were echoed by other stakeholders. It was evident that MSIP inputs had significantly improved the capacity of TAC tutors to deliver teacher and school support services through the TACs.

We were frequently reminded that, while the roles of TAC tutors and POs were similar, there were major differences in scope. POs focused their in-service training efforts on the teaching of English, mathematics, and science in Standards I to VI. TAC tutors were responsible for helping teachers in all subject areas from Standards I to VIII. POs provided in-school assistance intensively to three schools at a time. TAC tutors were expected to be available for in-school assistance to all schools within their zone at all times. Finally, whereas POs confined their work to teacher development (workshops, in-school assistance), TAC tutors had a wider range of expected duties, including workshops and school visits, facilitating the work of subject panels, assisting with the preparation of examinations and analysis of examination results, managing TAC libraries, providing support for district school inspections, coordinating inter-school extra-curricular activities in their zone, and hosting meetings of head teachers, teachers, and TAC committees. We were told that some TAC tutors were also being used by district officers for administrative and supervisory duties due to the lack of sufficient zone inspectors.

Examination of monthly TAC tutor reports from 1997 and 1998 confirmed that TAC tutors were routinely involved in zonal and district-level 'other duties' as listed above. Some did carry out duties at odds with their official role, such as registering teacher college applicants, assessing teacher promotion applications, and serving on school inspection teams. TAC tutors organized their own workshops for teachers, in addition to participating in MSIP workshops. In monthly TAC reports, we noted examples of TAC tutor workshops on making and using teaching aids, developing schemes of work for Kiswahili and social studies, 'filing' systems, and science examination topics. TAC tutors, however, did not organize workshops on a regular basis beyond those offered jointly with MSIP project officers (see next section).

TAC tutors made school visits for varied purposes, such as advising head teachers about administrative matters related to TAC services, communicating information to heads and teachers about examinations, checking on school TAC fee payments, individual teacher observation and feedback, and meeting with groups of teachers about needs for improvement. Inspection of monthly TAC reports showed from 1 to 10 school visits per month per tutor. However, visits for the purpose of teacher coaching, or to meet with groups of teachers based on prior observations or examination results, only averaged about one a week (three to six schools a month). With their additional responsibilities and no independent transportation, TAC tutors were not able to provide in-school assistance to teachers at the same level of intensity as POs (see next section).

While we could confidently say that the capacity of TAC facilities, management, and tutors to deliver professional services to teachers had improved, the question remained whether teachers and head teachers were taking advantage of these services. Interview comments from heads and teachers in project and nonproject schools supported claims that teachers were making greater use of the centers than in the past.

> We used to not go there. Now it is known and being used as a resource center where teachers can borrow books, or get ideas for preparation of charts, or preparation of exams ... To see teaching materials and learn to prepare them ... I go there to borrow books for myself and other teachers every 2 weeks ... (SIP) reading workshop at the TAC ... Workshops at TAC different from SIP workshops ... TAC tutor and a primary school inspector deliver those. *(Composite quote from interviews with teachers and head teachers)*

The monthly TAC reports in MSIP files provided a sense of the scope and variation in TAC use. The March 1997 Mtongwe TAC report, for example, recorded visits from 25 teachers for material production: 10 for library services, 7 borrowing reference books, and 19 for consultations, plus 2 TAC-sponsored workshops attended by 19 teachers. A May 1997 report from the Ziwani TAC listed 5 visits by teachers for material development, 14 for moderation of examinations, 18 for proofreading exams, and 24 for borrowing books and equipment. In addition, 23 teachers attended workshops in science and mathematics. While individual teachers reported varying levels of use of TAC services ("once or twice a week … once a month … I haven't taken time to go"), the evidence overall confirmed that teacher utilization of TACs had, indeed, increased.

Continuing Development and Sustainability of Teacher Advisory Centers

The capacity-building efforts of the MSIP project were successful in reactivating and mobilizing the network of previously ill-equipped, ineffective, or nonexistent TACs and TAC services across the district. That said, the sustainability of TACs and TAC services remained uncertain. Key issues for sustainability were identified under four broad areas: funding, workload, staffing, and transportation. The progressive decline in school payments to fund the TACs after they were 'opened' was the most serious issue. The myriad and competing expectations of TAC tutors was a second.

We recommended a system-level review and reallocation of the duties assigned to TAC tutors in order to emphasize their teacher development functions. It was also clear that TAC tutors alone could not possibly provide the level of classroom-based support required to enable significant, widespread changes in teacher practice under existing staffing levels (one tutor per TAC). We recommended that each zonal TAC be staffed by three tutors, and that this could be accomplished within current funding levels by restructuring the roles and nomenclature of zonal personnel (i.e., TAC tutors, zonal inspectors, and MSIP project officers). We also cited the need for subordinate personnel to assist in keeping the centers open for teachers and to help with TAC housekeeping chores. We argued that sustainability of effective TACs would also require ongoing opportunities for professional development and networking among TAC tutors. Finally, distance and transportation to and from TACs for both consultants and teachers required further attention to sustainability beyond the 'project.'

The mobilization of TACs through the MSIP did not, in itself, yield evidence of significant changes and improvements in teaching practice. As explained in the next section, the MSIP impact on teaching at the classroom level was closely linked to the provision of in-class coaching by project officers.

Evaluation of MSIP In-Service Activities

The MSIP in-service training game plan combined ongoing workshops for teachers from all primary schools delivered through the zonal teacher advisory centers and school-based in-service assistance to designated project schools. The project officers were the key agents for in-service training planning and delivery. Here we review the role of the project officers (POs) and examine the nature and quality of their in-service training activities in detail. Then we evaluate the impact of their in-service training activities on teaching and student learning.

The MSIP Project Officers

The MSIP project officers filled a new role in the district uniquely focused on school-based teacher development for school improvement. The original project design envisioned a team of professional development consultants with varying areas of subject expertise working jointly in project schools. Once the project shifted to a zonal delivery structure, with each PO assigned a particular zone, POs were challenged to assume the role of generalist teacher developers capable of providing in-service support in multiple subject areas (English, mathematics, science). The POs shifted increasingly from reliance on their own pedagogical expertise to a strategy of identifying other resource personnel (TAC tutors, teachers college tutors, education officers) and expert classroom teachers who could be recruited to assist with the planning and delivery of workshops at the TACs, and who could be encouraged to provide in-school leadership for teacher development once a school disengaged from the 'project school' phase of site visits.

The creation of the project officer teacher development role provoked discussion and decisions about a variety of key issues, including creating new roles versus developing existing roles; the need for subject expertise versus general teacher development expertise; the need for classroom teaching expertise relative to teacher development goals; the capacity of consultants to deal with management issues impinging on their teacher development activities; the importance of professional development for teacher developers; the need to work collaboratively at the school and system levels with other people involved in teacher leadership and development roles; and the challenges of creating shared understandings about innovative teacher development consultant roles.

The individuals selected for the project officer positions were well qualified for the job. Most had over 20 years of experience in the local education system. A majority had prior experience as primary school head teachers and as district education officers. Over half had in-service teacher training experience as TAC tutors, or in training positions with other teacher development projects. The POs were seconded from their existing district education officer positions (in which they were school advisors) in order to increase potential sustainability of the PO role beyond the project. This strategy circumvented the bureaucratic complexities of formally establishing a new position category within the education bureaucracy. On the other hand, the positions were defined at levels of seniority and salary that hindered the recruitment of expert teachers directly from the classroom. Seven POs had been out of the classroom for 10 years or more. While as teacher trainers and supervisors (e.g., head teachers, inspectors) they were familiar with contemporary teaching methods, they may not have had extensive classroom experience implementing those methods themselves.

The MSIP project included staff development for its project officers, as well as change facilitator training for district officers, head teachers, and TAC tutors (see preceding sections). The professional development program for POs included workshops with consultants from the Kisumu SIP-Aga Khan, visits to other SIPs in Uganda and Tanzania, weekly team meetings (for joint planning, sharing, and problem-solving), and the opportunity to pursue a modular program leading to a master's degree in education. The degree program was organized with the School of Education of the University of Leeds, England, was tailored to the Kenyan context and MSIP goals, and was delivered by visiting professors in a series of modules, with field-oriented reading and research assignments. The professional development program for POs included access to information about teaching methods advocated in the project, as well as training in change process skills (e.g., consultant role, teacher observation, action research).

Teacher Workshops

During Phase 1, in-service workshops were offered to teachers from the original four project schools and a selection of nonproject schools, at the project office/training facility. Once the Teacher Advisory Centers were reactivated, all teacher in-service workshops were delivered at the TACs, and were open to teacher representatives from all project and nonproject schools within each zone. During Phases 1 and 2, most of the workshops were delivered by POs, in accordance with their strongest areas of subject expertise. Toward the end of Phase 2 and into Phase 3, each PO became responsible for arranging for local workshop facilitators, in collaboration with zonal TAC tutor and education officers. During Phase 3, the project settled into a cyclical pattern of workshops. MSIP-sponsored workshops were jointly planned and offered across all zones each term, alternating the subject emphasis by term from English to mathematics to science. TAC tutors attended to other subject areas, such as social studies and agriculture.

During Phase 1 of the project, baseline surveys were developed and administered in project schools. These included a battery of survey instruments designed to assess the attitudes and behaviors of pupils, teachers, head teachers, and parents. Achievement tests were also developed by project staff and administered to pupils in Standards III and V, from a sample of project and nonproject (control) schools. Analysis of the test results enabled the POs to identify problematic areas of student performance and to use these for planning future workshops. Other MSIP teacher surveys were administered to assess needs for in-service assistance to aid teachers with student difficulties in reading skills and in science topics. Workshops on oral communication in English in Phase 3 resulted from PO observation and discussion about confusion in project school classrooms between oral language development, reading, and writing. In sum, the selection of topics for MSIP workshops was data based, involving a combination of teacher surveys, PO observations, and analysis and comparisons of student achievement test outcomes.

Workshop attendance from mid-1995 to mid-1998 numbered 1,746 teachers. It was not possible to determine the proportion of project versus nonproject school teachers attending, nor how many teachers attended multiple workshops. Teacher attendance in Phase 3 hovered around 385 teachers per set of workshops (about 40 teachers per zone).

The descriptions of workshop activities provided in teacher and PO interviews were consistent, and led us to conclude that the MSIP workshops had incorporated many key elements of effective in-service training. The workshops typically incorporated a combination of theory, demonstration, and practice, which provides a powerful basis for the transfer of training when combined with in-class follow-up coaching (see next section). Other positive features of the MSIP workshop activities included (a) the scheduling of workshops during the school day (valuing teacher time and professional development), (b) the intensity of the workshops over several days (increasing potential for teacher learning), (c) the grounding of the workshop in data on student performance (achievement tests) and teacher needs (surveys, observations; increasing perceived relevance), (d) the efforts to get schools to send small groups rather than individual teachers to workshops (increasing potential for mutual support and implementation at the classroom level), (e) the use of strategies and materials that are feasible within the Kenyan classroom context (increasing perceived practicality), (f) the involvement of practicing teachers as co-trainers (increasing credibility of the practices), and (g) the use of external incentives (e.g., money for transportation, meals, certificates, free materials). The overall power of the in-service model, however, depended on the complementary provision of in-school assistance, and on other supporting features of the organizational context, such as head teacher management, parental support, and the active presence of Teacher Advisory Centers and tutors.

School-Based In-Service Training

School-based in-service training through the MSIP project was intended to follow up implementation of teaching methods introduced in external workshops and to address school-specific in-service needs. The MSIP implementation model provided for intensive in-school assistance from POs for up to 1 full school year for Standards I to VI teachers in a school. By then, teachers were expected to have a basic level of competence in child-centered teaching methods, and to have the capacity to continue developing their expertise by helping one another with ongoing, but less frequent, support from external experts. This section illuminates the scope and nature of school-based in-service assistance provided by MSIP project officers.

Scheduling School and Classroom Visits

During Phases 2 and 3, the POs were assigned zones and were expected to provide in-school aid to three schools per zone for a full school year, after which they would move to three more schools (taking 4 to 5 years to reach the 12 to 15 schools in most zones). In-school help was targeted to Standards I to VI teachers of English, mathematics, and science.

The POs experimented with various scheduling plans for school visits, including weekly 1-day visits to each of their three schools on the same day each week and on different days each week, as well as spending 2 or 3 consecutive days in a single school and rotating among their three schools on a 3-week cycle. The availability of the two project vehicles posed constraints on the scheduling of school visits. At the time of our visit, the POs were working with different schools on alternate days of the week.

Exact records of the frequency of classroom and school visits were not on file. By examining a sample of the POs' daily journals, we were able to estimate that they made an average of 3.5 observations per school visit. The observations were evenly distributed between lower primary (48%) and upper primary classes (52%). English (41%) and mathematics (33%) were observed more frequently than science (18%). A few POs reported some observations of teaching in other subjects, such as Kiswahili.

The figure of 3.5 classroom observations per visit enabled us to appreciate practically the scope of PO input at the classroom level. The average number of teachers per school among the 50 project schools at the time of our study was 21. Assuming one school visit per week, a PO could see every teacher at least once over a 10-week term, and could reach about two thirds of the teachers for multiple visits. This supported the Phase 2 decision to extend the time for intensive work in project schools from 1 to 3 terms (a full academic year). One or two visits over a single term is unlikely to have much impact on teacher implementation. When the time frame increases to 3 terms (30 weeks), the scenario expands to five or more in-class visits per teacher. Research on in-service training and teacher change suggests that the degree of implementation and continuation of new teaching practices is much higher in situations of sustained classroom help (Joyce & Showers, 1980; Huberman & Miles, 1984; Ingvarson & MacKenzie, 1988).

Classroom Visits

The provisions for in-class assistance set the MSIP model for in-service teacher development apart from the external workshop and 'train-the-trainer' cascade models so typical of system-level in-service training approaches around the world. We were interested, therefore, in understanding what actually happens during a classroom visit that might contribute to teacher change. While we were not able to observe the POs in action, we did obtain samples of school and classroom visits from their work diaries. Inspection of these diary samples (covering eight POs, nine schools, 52 school visits, 180 class observations, and 96 different teachers) revealed a high consistency between PO observations and suggestions to teachers and the pedagogy advocated in project goals and workshops.

The PO diaries revealed five classroom observation scenarios, as follows:

1. PO observes and provides feedback with suggestions for practice.
2. PO does demonstration lesson (or observes and arranges for a demonstration).
3. PO observes and suggests or arranges for teacher to observe another teacher.
4. PO co-plans lessons (or observes and arranges for joint planning).
5. PO observes and consults with teacher on progress of nonreaders.

The vast majority of the 180 classroom observation visits sampled adhered to the first scenario: observe, make suggestions for changes in practice, agree on next steps. We found occasional examples of demonstration teaching and co-planning of lessons, but these amounted to less than 5% of the documented observations. In a few episodes (less than 5%), the PO recommended that the observed teacher visit and observe another teacher, or allow other teachers to observe the teacher's classroom.

Teacher Meetings with Project Officers

In addition to their classroom visits, Project Officers occasionally met in the schools with teachers grouped by class, subject, or division (e.g., lower primary, upper primary). Sometimes they held meetings with an entire school staff. Teacher meetings included

- staff meetings: orientation to MSIP project;
- staff meetings: feedback on PO observations and action planning with teachers and head teacher;
- materials development workshops; and
- school-based follow-up workshops (after zonal TAC-based workshops).

School-specific advice on the removal of barriers to the implementation of changes in teaching methods appeared to be an integral feature of the MSIP in-service approach.

Teachers attending MSIP- and TAC-sponsored external workshops were encouraged to hold in-house workshops or presentations at the school level to share their learning with other colleagues. The records at our disposal did not enable us to assess how frequently and consistently this sharing happened. A March 1998 6-month report identified eight schools in five zones where "after the main workshops at the zonal level, some teachers then cascaded the training in their schools." In our teacher interviews in project and nonproject schools, we had mixed reports of teacher follow-through with in-house presentations (with or without the assistance of POs). A senior English teacher from one project school reported organizing a workshop follow-up session with the other English teachers in her school on her own, without the PO. A nonproject school teacher gave a contrasting report. "We are supposed to hold a seminar here for people who go to (TAC seminars) to teach each other, but we haven't done that." We concluded that school-based workshops organized by selected teachers to extend the learning acquired in TAC-based workshops were not necessarily routine events, particularly in nonproject schools. Furthermore, the professional development power of a brief sharing session by teachers who may report what they learned at a workshop without having mastered the new practices themselves cannot be fairly compared to the power of the original 2- or 3-day workshop given by with experts in the workshop topics (Corbett & Rossman, 1989).

Materials development sessions were a second type of PO-facilitated in-school workshop. At one project school, for example, lower primary teachers reported that the PO arranged for them to meet as a group in the afternoons, on a regular basis, to make materials. Each week, the group made teaching aids and charts for one of their colleagues in accordance with the colleagues' needs. The PO provided some of the materials; others were provided by the head teacher in response to teacher requests. These teachers reported that they "began with in-school workshops like at the TAC center. Then it evolved to teachers saying what they needed help with and teachers working together." This comment was significant, because it was suggestive of the developmental nature of teachers becoming comfortable learning together and helping one another. In the evaluation of the Mzizima SIP in Tanzania, participants reported that it took as long as a year for new teachers to fully adopt norms and practices of peer collaboration on matters of teaching and learning (Anderson & Sumra, 1995; see also Chapter 3 of this book).

Conclusions About MSIP In-Service Assistance

From our analysis, we concluded that project staff had successfully carried out external workshops on child-centered methodologies through the zonal TACs, and that these workshops had been accessible to teacher representatives from across the district during Phases 2 and 3 of the project. The project succeeded in demonstrating the feasibility of delivering school-based in-service assistance to teachers at the classroom and school levels, in successive waves of schools, for sustained periods of time (up to a full school year), through the existing decentralized organizational structure of the local school district. The in-service model enacted through the MSIP exemplified many features of effective staff development design, as previously noted (Loucks-Horsley et al., 1987; Andrews et al., 1990; Joyce & Showers, 1980).

Our evaluation also suggested some areas for concern and future development of the in-service model. Chief among these was the individualistic model of school-based support that appeared to have evolved. The overwhelming majority of PO interactions with teachers in the schools occurred in one-on-one classroom observation and coaching episodes. This is a powerful mode of in-service assistance, but is limited in terms of the number of teachers who can be reached and supported on an ongoing basis, particularly when intensive access to consultant help is removed. While we found isolated examples of POs encouraging and facilitating teacher intervisitation and peer assistance, the possibility of developing self-sustaining, school-based, professional learning communities of teachers did not appear to have been fully taken on (Kruse et al., 1995). We return to this argument in the concluding discussion of this chapter.

Teacher Implementation Outcomes

There were no baseline descriptions of primary school teaching practices prior to the MSIP project. The 1993 project proposals, however, provided some images of teaching and learning in typical pre-SIP Mombasa primary school classrooms.

> For the most part present classroom practice is teacher dominated, where lessons consist of presentation of textbook material forcing children to memorize information with little understanding; children remain passive and do not ask questions and are not accorded the opportunity for discovery or problem solving; children lack stimulation due to inadequate or nonexistent use of teaching aids; there is an overemphasis on testing all the way from Standards I to VIII.

MSIP-sponsored workshops and in-school assistance were designed to support the adoption and implementation of more activity-based and child-centered teaching methods, including the use of low-cost teaching and learning aids. Our evaluation attempted to judge the degree to which current teaching practices conformed to these ideals, and to explore whether current practices reflected change from the ways teachers taught prior to coming into contact with MSIP in-service activities and support.

Although baseline survey data on teacher beliefs, opinions, and attitudes were gathered by project officers at the point of entry in project schools, the surveys were not readministered in those schools in an effort to gauge the impact of project interventions. None of the teacher surveys measured the kinds of teaching methods teachers were using in the classroom; hence, they were of minimal value as a baseline or post-intervention tool for evaluating teacher implementation of child-centered teaching methods.

Our evaluation of teacher outcomes focused largely on the analysis of teacher implementation findings from our interviews and observations in a sample of four project and two nonproject schools. We developed an observation checklist that described a continuum of teacher and pupil behaviors, from teacher centered to child centered, across multiple dimensions of teaching and learning. These behaviors included: lesson planning, instructional approach, teacher questioning practices (quality of questions, distribution of questions, responses to pupil answers), latitude for pupil input and choice in lesson activities, instructional grouping, pupil interaction in small groups, assessment of pupil performance, remedial interventions, availability and use of teaching and learning aids by teachers and pupils, orientation to textbook content, availability and use of learning centers, pupil engagement in activity-based learning activities, and pupil involvement in inquiry-oriented learning activities. The checklist items were based on project documents describing ideal teaching methods and on consultations with project personnel. Since the aim was to determine the degree of teacher use of teaching methods advocated in MSIP in-service activities, it was important to ensure that the evaluation tool mirrored as closely as possible the project goals for teaching. The observation checklist was developed using checklist creation procedures to assess the fidelity of implementation of multidimensional classroom innovations, while allowing for variations in use and proximity to ideal use within each dimension of practice assessed (Heck et al., 1981; Hall & Hord, 1987; Anderson, 1997). Checklists were completed immediately after each observation. As we observed we made anecdotal notes, which we later wrote up as narrative profiles of teaching practice.

Teacher Understanding

In our interviews, we asked teachers from project and nonproject schools to describe what child-centered teaching looked like in practice. We found a high degree of consensus in the way the project school teachers we interviewed defined child-centered teaching. These teachers emphasized (a) student learning through active participation in learning activities, (b) the teacher's role as planner, guide, monitor, and facilitator of student participation and learning in those activities, (c) the use of varied activities and instructional groupings in the course of a lesson, (d) the use of practical teaching and learning aids by teachers and pupils, and (e) children learning at their own pace. Teachers from nonproject schools offered somewhat less multi-dimensional descriptions.

It would be a mistake to interpret teachers' ability to describe the MSIP approach to teaching as an accurate measure of their implementation and skill in the use of these teaching methods. It is significant, however, that the project successfully communicated a common vision of desirable teaching and learning practices. Several teachers declared that they were first introduced to this pedagogical approach during their initial teacher training, but had not actively put its methods into practice prior to MSIP.

Change in Teaching Practices

We asked MSIP project officers and head teachers from the sample schools what changes in teaching practices they had observed and would attribute to contact with the MSIP. We asked the teachers how contact with the MSIP had affected their teaching. Responses were similar across all three groups. They emphasized an overall reduction of teacher dominance in the teaching and learning process, and an increase in student participation, learning activities, and interaction with teachers, as a result of MSIP contact. Instructional grouping and small-group learning activities were reported as more common. Members from each group commented on the increased presence and use of teaching aids and centers. More positive teacher-student relationships and communication were reported, as well as more individualization of instruction by teachers. All these changes were consistent with the pedagogical methods advocated in MSIP documents and training. With this as a foreground, we compare the interview portrayals of change with observations of teaching in project and nonproject schools.

Teaching Practices

Observation Checklist

Tables 1 and 2 display observation checklist judgments recorded for teachers in four project schools (16 teachers) and two nonproject schools (8 teachers), for each teaching practice component and variation.[11] We found that project school teachers were more likely to implement the kind of teaching methods emphasized in MSIP in-service activities in virtually all the dimensions of teaching and learning we assessed. Of a total of 340 teacher and pupil behaviors assessed in 17 project-school classes (1 teacher was observed twice), a total of 32% scored in the most child-centered options on the rating scale, and a total of 29% in the most teacher-centered option (Table 1). This overall finding contrasted strikingly with the two nonproject schools, where only 6% of 160 teacher and pupil behaviors assessed in eight classrooms were judged highly child centered, and 55% were judged strongly teacher centered (Table 2). These findings support the premise of the project's in-service training strategy: that regular, in-class follow-up assistance over a sustained period of time (6 months to a year) has a significant positive impact on teacher commitment and capacity to put new ideas and methods into practice, when comparisons are made with teachers who have only benefited from access to external workshops and TAC resources.

Table 1: Percentage of Teachers Implementing Child-Centered Teaching Methods in Project Schools

TEACHING PRACTICE COMPONENTS ASSESSED	TEACHER PRACTICE COMPONENT VARIATIONS[a]					
	N/A[b]	1 Teacher-centered	2	3	4	5 Child-centered
Lesson plans			***** 29%	********* 53%		*** 18%
Instructional approach		*****[c] 29%	********* 53%			*** 18%
Quality of questions			********** 59%	*** 18%		**** 24%
Distribution of questions				***** 29%		************ 71%
Responding to pupil answers				************ 71%		***** 29%
Dealing with incorrect answers	* 6%	* 6%	**** 24%	****** 35%		***** 29%
Decision-making for activities		*************** 88%				** 12%
Grouping of pupils		***** 29%	****** 35%	** 12%		**** 24%

<table>
<tr><th rowspan="2">TEACHING PRACTICE COMPONENTS ASSESSED</th><th colspan="6">TEACHER PRACTICE COMPONENT VARIATIONS[a]</th></tr>
<tr><th>N/A[b]</th><th>1
Teacher-centered</th><th>2</th><th>3</th><th>4</th><th>5
Child-centered</th></tr>
<tr><td>Observation and interaction</td><td></td><td>*
6%</td><td>***
18%</td><td>*****
****
53%</td><td colspan="2">****
24%</td></tr>
<tr><td>Written assignments</td><td>*****
****
53%</td><td></td><td>****
24%</td><td>***
18%</td><td colspan="2">*
6%</td></tr>
<tr><td>Accessibility of teaching aids</td><td></td><td></td><td colspan="2">****
24%</td><td colspan="2">*****
*****
***
76%</td></tr>
<tr><td>Source of teaching aids</td><td>*
6%</td><td>***
18%</td><td colspan="2">****
****
47%</td><td colspan="2">*****
29%</td></tr>
<tr><td>Teacher use of teaching aids</td><td>*
6%</td><td>*
6%</td><td>***
18%</td><td></td><td colspan="2">******
******
71%</td></tr>
<tr><td>Textbook use</td><td>***
18%</td><td>***
18%</td><td></td><td>*****
29%</td><td colspan="2">******
35%</td></tr>
<tr><td>Availability of centers of interest</td><td>*
6%</td><td>*****
29%</td><td>*
6%</td><td>*****
29%</td><td>***
18%</td><td>**
12%</td></tr>
<tr><td>Use of centers of interest</td><td>******
35%</td><td>*****
29%</td><td>*
6%</td><td>***
18%</td><td colspan="2">**
12%</td></tr>
<tr><td>Pupil engagement in active learning</td><td></td><td>*****
*****
59%</td><td colspan="2">*
6%</td><td colspan="2">******
35%</td></tr>
<tr><td>Inquiry-oriented activity</td><td></td><td>****
***
41%</td><td colspan="2">******
35%</td><td colspan="2">****
24%</td></tr>
</table>

TEACHING PRACTICE COMPONENTS ASSESSED	TEACHER PRACTICE COMPONENT VARIATIONS[a]					
	N/A[b]	1 Teacher-centered	2	3	4	5 Child-centered
Interaction in small groups		**** *** 41%	*** 18%	*** 18%		**** 24%
Use of supplementary learning materials		***** **** 53%				**** **** 47%
TOTAL BEHAVIORS ASSESSED $N = 340$	22 *(6%)*	77 *(23%)*	133 *(39%)*		108 *(32%)*	

[a] The number of variations per component identified in the observation checklist ranges from three to six. To enable a summary of findings, we combined some variations to highlight overall tendencies in the figures reported under Total Behaviors Assessed at the bottom of Tables 1 and 2.
[b] For gross comparisons, we combined the figures in the N/A (non-applicable) column with those in the Teacher-Centered column. Many of the N/A-column figures were linked to previous responses. For example, seven nonproject school teachers had no centers of interest set up in their rooms, and so we gave them a score of "1" for Availability of Centers of Interest. If there were no centers available, we did not score teachers on use of Centers of Interest.
[c] Asterisks signify the actual frequency count of observations recorded for each checklist component variation.

Table 2: Implementation of Child-Centered Teaching Methods: Nonproject Schools

TEACHING PRACTICE COMPONENTS ASSESSED	TEACHER PRACTICE COMPONENT VARIATIONS[a]					
	N/A[b]	1 Teacher-centered	2	3	4	5 Child-centered
Lesson plans		*** [c] 37.5%	***** 62.5%			

TEACHING PRACTICE COMPONENTS ASSESSED	TEACHER PRACTICE COMPONENT VARIATIONS[a]					
	N/A[b]	1 Teacher-centered	2	3	4	5 Child-centered
Instructional approach		****** 75%	** 25%			
Quality of questions			****** 75%	** 25%		
Distribution of questions		*** 37.5%	* 12.5%	* 12.5%		*** 37.5%
Responding to pupil answers		* 12.5%	* 12.5%	****** 75%		
Dealing with incorrect answers	* 12.5%	** 25%	* 12.5%	**** 50%		
Decision-making for activities		**** **** 100%				
Grouping of pupils		**** 50%	** 25%	** 25%		
Observation and interaction		* 12.5%	****** 75%	* 12.5%		
Written assignments	** 25%		****** 75%			
Accessibility of teaching aids		** 25%	***** 62.5%			* 12.5%

TEACHING PRACTICE COMPONENTS ASSESSED	TEACHER PRACTICE COMPONENT VARIATIONS[a]					
	N/A[b]	1 Teacher-centered	2	3	4	5 Child-centered
Source of teaching aids	* 12.5%	****** 75%	* 12.5%			
Teacher use of teaching aids	* 12.5%	**** 50%		* 12.5		** 25%
Textbook use	** 25%	** 25%	* 12.5%	* 12.5%		** 25%
Availability of centers of interest		**** *** 87.5%	* 12.5%			
Use of centers of interest	**** *** 87.5%	* 12.5%				
Pupil engagement in active learning		**** **** 100%				
Inquiry-oriented activity		***** 62.5%	*** 37.5%			
Interaction in small groups		****** 75%	* 12.5%	* 12.5%		
Use of supplementary learning materials		***** 62.5%	* 12.5%			** 25%
TOTAL BEHAVIORS ASSESSED $N = 160$	14 *(9%)*	74 *(46%)*	62 *(39%)*		10 *(6%)*	

[a] The number of variations per component identified in the observation checklist ranges from three to six. To enable a summary of findings, we combined some variations to highlight overall tendencies in the figures reported under Total Behaviors Assessed at the bottom of Tables 1 and 2.
[b] For gross comparisons, we combined the figures in the N/A (non-applicable) column with those in the Teacher-Centered column. Many of the N/A-column figures were linked to previous responses. For example, seven nonproject school teachers had no centers of interest set up in their rooms, and so we gave them a score of "1" for Availability of Centers of Interest. If there were no centers available, we did not score teachers on use of Centers of Interest.
[c] Asterisks signify the actual frequency count of observations recorded for each checklist component variation.

Item analysis of the checklist findings strongly supported the overall finding that teachers and pupils in project schools demonstrated more of the child-centered teaching and learning behaviors than teachers in nonproject schools, for all dimensions of teaching practice assessed. Analysis of variations within different components of instructional practice, however, indicated that the fidelity (closeness to 'ideal' child-centered practice) of teacher implementation remained highly variable in project schools. A majority of project school teachers were judged still highly teacher centered for various components of practice. The use of short rote-answer questions, for example, remained the norm in many project school classrooms (59%). Student involvement in inquiry-oriented learning activities was not evident in a substantial number (41%) of project school classrooms. Appropriate use of small-group learning methods was limited to about a quarter of the project school teachers observed. Many teachers provided little remedial attention to students as they worked. While project school teachers made frequent use of teaching aids as they taught, more than half (53%) of those observed did not provide opportunities for students to interact with hands-on learning materials. The proportion of project school teachers who did not engage pupils in activity-based learning tasks was also high (59%). For one area, providing opportunities for pupils to participate in decisions about lesson content and activities, virtually all project school (88%) and nonproject school teachers (100%) were highly teacher centered. There are limitations to these findings. Teachers do not use all their teaching methods in every lesson (our observations were limited to one lesson per teacher). Nonetheless, the aggregate findings revealed distinct patterns of difference and similarity between teachers in project and nonproject schools.

These findings led us to conclude that, while an intensive dose of in-class follow-up support to initial workshops definitely carries teachers down the road to more child-centered teaching methods, the teachers reached will not all attain 'ideal' practice in a 6- to 12-month period of intensive, school-based support. While such findings may disappoint project developers and policy-makers, the finding is normal, given what is known about the developmental nature of teacher learning of new teaching methods.

Teachers, head teachers, project officers, and TAC tutors presented a common view of key factors influencing teacher progress with implementation of child-centered, activity-oriented teaching methods. The factors cited are consistent with past research on the implementation of changes in classroom practices. *Access to materials* required for implementation is a basic requirement. Effective *introductory and follow-up in-service assistance* is a close second. Teacher willingness to commit to the use of new methods is highly dependent on *evidence of immediate benefits* for students, particularly when *teacher workload* for preparation is intensified in the early stages of implementation. *Interaction (sharing, problem-solving) with other teachers* implementing the same changes helps teachers transfer new practices to their specific settings. It also provides social and emotional support for change. Finally, continuous *active support from head teachers and from parents* has important effects on teachers' capacity to implement new methods and on teachers' sense of priority and commitment to those changes.

Teacher Observation Profiles

The checklist data enabled us to judge the degree to which the kinds of teaching and learning behaviors advocated in MSIP goals and in-service activities were being put into practice. Our anecdotal notes and narrative profiles of classroom practice provided a basis on which to judge teacher effectiveness in the use of those methods (Avalos, 1990). While space precludes presentation of all the profiles, the following sample of classroom vignettes draws attention to issues of critical interest to this and similar teacher development initiatives. We drew on our experience as teacher educators in judging the quality of teaching displayed in these qualitative accounts of classroom practice.

Some teachers implement 'activities' without obvious connections to 'learning.' One of the most troubling findings was that some teachers appeared to have adopted the idea that simply involving students in activities constituted effective activity-based learning. These vignettes from project and nonproject schools illustrate this tendency.

Project School: Standard II Science

Forty-two students seated around four sets of tables. The teacher began by telling the children that it was a science lesson and wrote 'science' and 'paper toys' on the board. "Today we are looking at paper toys." The class repeated "Paper toys." The teacher introduced the lesson asking the children what a pile of art papers was made from. The children answered when asked, "Paper." Then she asked individual children to name kinds of toys. They responded with short phrases, such as "A car." "An airplane." Then she pulled several toys (e.g., a toy car, a toy camel, a toy paper boat) from a bag and asked closed short-answer questions of individual pupils around the room. "What is it? Is it real? Can you get in it? What is it made from?" She provided paper to two children and asked them to demonstrate how to construct a paper rocket and a paper fan. Then she herself showed the class how to make a paper boat. Next she told the group leaders to give out paper and told each group to make a fan, a boat, and so on. As they worked, she circulated from group to group, helping those who were stuck, praising good work, and instructing the children to color their toys when finished. The teacher closed the lesson by writing a list of toy materials on the board and a list of the paper toys the pupils had made, and by telling them to copy the lists in their exercise books. She asked the children to leave their toys on her desk to be displayed.

Nonproject School: Standard III Mathematics

Thirty-two pupils seated in individual, paired desks, arranged in three rows. The teacher introduced the lesson, asking the children to read the word 'small,' which she had written on the board. They responded chorally, and she had them repeat in chorus two times. She repeated this procedure with the word 'smallest.' Then she asked one child to state the opposite of 'small' and another the opposite of 'smallest,' which they did without difficulty. She wrote the lesson heading on the board and had the children read it and repeat it chorally two times: "arranging numbers from smallest to biggest." She began the actual lesson with a physical activity. She took the children outside and had them run a 100-meter race. She noted who came in first, second, and so on, to tenth. They returned to the classroom, she wrote the names of these children on the board and awarded the first-place winner 20 shillings. She then asked the children to award marks (100, 90, 80, etc.) to the ten winners, in descending order. She pointed out that the largest number was 100, followed by 80, 70, and so on, down to 10, the smallest number. She asked one child to come to the board and rewrite the same numbers from smallest to biggest, which the pupil did without difficulty. Next, she wrote three sets of numbers on the board and told the children to arrange them from smallest to biggest (e.g., 85, 106, 60, 95, 120, 72). She circulated, marking the pupils' problems and offering no help to weak pupils. There was no clear-cut conclusion to the lesson.

The first vignette was a 'science' lesson, though it was not evident what science concepts the teacher was trying to present, nor what learning she expected to result from paper-toy construction and copying word lists of toy materials and objects. The science activity involved no inquiry, prediction, or analysis, and the teacher made no attempt to explore what students might be getting out of the lesson, other than the fun experience of building paper toys. The second example is from a nonproject school; therefore, the teacher's superficial implementation of activity-based learning strategies was less unexpected. The mathematics lesson was about sequencing numbers. The 100-meter dash was entertaining, but contributed little to pupil learning. The rest of the lesson was conducted as a traditional chalk-and-talk, teacher-centered lesson. The following vignette is from another mathematics class in the same school.

Nonproject School: Standard VI Mathematics

Thirty-five pupils seated in individual, paired desks, arranged in three rows. The teacher (male) began the lesson with 4 minutes of quick mental division problems: "What is 8 divided by 2? What is 100 divided by 25?" The students responded quickly and correctly. He introduced the lesson, using a chart showing the formula "speed equals the distance covered divided by the time taken". He wrote an example on the board connected to everyday experience: a person traveling from Nairobi to Mombasa, 480 km in 5 hours. He asked the children to tell the distance, the time, and to calculate the speed, which they did positively. Most of his questioning was directed to the pupils seated toward the front of the room. Then he moved to the application part of the lesson, which was to have the children do ten speed problems from their mathematics textbook. He circulated among the desks, marking problems as the pupils worked, without helping the weak pupils.

The addition of a student 'activity' to the Standard III mathematics teacher's lesson did nothing to enhance the quality of student learning, when compared to the traditional mathematics lesson without 'activity-based learning' illustrated in the Standard VI example. These were not isolated examples. The frequency of skillfully implemented, activity-based learning activities observed in the 24 classrooms sampled was low. This raises a fundamental question for this and other in-service projects that promote a pedagogy of active learning. Is it is 'better' to invest training resources in the promotion of activity-based child-centered methods, which appear to be very challenging for teachers to internalize and skillfully apply, or might resources be better directed toward improvement in teacher use of more traditional didactic methods (Guthrie, 1990)? Given the developmental process of teacher change, it is possible that the teachers were in the early stages of mastering this type of teaching. Gardner (1990) studied the adoption and use of active learning in Indonesia. Initially, teachers implemented activities that merely kept students busy. Over time, the educational value of activities became a more important consideration.

Inquiry-oriented teaching and learning activities are the most difficult to master. Student participation in inquiry-oriented learning activities was rare in all classrooms we observed. While a few project school teachers involved students in activities that had potential for student inquiry and discovery, these teachers were either unaware, not skilled enough, or chose not to take their lessons to that level.

Project School: Standard I Science

Twenty-nine pupils seated in pairs, arranged in three groups of 9 to 10. Classroom rich in learning areas (science nature chart, nature table, curiosity chart, hanging wordcards, word tree, shop, charts in all subject areas, physical education corner). The lesson dealt with learning about animals. The teacher had the students begin by naming animals they knew (e.g., dog, lion, snake). Next, she gave each group a sheet of paper and directed the group leaders to write the names of all the animals that the group could think of. She told the children to consult the curiosity corner if they needed (there was information about animals there). When this task was completed, the teacher asked the pupils to pair up and to classify the animals into two groups: "animals found at home" and "not found at home." Children had difficulty classifying some animals, such as rats, pigs and birds, which could be grouped in either category. The teacher responded vaguely to children's questions about birds and pigs. She concluded the lesson, asking children to name, out loud, animals depicted on flash cards she had prepared.

Nonproject School: Standard III Science

Twenty-six students seated individually, in three rows. The teacher (female) began the lesson, asking the children to recap the previous lesson. The children were able to recall that air allows sound to travel from a radio to their ears. The teacher demonstrated this by having a child go outside and call a name. She told them that they were going to find out whether sound travels through solids. She asked the children to pair up and then asked one child to put her/his ear to the desk while the other knocked the desk. Teacher: "What do you hear?" Child: "Air." Teacher: "Air?" Child: "Sound." The teacher then announced, "The sound traveled through the desk to your ear." Then she sent a child outside to beat the wall with a stick and had three children put their ears to the wall. She asked whether they heard something, and the pupils responded affirmatively. The teacher concluded that the children could hear even though they couldn't see anything. Then she showed the pupils a telephone made from string and matchboxes and sent one child outside with one end of the telephone to talk to another child inside the classroom. The two pupils reported that they could hear each other and the teacher announced that they could hear sound. Next, she gave similar 'telephones' to groups of students and had them all go outside to try the experiment. Back in the class, the teacher concluded that sound traveled through the string, which is a solid, and related the experiment to sound traveling through telephone wires from Mombasa to Nairobi. She asked the pupils to draw the telephone in their exercise books.

Both lessons were reasonably effective in the use of teaching and learning aids, the integration of small-group work, and to a lesser extent, teacher questioning. Both lessons, however, had the potential to involve students in meaningful exploration, discovery, and analysis of their own conclusions. This potential was not exploited. So, while these were positive examples of implementation of some MSIP expectations, the teachers had not progressed to a high level of student-centered inquiry in lesson design and delivery.

Teacher mastery of child-centered teaching methods proceeds incrementally over time. The final vignettes are from nonproject schools. They depict two of the more teacher-centered lessons we observed.

Nonproject School: Standard I English

Twenty-nine children seated in groups of six to eight, around four rectangular tables. Back wall covered with vocabulary, spelling, and reading charts. Some had hand-drawn or magazine pictures for key words (e.g., "This is a table"). Wordcards were strung along one wall and a vocabulary fishing game, *Lucky Dip*, sat in a box on a small table in the back of room. The entire oral lesson was devoted to repetition of teacher-directed phrases in English. Often, the teacher directed the children to continue repeating a phrase several times, and to raise the volume of the voices with each repetition. Several times during the lesson, she interrupted choral recitation to do a whole-class round-robin with each student standing and reciting a specific phrase ("This is the top of the bottle." "I am a boy." "I am a girl."). She frequently asked the students to repeat phrases until she was satisfied with their pronunciation. Apart from English pronunciation and vocabulary, the lesson appeared to be about opposites: top and bottom, inside and outside, boy and girl. The teacher attempted to use teaching and learning aids, such as a plastic bottle passed around, children pointing to the top or bottom of the bottle as the spoke the key phrase, and children pointing and gesturing to different objects (e.g., top of door, top of table). The children mimicked each gesture, as well as the teacher's words. The lesson had no clear introduction, development, or closure. At the end, the teacher instructed the pupils to repeat the ten times table while she wrote instructions for the next lesson on the board. They chanted, "One 10 is 10, two 10s are 20, three 10s," up to 100, over and over until she signaled them to stop after about ten repetitions.

Nonproject School: Standard VI Mathematics

This teacher (male) had no lesson plan. The lesson was on rounding off to the nearest ten and was conducted in a highly teacher-centered manner. During the introduction, for example, he made statements such as, "I told you that 5 becomes 10, 4 becomes 0 because it is smaller than 5, and 8 becomes 10 because it is greater than 5. I have told you this as an example." Through closed questions, he developed the following summary: 3→0, 1→0, 7→10, 11→10, 17→20, 10→10. He explained everything most of the time. For example, for the number 17, "This one here is more than 15, so it goes to 20." He did not seem to know many of his pupils' names. Quite a few times, he pointed to them with statements such as, "Yes, you at the back." Sometimes, he posed a question and answered it himself.

These two examples are important reminders of the incremental process of classroom change. The first example is actually an MSIP success story, in terms of teacher change. According to the teacher, before attending an MSIP oral language instruction workshop, she would only have been observed 'saying' and the children in chorus 'repeating', with no instructional aids or supplementary vocabulary activities (the word game *Lucky Dip*). She attributed some of her tactics, including her use of teaching and learning aids (the bottle) and having the children individually recite the vocabulary words and phrases, to the SIP workshop. In her mind, the fact that she was now listening to individual responses and correcting when necessary made her more aware of and responsive to pupil needs and progress than before, which therefore meant that the lesson was more student centered. Using objects and gestures to illustrate the meaning of words and phrases made the learning more activity based. The second teacher judged his mathematics lesson acceptable, because (a) most of the children participated, answering questions, (b) a few worked on the board and gave correct answers, (c) the answers in their exercise books were mostly correct, and (d) the children behaved. The observer, an experienced mathematics educator, had noted from the pupil's exercise books that a few had no idea what 'rounding off' was about (e.g., rounding of 46 to 70). After the observation and interview, she suggested a simple visual aid to the teacher: a number line, shaded from 5 to 10 and unshaded from 0 to 4, with single-digit, double-digit, and three-digit examples. Later that day, the teacher reported that he had used the learning aid with another class and that it had been useful for 'slow learners.'

These two last examples illustrate the incremental and content-specific nature of teacher learning. The examples can be viewed optimistically as demonstrating the potential for change. On the other hand, they reinforce the reality that the MSIP strategy of periodic subject-specific workshops and follow-up visits by external helpers is likely to produce small gains that may or may not accumulate over time. The last example also illustrates the value of consultant subject expertise. Another member of the evaluation team was an experienced user of cooperative learning, but it would not have occurred to him to suggest an instructional aid in teaching 'rounding off' in mathematics.

Some further key findings arising from our analysis of classroom observation profiles were as follows. The quality of implementation of child-centered teaching methods varied for teachers within project schools, as well as between project and nonproject schools. This finding was a significant reminder that the degree and quality of implementation of new teaching methods cannot simply be predicted from knowledge of in-service interventions delivered. The degree and quality of implementation of child-centered teaching methods was also affected by teachers' subject expertise and interest. One project teacher was observed teaching lessons in different subject areas. Her use of child-centered methods was much stronger in mathematics (her special interest) than in English, despite her having attending MSIP in-service workshops in both subjects. This teacher's case also highlighted the subject-based pedagogical orientation of many of the teachers we contacted. It may be easier for teachers to adopt subject-specific teaching methods than to learn generic methods and then apply those methods to a variety of subjects. Finally, most teachers demonstrated a novice understanding of small-group learning methods. Some organized pupils in groups, but used whole-class methods of instruction. Some involved pupils in small-group activities that could have been more effectively carried out with individuals. Evidence that teachers expected both individual and group accountability for task completion and learning outcomes was not evident in most instances. Developing expertise in using small-group learning strategies is a long-term professional learning goal (Anderson, Rolheiser, & Bennett, 1995). We would not expect a high level of mastery over the short life of the MSIP.

Synthesis of Classroom Observation Profile Findings

Our interview and observation checklist data supported the claim that significant changes in teaching practice had occurred in schools and classrooms influenced by MSIP in-service activities, and that the greatest changes had occurred in schools that had benefited from in-class assistance by MSIP project officers. The findings suggested that teacher adoption of some teaching methods advocated through the MSIP was attainable with adequate material and in-service support, such as training in preparation and use of teaching aids, activity-oriented lessons, and small-group learning. However, teacher skill in the use of these and other methods associated with child-centered activity-based instruction remained highly variable and often in the early stages of development, despite the intensive in-service training and material support given to project schools.

This was not a negative finding in terms of the MSIP's impact on teaching; rather, it was a realistic finding about the complexity of attempting fundamental changes in teaching practice. The MSIP design proved successful in getting teachers to adopt and begin to use a wide variety of nontraditional teaching methods in the classroom. We concluded, however, that teacher access to *continuous*, external and classroom-based in-service training and material support would be needed to enable teacher mastery of the kinds of teaching methods advocated in the MSIP in-service activities.

Student Outcomes

The absence of reliable pupil outcome data limited our ability to fairly judge the ultimate effects of the MSIP in-service and school improvement activities on the quality of student learning. Following on recommendations from previous AKF- and AKES-sponsored SIP projects in Kisumu, Kenya, the project staff developed its own standardized tests to track student performance in English, mathematics, and science and administered them to students in seven project and six nonproject 'control' schools in 1996, and again in 1997. Comparisons of the aggregate mean scores of project and nonproject schools revealed little difference (e.g., no more than a one-point difference in each subject). This was not surprising, since MSIP in-service work in Phase 2 schools did not begin until the first term of 1997. Significant effects would not likely have shown up so quickly. More important, the project staff identified shortcomings in their evaluation design that called into serious question the reliability and validity of the evaluation results (e.g., lack of data on student retention and transfers, variability in the conditions of testing). As a result, they abandoned the MSIP testing program and shifted their attention to student results in district and national testing programs (i.e., the Standard VIII primary school leaving examination, district mock-testing programs leading up to the Standard VIII examination). From the perspective of local stakeholders (teachers, students, parents, supervisors), student performance on independent MSIP tests held little meaning, in comparison with student performance on high-stakes district and national examination systems' associated with student promotion and with school and teacher reputations. We concluded that the expertise and resources required to carry out credible independent testing of pupil outcomes exceeded the scope of the MSIP. We recommended that project efforts could be more productively directed toward influencing the quality of local and national examination systems, and lending technical support to school and district personnel to enhance their assessment literacy and capacity to use examination results for in-service training and curriculum development plans.

In 1995, Mombasa District ranked 60th out of 63 primary school districts across Kenya, in terms of average student results on the Standard VIII primary school leaving exam. By 1997, Mombasa District's ranking had improved to 53rd, and by 1997 to 37th. In 1996, the district was recognized as the second most improved in the country. The MSIP project leaders were appropriately modest about the project's contribution to that accomplishment: "SIP is proud of this improvement, though we can't directly say we are wholly responsible for the good work. It's a collaborative issue." (1998 project report) Since MSIP in-service activity targeting teachers in Standards I to VI across the system did not seriously get underway until the first term of 1997, it would be difficult to argue that in-service activities in 1996 and 1997 would have had a system-wide impact on Standards VII and VIII teachers and students. MSIP, however, was a partnership with the municipality that fitted into a broader district and provincial commitment to improving the quality of teaching and learning in Mombasa primary schools. It represented one thrust of this effort, as did the ongoing, national Strengthening Primary Education program (SPRED) and the management training program for primary school head teachers (PRISM). Improvement in student results was more likely attributable to the combined effects of these initiatives than to any one input. We surmised that the MSIP contributed to the improvement in student results through its system-capacity-building initiatives after 1995, which included reactivation of the TACs, increased parental support and funding for schools and TACs, and activities sensitize to stakeholders at all levels, in addition to the in-service work with teachers and heads teachers.

Ultimately, we were left with weak measures of project impact on students: the perceptions of teachers and project officers interviewed during our site visit. Comments from these groups were consistent: They highlighted improvements in student attitudes toward school and learning, more active participation in learning, improved reading performance and oral language skills, increased attendance, and, to a lesser extent, improved academic results. Further confirmation of improvement in student attitudes toward school and learning came from a small group of parents sampled in an MSIP school committee focus group conducted as part of the MSIP evaluation.

Sustaining Improvement

The Mombasa School Improvement Program demonstrated the feasibility of implementing a consultant-centered professional development system in an urban African primary school district, capable of delivering a coordinated combination of external workshops and intensive school-based assistance consistent with research on the characteristics of effective in-service teacher development. The in-service approach yielded evidence of positive results for change in teaching practices, for teachers who benefited from access to both the school-based assistance and external workshops, in comparison with teachers who benefited directly or indirectly from only the external workshops. This finding confirms past research on the significant effects of follow-up coaching in the classroom, to support the implementation of pedagogical innovations introduced through workshops and similar in-service events outside the classroom (e.g., Avalos, 1985; Joyce & Showers, 1980; Djalil & Anderson, 1989; Harvey, 1999). This occurred in the context of an increasingly decentralized system of school governance, funding, and teacher and school development.

The in-service design training was implemented as part of a comprehensive and integrated initiative aimed at developing a sustainable and effective infrastructure for continuous development for primary school teachers across the district. Implementation of the MSIP in-service model was supported by the following key capacity-building initiatives: (a) sensitization of parents and mobilization of school committees to take on greater responsibility for school management and funding along with the head teacher, (b) mobilization of zonal Teacher Advisory Centers (refurbishing and equipping TAC facilities, establishing TAC management committees, securing funding from school committees for TAC services, appointment of competent TAC tutors), (c) basic management training for head teachers, school committees, and TAC management committees, and (d) change facilitator training for project consultants, TAC tutors, and head teachers. External funding was provided for (a) the purchase of equipment (e.g., typewriters, overhead projectors) to aid in mobilizing the TACs, (b) the delivery of MSIP in-service activities, (c) transportation for MSIP project officers, (d) community development interventions (the community development officer), and (e) project direction. The expectation was that start-up costs would be sustained through the district budget for education personnel, and through the financial contributions and fundraising activities of parents.

While evidence of the positive effects of the in-service training strategy was documented through our evaluation, challenges remained for continuation and refinement of the teacher development goals of the project, and for sustainability of the professional development infrastructure. The teachers we observed in MSIP schools exhibited variable levels of expertise across the 12 dimensions of teaching and learning behavior assessed. We concluded that widespread mastery of the kinds of teaching and learning methods introduced and supported through the MSIP would require continuing school-based in-service assistance, beyond the initial period of intensive regular visits from teacher development consultants provided through the MSIP in-service training design. In our recommendations, we highlighted two directions for development of the support system to further strengthen the local capacity for continuous improvement.

The future status of the MSIP officers remained in question. While the project officers were District employees seconded from established positions, it remained uncertain whether the teacher development activities of the POs could continue at the same level of intensity when the 'project' ended and the people in those roles were reassigned to traditional education officer roles. In addition to the MSIP officers, the District was only able to fund about half the zonal inspector and divisional school advisor positions during the project years (These jobs are funded through the civil service, not the Teacher Service Commission, as the TAC tutor jobs were.) This raised doubts about the capacity of the District to sustain the appointment of a full complement of project officers, inspectors, and school advisors. This was also a key factor underlying the diversion of TAC tutor time away from teacher development activities, into supervisory and administrative duties normally carried out by inspectors.

Since the evidence of teacher development was greatest in schools that benefited from the POs' school-based in-service assistance, it seems worthwhile to consider ways of increasing both the scope and duration of assistance to teachers at the school level. As a partial solution, we argued for a restructuring of supervisory and teacher development roles across the district. The aim would be to increase the number of teacher development consultants per zone to three, to shift many of the administrative and supervisory responsibilities of inspectors and TAC tutors to other personnel (e.g., zonal head teacher associations, subject panels, TAC support personnel), and to increase the school-based teacher development activity of people previously in TAC tutor and inspector roles. It was evident, however, that project officers, even in closer collaboration with TAC tutors and inspectors, could not possibly provide the frequency and intensity of classroom-based in-service assistance that would be required to genuinely carry all primary teachers in the district to higher levels of instructional expertise. We concluded that continued development of instructional expertise would require greater attention by consultants, supervisors, and head teachers to the development of more self-sustaining, school-level professional communities, more networking among schools, and more strategic use of (without overdependence on) external expertise at the school level.

The benefits for teacher and student learning of strong, school-based professional communities of teachers are well documented in the school improvement literature (Kruse & Louis, 1995; Joyce, Calhoun & Hopkins, 1999; Thiessen & Anderson, 1999). Kruse and Louis (1995), for example, identify five critical elements of professional community in schools: reflective dialogue, deprivatization of practice, collective focus on student learning, collaboration, and shared norms and values. The MSIP had contributed to awareness of the importance of these elements across the district, especially in project schools. There remained significant work at the school level, however, to develop the structural conditions (time to meet and discuss, physical proximity, interdependent teaching roles, communication structures, teacher empowerment and school autonomy) and the social and human resources (openness to improvement, trust and respect, cognitive and skill bases, supportive leadership, teacher socialization practices) necessary to the establishment of vigorous, self-sustaining professional communities in Mombasa schools.

Interpretation of the teacher development outcomes of the Mombasa SIP is contingent in part on the nature of the 'innovation' in teaching practices promoted through the project. The major focus of change was the acquisition and improvement of instructional methods, not the implementation of a specific program and materials, with prescribed teaching methods for teachers and prescribed learning activities for pupils. The mastery of generic teaching methods and skills, we would argue, is a more difficult and longer term goal than helping teachers learn to carry out lesson-specific teaching and learning methods. Many of the teachers we interviewed in MSIP project and nonproject schools reported that the teaching methods advocated through the MSIP were not 'new' to them. They had been exposed to knowledge about similar methods during their pre-service teacher training. The working conditions, instructional norms and expectations they encountered as practicing teachers, however, had never been conducive to putting those ideas and methods into practice. The MSIP significantly altered their conditions of practice, thereby encouraging and enabling them to begin implementing teaching methods that they had previously regarded as impractical, if not impossible. These conditions included (a) ongoing access to in-service assistance from consultants in both workshop settings and in the classroom, (b) the ready availability of low-cost materials for teaching and learning aids at the school and zonal TACs, (c) a shared understanding and commitment to implementation across the school and district, (d) a high degree of instructionally focused teacher collaboration in the school, (e) instructional leadership and support from head teachers and supervisory officers, and (f) demonstrations of parental support for the implementation of nontraditional teaching and learning methods. Teacher training in the use of progressive teaching and learning methods stands little chance of being put into practice on a wide scale without an effective support system for implementation and ongoing teacher learning.

Beyond the need for strengthening access to ongoing in-service training and consultative services for teachers, the continuation of high levels of parental involvement in school and TAC financing and governance was clearly essential to sustaining the MSIP-inspired support system for continuous teacher and school development. School- and TAC-based control of funding made it possible for head teachers and TAC-based teacher development personnel to respond to the material and in-service support needs of teachers, without getting caught up in centralized control of funds at the district level. It seemed clear, however, that the sustainability of parental involvement of this sort would require the continuing intervention and support from someone, ideally a team, acting full-time in a community development role. In our evaluation, we recommended that steps be taken to institutionalize the role of community development officer to aid parental involvement in site-based management of schools and TACs, in addition to the teacher development work of district consultants such as the POs and TAC tutors.

In an examination-conscious school system such as Kenya, the sustainability of nontraditional teaching methods and of innovative support systems for continuous school and teacher development is inevitably and inextricably linked to the short- and long-term evidence of impact on student learning. While the national standing of Mombasa primary schools on primary school leaving examinations improved dramatically from 1995 to 1997, neither we nor project staff were able to attribute this improvement directly to the MSIP and to the changes in teaching methods observed and reported, particularly in project schools. The original design of the MSIP called for the development of independent baseline measures of teacher performance, school management, parental involvement, and student learning. The intent was that such data would be useful, not only for planning project interventions, but also for tracking project impact through the use of repeated measurements. Despite genuine effort by the project personnel, the technical expertise and logistics of developing and carrying out a such a complex monitoring and evaluation system proved beyond the scope of the project's human and material resources, and diverted attention and energy from the project's in-service and parental involvement activities. The common expectation, by external funding agencies, that education improvement initiatives such as the Mombasa School Improvement Project create and carry out their own results-oriented evaluations is understandable from the donor-agency perspective. In the long run, however, this kind of within-project monitoring and evaluation research may contribute little to the capacity of local school systems and schools to critically appraise the worth of educational innovations. In our view, future projects of this genre might well devote more attention to furthering the capacity of local school systems to become critical assessors of educational innovations. A key function of the central support system would be to organize and technically support evaluations of innovative projects such as the Mombasa School Improvement Program, rather than to encumber the project personnel with daunting expectations for development and use of highly technical monitoring and evaluation systems that may duplicate, stand apart from, or even conflict with existing evaluation systems, and with those of other externally funded school improvement projects.

1. Children in school uniforms standing outside Union Primary School. Kisumu School Improvement Project, Kenya

2. Teacher assists pupils seated in groups of six. The classroom walls are covered with children's work and visual aids.Kodiaga Primary School, Kisumu School Improvement Project, Kenya

3. Despite crowded classroom conditions teachers do manage to organize the class for small co-operative group work. Ezra Gumbe Primary School, Kisumu School Improvement Project, Kenya

4. SIP Team members help teachers make teaching aids at Teachers' Resource Center. Kampala School Improvement Project, Uganda

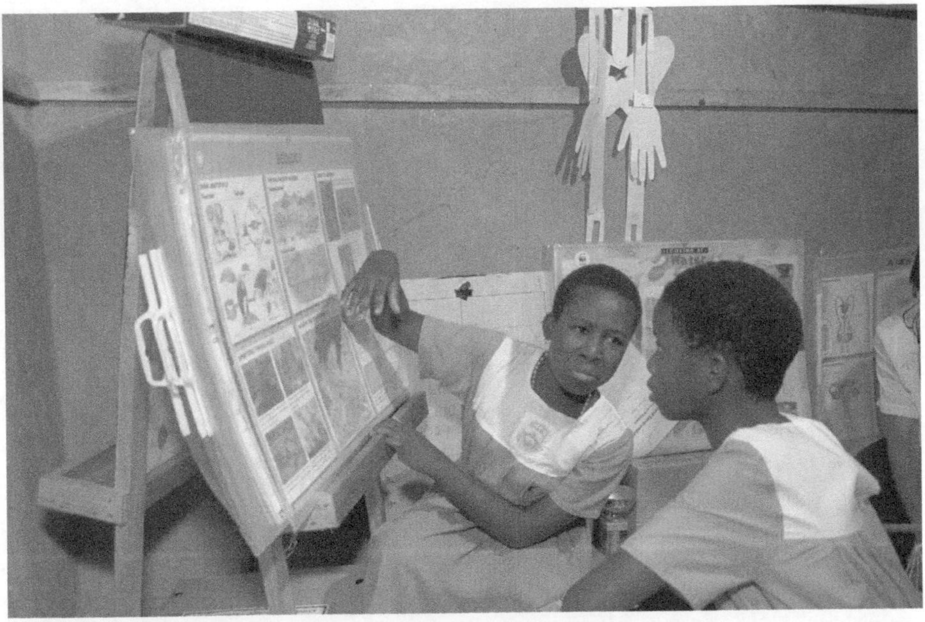

5. *Students discuss learning charts created by teachers. Makerere Church of Uganda Primary School, Kampala School Improvement Project, Uganda*

6. *Children seated on wood benches with desks use learning aids created by teachers. Makerere Church of Uganda Primary School, Kampala School Improvement Project, Uganda*

7. *Outdoor morning assembly at Aga Khan Mzizima Secondary School, Dar-es-Salaam. Mzizima School Improvement Project, Tanzania.*

8. *Teachers jointly prepare class library materials during weekly teachers' workshop at Mzizima Secondary School, Dar-es-Salaam. Mzizima School Improvement Project, Tanzania.*

9. Panorama of grounds of Mzimuni Government Primary School, a large urban school serving over 3000 pupils. Site of project office for the Dar-es-Salaam Primary Schools Project, Tanzania.

10. Teacher conducts a class of 100 in a typically crowded, resource-poor primary school prior to project intervention. Dar-es-Salaam Primary Schools Project, Tanzania.

11. Teachers working together to prepare teaching and learning materials at a Teacher Resource Center. Dar-es-Salaam Primary Schools Project, Tanzania.

12. Kinondoni Teacher Resource Center at Mzimuni Government Primary School. Dar-es-Salaam Primary Schools Project, Tanzania.

13. *Bondeni Girls Primary School, site of project headquarters. Mombasa School Improvement Project, Kenya.*

14. *Project Director visits a SIP class. Children are divided into groups to use learning aids made from everyday materials. Central Girls Primary School, Mombasa School Improvement Project, Kenya.*

15. Teacher involves two children in acting out a story in English. Secondary English Language Orientation Project, Zanzibar.

16. A group of students share a textbook produced by the project team. Secondary English Language Orientation Project, Zanzibar.

7

Incentive Grants, Management Training, and Teacher Change in Dar-es-Salaam, Tanzania

Geoff Welford and Herme Mosha

This chapter describes the Aga Khan Foundation-Tanzania (AKF-T)'s Dar-es-Salaam Primary Schools Project (DPSP), based on the full-term external evaluation of the project. We start with a description of the project, providing background to its establishment and setting it in its local context. The project's design and implementation strategies are summarized before examining, in some detail, the two main strands of the project's work: school management and teaching in the core subjects (mathematics, science, and English). The DPSP has introduced an innovative incentive grant to stimulate and encourage participation and full implementation of project training. The effectiveness of this innovation is of interest and is probed. The DPSP was intended to develop the Teacher Resource Center in which it was based, and the TRC's role here is evaluated. The DPSP has operated in a wider context of development, and this wider context is the subject of discussion leading to proposals for the future. The chapter concludes with consideration of the lessons learned in this intervention.

Background

At the time of the evaluation, November 1999, the Dar-es-Salaam Primary Schools project was working with 15 schools in the Kinondoni District of Tanzania's capital city, Dar-es-Salaam.

The DPSP interventions started in 1996. The project was implemented through the Kinondoni Teacher Resource Center (TRC), in a phased approach over the 4-year period from 1996 to March 2000, in 15 of the district's 73 public primary schools. The project was thus in its final year during the evaluation, having included five schools in each of three cycles of training and follow-up activity. The DPSP was funded by the Canadian International Development Agency (CIDA), the British Government's Department for International Development (DFID), and the Aga Khan Foundation (AKF).

The Genesis of the Project

The project was conceived following an analysis of local primary education provision by AKF-T, carried out using its own funds between July 1994 and June 1995. The survey described constraints on the provision of quality primary education in Tanzania and prompted the development of a project proposal concentrating on the following goals.

- Improvement of school management by providing head teachers and school committees with knowledge and skills needed to prepare good annual plans, maintain effective school-community relations, and manage school finances
- Help to build the capacity of government-run Teacher Resource Centers to provide support for teachers of mathematics, English, and science
- Development of a whole-school improvement approach, intended to enhance the accountability of schools to the school committees and parents, in the light of a government decision to decentralize the management of primary education. (The government had decided that primary schools should assume responsibility for managing themselves, with head teachers, teachers, school committees, and parents making decisions together.) (AKF-T, 1995)

The Tanzanian Education Policy Context

Tanzania has a 2-7-4-2-3+ system of education (2 years of pre-primary education, 7 years of compulsory primary education, 4 years of secondary ordinary level, 2 years of secondary advanced level, and a minimum of 3 years of university education). It is intended that children enroll in primary school at the age of 7 years, although in practice children both younger and older (up to 12) start Standard I. Ownership, management, and administration of primary education involve central government, local authorities, and sometimes nongovernmental agencies. Local authorities administer primary schools, while the Ministry of Education and Culture (MOEC) lays down the curriculum, prescribes and provides textbooks, and employs teachers and inspectors of schools (URT, 1993).

The Government of Tanzania has provided explicit objectives for primary school education (URT, 1995). In order to implement these policy objectives for primary and other levels of education, the Tanzanian government, in 1999, launched the Education Sector Development Program (ESDP). The ESDP is a 15-year phased program. The first phase is the Medium-Term ESDP Plan, which runs from 1999/2000 to 2003/2004. The Plan is predicated on Tanzania's long-term development policy, Vision 2025, which emphasizes the development of a high-quality, relevant, and equitable education system. A well-educated Tanzanian, in turn, is expected to be able to use accumulated knowledge, skills, and aptitude to eradicate poverty, one of the key objectives of Vision 2025.

The aim of the first phase is to provide a high standard of education to all children between the ages of 6 and 15. Basic science and mathematics are to be accorded great importance in keeping with the demands of the modern technological age, while sight is not to be lost of humanities, especially those elements that are necessary for promoting democracy, entrepreneurship, and good governance. The Medium-Term ESDP Plan is described as child-centered. All planned reforms and interventions (inputs and processes) are directed at improving the teaching and learning capacity and the learning environment, as well as enhancing the efficiency and effectiveness of the teaching, learning and assessment processes, in order to ensure effective learning.

The Dar-es-Salaam Primary Schools Project is one of the collaborative ventures between the Tanzanian government and its partners that address problems inhibiting effective learning at the primary-school level. A brief account of primary education in Kinondoni District is provided in the following section in order to set the context for the DPSP intervention.

The Kinondoni District Education Context

Kinondoni is the one of three districts in Dar-es-Salaam. In 1996, it had 68 primary schools with a pupil population of 108,683; in 1999, it had 73 schools with a pupil population of 110,787. The official teacher-pupil ratios were 1:42.5 in 1996 and 1:44 in 1999. However, in our evaluation, we never saw classes with fewer than 50 children, and several of the classes observed had more than 120, one in excess of 160!

In 1994, of 2,882 primary teachers in Kinondoni, 1,356 (47.1%) were classified as Grade B/C (those who have completed Standard VII but failed to gain entry into secondary school). Most of these teachers have received little formal teacher training, and many are Universal Primary Education teachers trained at a distance. In 1994, 91.7% of all Grade B/C teachers were female. In 1999, 92.2% were female.

In 1994, there were 1,526 Grade A teachers in Kinondoni primary schools (53% of all primary teachers), 85.5% of them female. Grade A teachers have completed secondary school and received some formal teacher training. By 1999, the number of Grade A teachers had risen to 1,653 (59.93% of all primary teachers), with no change in the gender balance. Government policy is for Grade B/C teachers to upgrade to Grade A. The figures above show that there was progress on this policy between 1994 and 1999, although it will take many years to reach the target.

Teachers' salaries are low. With increased liberalization of the economy, the majority of primary school teachers in Tanzania live below the nominal poverty line of US$ 500 a year. Most teachers supplement their incomes through small businesses on school premises or elsewhere. Many run private-tuition classes, sometimes during normal school hours but mostly after school, on weekends and on public holidays, as well as during vacation. The priority for most teachers appears to be to raise their incomes to acceptable levels (GTZ, 1995).

Teacher motivation has been further eroded by lack of resources at school level. We found in our evaluation that primary schools in Kinondoni offered children a poor learning environment. In the suburbs many school buildings required rebuilding, while in the urban core nearly all schools needed some maintenance and rehabilitation. School buildings had leaking roofs, no ceilings, and rarely even wire mesh in window spaces. Most classrooms did not have lockable doors. There was an acute shortage of desks and space, as some classrooms housed over 150 students. In 1994, baseline data showed that there was an average of one textbook for every 13 pupils. School management was generally poor, as few head teachers had received training in management. Teaching materials, such as manila sheets for the production of teaching aids, were not available.

At the time of the inception of the DPSP, Kinondoni District had a Teacher Center (TC) that was financed by contributions from parents. It provided some refresher courses for teachers, but more important, was used to upgrade primary school teachers. It provided a program of tutoring for good Grade B/C teachers, allowing them to complete secondary education and satisfy entry requirements for enrollment in a longer teacher-training course.

The Dar-es-Salaam Primary Schools Project

Project Goals

The project's objectives were to:

- recruit, orient, and train a project team in the project's improvement model, and to carry out the training and supervision tasks necessary;
- train primary school head teachers to lead academic and financial planning, and management of change at school level, working with school committees, parents, and TC staff;
- support school head teachers and school committees to clarify their roles and to develop their ability to undertake results-based school management (of academic, financial, and human resources) and to monitor and upgrade the schools' academic programs;
- improve accountability for funds raised by parents, by training head teachers and school committees to disseminate easy-to-assimilate accounts (and

through this process, to further involve parents in decisions about how school funds should be spent, thereby motivating them to provide consistent, sustained financial and other support to schools);

- train Kinondoni TC staff to upgrade their knowledge and skills in order to strengthen their capabilities to train head teachers, school committees, and teachers;
- train school academic coordination committees to assess needs and to ensure that improved teaching methods, curriculum, materials, etc., are introduced into classrooms, and that their use is sustained and further developed with active support from head teachers and the TC;
- train teachers in child-activating pedagogy, teaching for individual differences, organization of large classes, student assessment, and upgrade teachers' subject knowledge and planning skills;
- set up simple, yet effective, monitoring systems to allow head teachers and school committees to assess changes in school management, teachers' abilities, and student performance;
- disseminate the results of school changes to policy-makers and other agencies and, through this dissemination, to set up a dialog with the government on improvements to the system; and
- set up processes to conduct and use research to develop the SIP model and to further its replication in other schools and TCs. (AKF-T, 1995)

Overall, the project approach combines Teacher-Center-based training using a modular approach with school-based support, follow-up, and monitoring.

At the end of the DPSP, the following outputs were expected.

At School Level

- Fifteen head teachers and up to 50 assistant head teachers better equipped to academically, financially, and administratively manage their schools
- Fifteen school committees with clearly defined roles and management skills, more actively involved in improving academic and financial performance
- Up to 150 teachers trained in the use of child-centered and activity-based teaching techniques
- Fifteen school coordinators with the skills and credibility to sustain good teaching practice and provide school-based support to teachers

At Teacher-Center Level

- TC trainers with 'master-trainer' skills in adult teaching, child-centered teaching techniques, skills for monitoring teacher performance, and management/organization
- An in-service teacher upgrading program for mathematics, science, and English successfully implemented

- A management and planning training program for head teachers and school committees
- A strengthened TC, with basic materials and essential equipment, capable of continuing support work for all schools in the district
- TCs in the other districts of Dar-es-Salaam having had the benefit of some training and sharing of resources and materials with the Kinondoni TC, thereby preparing them to embark on management and training programs within their own schools

The project is administered by the Aga Khan Foundation-Tanzania. An agreement signed between the AKF-T and the City Council specifies roles for the City Council, Zonal Inspectorate and the district education officer (DEO) in the management of the project.

Project Management

Figure 1 shows the organization of the project team and its management as of mid-1999. The project director was appointed from a pool of experienced international and national educationalists. The British Voluntary Service Overseas (VSO) provided three volunteers: one to develop mathematics and science, one for English, and a third volunteer, who in addition to maintaining project accounts, had a major responsibility in training school committees and administrators in accounting and financial planning (AKF-T, 1995).

Three members were to be recruited from the Kinondoni Teacher Center (KTC) and involved as trainers in the project team. They were to ensure a degree of institutional capacity building for the TC. On the basis of advice received from the District authorities and the head of KTC, it was decided to take only one teacher from the center who would be responsible for science, as other staff were judged not to have the necessary skills for the trainer posts. Through consultation with the district academic officer and the district chief inspector of schools, candidates for the mathematics and English trainer posts were identified from schools in the District. The staff were selected on the basis of merit, and formed the core of 'master trainers' for the project. The attachment of the three trainers to the project was approved by the City Council. The zonal chief inspector also approved the secondment of one of the district inspectors to the project team. She was to assist in liaison and coordination between the project and the inspectorate in the district.

Two support staff were employed: a driver who also acted as a messenger and an office attendant responsible for assisting with cleaning, catering, and some minor office duties. There was no secretarial support, all correspondence, materials writing, and other tasks requiring word processing being done by team members.

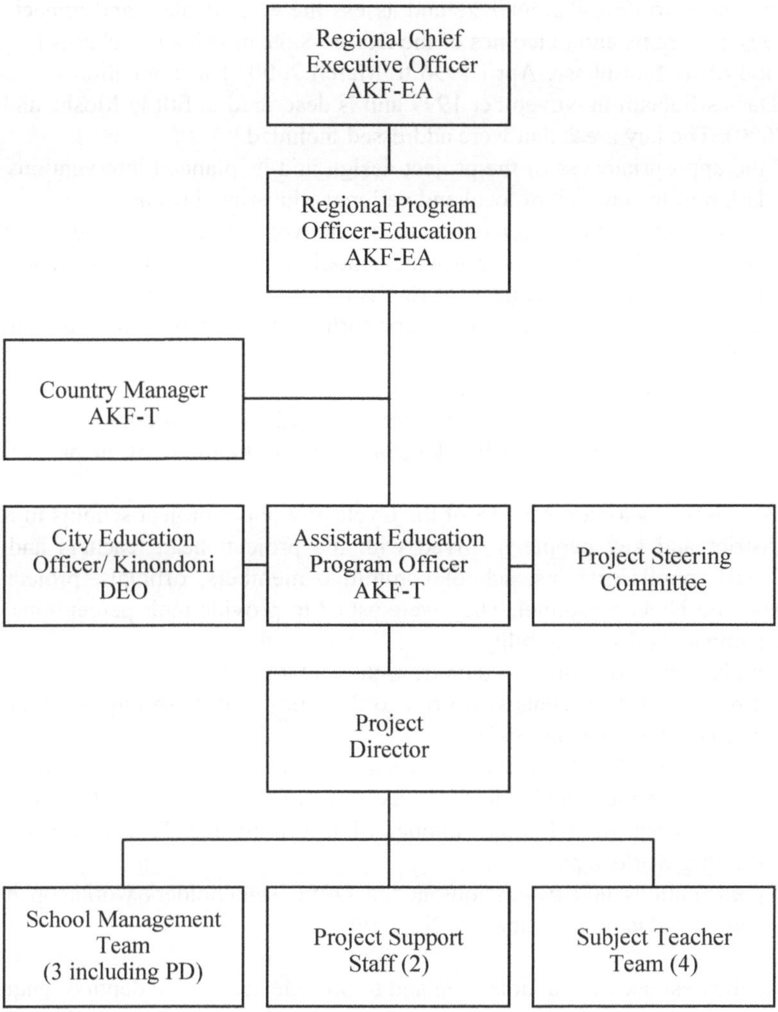

Note. AKF-EA = Aga Khan Foundation-East Africa, AKF-T = Aga Khan Foundation-Tanzania, DEO = district education officer, PD = Project Director

Figure 1: Organization of the Dar-es-Salaam Primary Schools Project

Evaluation Design and Methods

We were asked to describe, analyze and assess the effectiveness and impact of the objectives, programs and outcomes of the Dar-es-Salaam Primary Schools Project in the period of its first phase, April 1996 to March 2000. The evaluation was carried out in Dar-es-Salaam in November 1999 and is described in full in Mosha and Welford (2000). The key areas that were addressed included

- the appropriateness of the project design and its planned interventions in relation to the context of local and national educational needs;
- the efficiency and effectiveness of the delivery of programs and inputs, particularly with respect to improving school management and the teaching and learning of the core subjects of English, mathematics, and science;
- the mechanisms for effective monitoring of programs, interventions, and outcomes;
- the governance of the project;
- the influence of the project on other local and national programs;
- recommendations for further development and the future of the project.

The evaluation was to target all 15 of the Cycle 1, 2 and 3 project schools in Kinondoni District and key people involved with the project: head teachers and other school staff, pupils, parents, school committee members, officials, project team members, and NGO personnel. They were asked to provide their perceptions of the project's impact and sustainability.

The study gathered a range of data from three main sources:

- a review of documents sent prior to the study and those requested and supplied during the exercise;
- detailed fieldwork involving individual interviews and focus group interviews with the stakeholders, observations of project-school classrooms, and observations of a limited number of in-school and Teacher-Center-based training workshops;
- presentations and discussions at the DPSP stakeholders workshop held in Dar-es-Salaam November 19-20, 1999.

To make the best use of available time and to provide acceptable depth of enquiry, it was agreed to focus only on a sample of project schools, and not to include as a 'control' measure any non-project schools. We drew on our extensive experience of Tanzanian primary schools when making comparative judgments of project schools.

The project had data showing estimates of the performance of the schools on the two principal dimensions of the project's interventions: management and core subject teaching. The evaluation sampled 7 of the 15 project schools. The sample included three schools from Cycle 1 where the project had been operating longest, and two from each of Cycles 2 and 3, the schools performing best and poorest against the dimensions. Table 1 provides background information about the sample schools.

Semi-structured interview schedules were developed and approved by the evaluation team to guide the interviews with stakeholder group and individuals. Each covered the main concerns of the terms of reference (TOR), but differed slightly to address the differing interests of the people interviewed relative to their role in the project. The evaluation team interviewed 212 representatives from all the stakeholder groups in the seven study schools, and 33 people from the Ministry, the City Council, and other officials associated with the DPSP.

Classroom observations were made using a simple checklist, which could be completed in situ without distracting the observer from the key points of the lesson under observation. This checklist was used to observe 17 classes in the seven evaluation sample schools. As with the interview schedules described above, the checklist was devised to gather data relating to various concerns of the evaluation's TOR, specifically the physical state of classrooms, the use of displays, child-grouping strategies, child behavior, teacher activity, curriculum, and teacher-pupil interaction. The instruments, interview schedules and checklist, were unique to this evaluation, although they naturally drew on other published materials and our previous experience of using and developing such instruments. In particular, we were attracted to the instruments used by Siraj-Blatchford et al. in their 1997 evaluation in Kampala (AKF, 1997) as a way of gathering appropriate data.

Table 1: Background Data for the Seven Project Schools Used in the Evaluation

Cycle	School	Pupil total	Girls	Boys	Teachers established	Teachers at post	Female teachers	Teacher-pupil ratio
1	MA	1,803	900	903	38	37	35	1:49
	ML	2,439	1,223	1,216	79	72	74	1:34
	UK	2,843	1,392	1,451	71	69	65	1:41
2	MI	2,786	1,368	1,418	65	57	60	1:49
	TA	2,229	1,008	1,221	50	46	40	1:48
3	TU	2,594	1,280	1,314	61	57	50	1:46
	KI	2,494	1,175	1,320	56	55	49	1.45
		17,188	8,346	8,843	420	393	373	1:45

Note. From DPSP Schools Basic Data, 1999. Ratio calculated using number of teachers at post at the time of data collection. Class sizes in Dar-es-Salaam are large by any standard: 50 to 80 in most schools, with 100 to 180 per class in some schools. The double shift system, introduced in 1997, has served to reduce class sizes and the official teacher-pupil ratio.

The information and data collected were jointly interpreted and analyzed by the evaluation team members. Preliminary findings were presented at a workshop organized by the AKF-T for initial feedback as well as for dissemination to stakeholders. The final evaluation report drew on discussions from that workshop offering advice on future directions for the project.

Evaluation of the Effectiveness of Project Management

The project design and intervention strategies sought to enhance learning through improvement of the teachers' knowledge and pedagogical skills. The project increased the resource base essential for effective teaching and strengthened the capacity at community and school level for effective management, leading to quality improvement in schools. At the local level, the project design met the challenge of decentralization of control. The project team took responsibility for diagnosing school-specific needs through the collection of data on which interventions were to be based. The project design emphasized greater participation by stakeholders, enhanced accountability, and management for results.

The project was well-managed and operated effectively. Team roles and lines of accountability were clearly spelled out in available documentation (AKF-T, 1995). The project director was accountable to the Aga Khan Foundation-Tanzania, with project decisions guided by a thoughtfully constituted and active steering committee.[12] The framework agreement with the City Council was functional and clear in its conception and its usefulness.

Resources were used effectively and efficiently. Team time was used in a variety of activities. Team members had individual schedules of work they agreed on and apportioned among themselves frequently and regularly. They worked together, when appropriate, for planning and development and individually, in delivering seminars, monitoring workshops, visiting schools, observing lessons, and so on.

We judged project staffing to have appropriately responded to changes in the objectives as the project developed. The project was well led and organized into appropriate teams for delivery of project objectives. The school management and core subject teams were well respected and liked, efficient and effective, as individuals and as teams. The input of the VSO volunteers was well used; volunteers supplied useful experience to a developing project.

The TOR for the evaluation asked us to carry out a cost-benefit analysis of project funds dedicated to the incentive grant and funds dedicated to materials development at the school and project levels. The costs in the project related to the inputs, while the benefits were ultimate improvements in learning that resulted from the investment in the interventions. The absence of quantitative measures of the growth in pupil learning made such a cost-benefit analysis impossible. Only informed (but nevertheless speculative) judgments of the gains were feasible. There appeared to be general satisfaction among all the stakeholders, both those in the school communities and those more loosely associated with the project, that there had been gains through improved management and better teaching.

Project Design and Baseline Data Collection

A good battery of baseline data had been collected, although more information about the personal background of some stakeholder groups would have aided the targeting of training. Baseline data were appropriate, both to the national and local context and to the project objectives. They were collected systematically and the analyses accurately performed. Good use was made of these data in the design of training courses.

The concentration on qualitative data was driven by pragmatic concerns and was, for the most part, appropriate. An exception was in addressing the effect on pupil performance of the intervention, given that there was pressure to measure the effect of the intervention on learning, and further, to attempt a cost-benefit analysis of the intervention. A targeted pre-test/post-test strategy could have been used if the desire was to measure more accurately the returns of project investment in terms of pupil scores. It would have necessitated the expensive design and trial of a standardized test with established parameters. It is likely that such a strategy would have been beyond the capacity of the project and perhaps even the Aga Khan Foundation itself. Other priorities were more pressing. Quantitatively measuring the impact of the project on pupil performance would have required the used of a standardized test. We comment later in the chapter on the development of quantitative pupil performance measures.

We judged program strategies to have been both appropriate and effective for a well designed, well implemented project that achieved considerable success. Design of guidelines and training materials was strong. The basic methodology adopted for training, seminars and follow-up workshops was appropriate and largely effective. According to our observations, the seminars for the targeted head teachers and teachers were well conceived, well carried out and met their objectives. Workshops in schools cascaded the training methods and the use of project materials. They were conducted and monitored professionally. We commented on a lack of formal 'training-the-trainer' inputs and, while the team structure was appropriate for the discharge of the objectives, there was a place for the explicit identification of key head teachers and teachers to take active training roles.

Finally, in project design, there was a case for the project to observe, monitor and record a 'control' group of similar schools. It would have been appropriate for project managers to be able to distinguish changes in the project's primary schools due to the intervention from those which may have taken place anyway.

The Project Incentive Grant

The project director described the project incentive grant as "an innovative element developed by the project and currently under test." The idea was to motivate primary schools to improve their performance by providing cash based on performance, measured against an agreed-upon set of performance indicators. The most recent indicators are shown in full in Mosha and Welford (2000, Appendix 7) and are summarized below in Tables 2 and 3. Schools are judged and rewarded on the effectiveness of their financial management, institutional development and curriculum improvement, with a matching award for fundraising.

The indicators against which performance of schools has been measured, and the maximum grant awarded to any school in each main indicator area, are shown in Table 2. Schools were judged, on the project team's 6-month evaluations done over the period of the award, against an extensive list of subindicators within the four main areas (see Table 2) and rated on a 5-point scale. The mean indicator score was then computed and, depending upon the rating achieved, all or part of the funds allocated to a particular indicator was awarded in the proportions shown in Table 3.

Table 2: Performance Indicators and Incentive Grant Awards

Performance indicator	Maximum award
1. Financial management	TSh[a] 450,000
Cash book preparation	
Budgeting	
Financial supervision	
Accounting	
2. Institutional development	600,000
School management and organization	
School development planning	
School-community links development	
3. Curriculum improvement	750,000
Organization and supervision	
Teacher development program	
Resources support	
Assessment and monitoring	
4. Matching grant for fundraising	500,000
Total	TSh 2,300,000[b]

[a] Tanzanian shillings, [b] Equals US $ 3,750

Table 3: School Performance Ratings and Resultant Incentive Grant Awards

Rating	Indicator award			
	FM	ID	CI	MG
< 3.0 points	0%	0%	0%	
3.0-3.4 points	50%	50%	50%	Funds raised matched to a maximum of TSh 500,000[a]
3.5-3.9 points	100%	100%	80%	
> 4.0 points	100%	100%	100%	

Note. FM = Financial Management, ID = Institutional Development, CI = Curriculum Improvement, MG = Matching Grant.
[a] TSh = Tanzanian Shillings

Implementation of the Incentive Grant

At the time of the evaluation, the first two of three phases of the incentive grant had been allocated. The Phase 3 focus is to be pupil attainment, which will be compared with evaluation criteria still to be agreed upon by the project and the schools.

While this incentive grant strategy is complex and problems remain, it is very worth pursuing. The performance indicators, and how schools perform against each subindicator, have to be shared with each school. These indicators will take time to understand, based as they are on impressionistic criteria, and they are likely to provoke argument and conflict. The project team made every effort to communicate the performance indicators to head teachers, assistant head teachers and school committees, but true understanding of how judgments are made and justified will be slow. Because ratings on a 5-point scale inevitably bunch around 3 and 4, we suggested a 10-point scale to spread the scores and produce wider differences on which to base the awarding of grants.

The sums involved in the incentive grant have been a relatively modest proportion of the project's funding and, once this pilot is fully evaluated (it is too early to fully judge its effectiveness), more substantial sums might be possible. Schools were fully involved in the awarding of the incentive grant, with encouraging involvement of the school committees, teachers and even pupils. All spoke of the need to publish transparent accounts of the grant awards and spending to counter suspicion of teachers by parents, which can arise when monies awarded to schools are not properly accounted for.

Schools need to determine priorities and be able to budget effectively. For the Phase 1 awards, the amount that could be spent, and what it could be spent on, was largely fixed by the project, and schools had relatively few decisions to make. The Phase 2 awards had just been made at the time of the evaluation, and schools appeared a little wary of committing themselves to setting budgets. They seemed to lack confidence in their ability to do this effectively. This learning process should be valuable and the ease with which subsequent budgets are drawn up will be straightforward to measure.

The Impact of the Incentive Grant

The effect on schools of having extra money to spend on improvements was immediately visible: new textbooks, furniture, storage cupboards, student writing materials and raw materials for teaching.

Schools described the awards as motivating rewards of merit. We sensed disappointment rather than anger when they failed to achieve the 100% award. We sensed a feeling of defeat in two schools that had not carried out fundraising activities. The feeling of defeat stemmed from the perception that parents at the school were too poor to approach for fundraising and that community poverty was beyond the control of the school. Parents, school committees, head teachers and teachers felt that, because their school was in a poorer area, they would receive a lower income from UPE contributions and that they would be further penalized by an award lacking the matching grant (MG). We suggested that the project address this demotivating effect, perhaps by considering mitigating criteria, in order not to further contribute to the gap between poor-parent schools and those with better off parents.

In another project school, there was a sense of the fairness of the process but a feeling that the school's case was unusual, and one in which more flexibility could have been applied. They had scored 0 on the School Management and Institutional Development performance indicators, but had been appraised under a head teacher who had since been replaced. The new head teacher had immediately implemented training, mobilized her staff, set up a resource area and improved the school environment. However, the judgment had been made and the award based on the achievements of a previous management team. We speculated on the use of a response fund to reward such an improvement.

The Capacity Within the Project for Effective Monitoring and Evaluation

Interviews with the project team revealed that this capacity was good, well understood and developing with experience. The mechanisms were well defined and working effectively. A wide range of indicators was established at the outset and these were refined over time. The instruments developed as part of the baseline surveys were appropriate and facilitated monitoring of the progress on indicators against starting points. Subsequently developed monitoring instruments were working well too and were used efficiently in developing all aspects of the program. The instruments picked up key information from schools, and from the school system and allowed for proactive development of strategic work to further effectiveness of training. For instance, an observation checklist developed by the core subject team identified difficulties teachers were having with planning and implementing groupwork. Subject team analysis of groupwork checklists led to remedial strategies for teachers and identified areas where school management could be helpful.

The SIP interventions took the form of seminars (including refresher courses), workshops, observations and teacher support. They were effectively monitored. Seminars were often peer observed and participants were required to complete evaluations. Project team members observed teachers who had been trained, using standardized pro-formas, as they ran the in-school cascade training workshops. Evaluations were discussed with the appropriate people and led to action and modification of materials or approaches. Classroom teachers were observed from time to time, and project personnel offered advice aimed at improvement.

Computer-based data storage and analysis was used efficiently to identify key trends. The data allowed the team to identify trends in performance indicators that informed action quickly. Data presentation systems were effective and well used. If there was a weakness, it was that too much technical analysis expertise resided in the project director, despite conscious efforts to provide information technology training to team members.

The team might be in a position now to design and monitor other ways of judging effectiveness. We have mentioned the use of data from a parallel group of non-project schools. Since much depends upon the affective domain, the attitudes of head teachers and teachers to both training inputs and training objectives, the measurement of attitudes and attitudinal change might also be attempted.

Management Training

The project concentrated its efforts on management training and on training for teachers of the core subjects, mathematics, science and English. The focus on management and the range of people to whom training was directed were fairly ambitious, especially for Tanzania, where schools are used to doing as they are told, with little real decision-making left to managers.

Target Groups for Management Training

The number of people who underwent training as part of the project, up to November 1999, is shown in Table 4. The project trained head teachers and assistant head teachers to make them better equipped to manage their schools academically, financially and administratively. For example, head teachers and school committee members were trained to prepare annual plans and budgets, which were monitored by the project team throughout the year. The roles of school committees in project schools were more clearly defined, and they too were provided with training to help them develop their management skills. The project encouraged school committees to oversee financial operations carried out by their head teachers. Management training was further broadened in 1999 to include ward education coordinators from Kinondoni District.

Table 4: Participants in the Project Management Training

Target group	Number trained
Head teachers	19[a]
Assistant head teachers	51
School committee members	50
Ward coordinators	14

Note. From SIP project director.
[a] includes replacements for head teachers posted away from project schools.

The range of target groups meant that all those connected with school management not only received training but were also kept abreast of the detailed implementation of project objectives. Transfer of key personnel, however, hampered the team and led to a perception of effort wasted in training head teachers. Four of the five original project-school head teachers were transferred in the first year, necessitating accommodation of the training needs of new head teachers in the program. That was achieved with a minimum of delay and fuss. The team provided appropriate and sensitive follow-up. (Head teachers spoke warmly of the way the project team approached them in school when following up observed inadequacies in implementation of training inputs.) The team saw its efforts further stimulated by the incentive grant.

Management training was based firmly on the project objectives and was informed in detail by the results of the baseline study. The project team used the comparisons with the baseline data to judge both the extent of objective coverage and the realities of the project's implementation. The team responded to needs not covered in the baseline study and included new content in the later courses and subsequent implementation monitoring. The management training team developed four training modules covering basic areas of school management. The manuals, two modules produced by November 1999[13] and two later in the project,[14] incorporated tested course input materials. The materials were widely distributed to other organizations and represent a valuable output of the project. Situated in the local context yet drawing widely on published theoretical material, they offered an opportunity for the enrichment of other school development programs.

Effects of the Management Training Program

School Development and Planning

We found strong evidence of transfer and application of the training content to schools. All schools visited had mission statements, school master plans and mottos.

When pressed, however, head teachers were not usually able to show more detailed, annual workplans, with identified priorities for action. They were able to describe the theory, but as yet perhaps lacked the confidence and practical skills to produce such plans for scrutiny. It could also be that they had yet to be convinced that to do so would be productive for their schools.

Improved Management

Visits to schools provided us with evidence of improved management as just described. Head teachers generally expressed confidence in themselves, confidence which arose from their training, support and monitoring. Leadership is difficult to describe and quantify. The effect of poor leadership is easier to see, perhaps, and the project, in concert with the district education officer (DEO), had taken action where appropriate. For instance, in one project school, an ineffective head teacher was transferred out and replaced. We also saw effective leadership in action in this school, where the new head teacher and her assistants had transformed things in 3 months. An atmosphere of enthusiasm prevailed in this school.

Financial and Resource Management

There was clear evidence of the project's effect in all project schools visited. Our review revealed that all schools issued receipts and maintained records through cashbooks. All heads involved their school committees in decision making regarding the disbursement of funds. They had annual budgets and presented financial reports to parents and to the DEO. This aspect of the training had made a real difference in the project schools. We pushed both head teachers and school committee members to describe how they published and displayed school fund accounts, and how they disseminated the information about accounting procedures to parents and to other members of the school community. Here there was uncertainty both about the mechanisms and the usefulness of such procedures.

Improved Curriculum Management

Significant improvement was evident. We noted delegation of responsibility for the school to assistant head teachers and the implementation of checking strategies in the management charts displayed. Lesson plans were being discussed with teachers, sometimes coordinated in subject areas, and checked by school management teams. There was variable practice here, however, and evidence of a recent falling off in maintenance of central lesson-planning records. In most of the classes visited, teachers did not refer to lesson plans and, in many cases, were unable to show their plans for the lessons being taught. There were schemes of work addressing the objectives of the National Curriculum. The translation of these into lesson plans and the use of these plans in teaching is regarded as part of the responsibility of the head teachers and assistant head teachers as they manage the curriculum and features in the project's management training. We concluded that this component of curriculum management was not yet well implemented in the project schools.

There was no evidence of teacher peer monitoring and assistance in classes. Ways in which teacher teams might work better in the classroom had not fully been taken on in project schools, despite being a deep concern of the core subject team. We did not see more than one active teacher in a class at any one time, often when 100 or more children were being taught and other teachers were seated outside classes. Team teaching is complicated to plan and implement but could be used where there are underutilized teachers and big classes. We wondered if this aspect of curriculum management was viewed as an issue for classroom teachers and thus as an objective of the core subject team alone. Team teaching does, however, depend on a school's senior management making decisions about the deployment of resources in the classroom, and we urged the management team to address this issue. Creative human resource management is a priority for effective curriculum management. It remained for the project team to work out a joint strategy, through both management and core subject training, to allow school managers to maximize their use of scarce resources.

Head Teacher and School Committee Support for Curriculum Delivery

This was evident in the project schools largely through the provision of material support. Teachers had been provided with tables and chairs. Textbooks and other books had been bought, though not yet in large enough quantities to suggest that teachers had sufficient textbooks to use them skillfully in class.

School committees in some schools had successfully mobilized parents to provide financial contributions, which had translated into more materials for classroom use. In two schools, the school committees helped management enforce attendance by mobilizing the community to send truants back to school. In several schools, the committee used some of their funds to provide incentives for teachers and for outstanding students.

Ownership and Participation

Ownership and participation by a cross-section of the school community in the financing and management of schools is evolving, but will take time. It was taking place at the time of our evaluation against contradictory and seemingly uninformed advice from the Ministry,[15] repeated in the press. (The capability of school committees to raise significant funds to support school improvement, however, varied.) All the project schools required a lot of money for rehabilitation, maintenance and purchase of additional teaching and learning materials.

Aside from the obvious participation of school committees and parents in school improvement through financial contributions or labor and materials to building projects, some schools reported increased interest among parents in supporting and monitoring learning quality. One of the school committees helped with weekend tutoring sessions, both by providing incentives to teachers to provide tutoring and by participating in tutoring where they had expertise to offer.

Parents are naturally concerned with overall school performance. Popular schools are those where the proportion of Standard VII graduates proceeding to secondary school is large and growing. We found parents proud of their schools and aware of the effect of poor leadership on their children's education. Parental pressure, for example, had been key to the transfer of two underperforming head teachers in the district.

School Capacity to Monitor Change

The project team expended considerable effort on following up its training. School visits were not haphazard events but had a clear agenda, and judgments about effective implementation of changes in practice were made against detailed set of indicators. These indicators had developed over the life of the project and had been usefully built into the incentive grant decision-making process. The criteria reflected the project's training components, and schools were clearly informed about how the various indicators worked. The project team's expectations were that changes in management practices would result in improvement on indicators and they monitored schools accordingly. Schools' capacity to monitor change in their management, teaching practices and pupil performance was developing. It was early in this complex process, and systems in most schools were in their infancy. Record-keeping is the first stage in the process. Test scores, for example, were kept in profusion, but the development of school capacity for making sense of the scores and other data, identifying causes of performance weakness and promoting mechanisms for remediation will take additional time.

Training in the Core Subjects: Mathematics, English, and Science

During the project an in-service teacher development program, focusing primarily on methodology, was developed for English, mathematics and science teachers.

Teachers were trained in the use of activity-based teaching techniques, with a concentration on stimulating pupil-initiated interactions in the classroom. Built into courses for teachers was training to organize and conduct workshops for other teachers in their schools, the idea being to empower teachers to provide school-based support for their peers. Because the project subject team members worked as an integrated team rather than separately as subject trainers, they produced methodology-training modules (published as a manual, Hardeman, 1999) that covered topics common to all three subjects. The modules, targeted at teachers, Teacher Resource Center coordinators and resource persons, should eventually prove useful for teachers of all subjects and may have use beyond the project, in other interventions. Most subject training materials are available in both Kiswahili and English. From our evaluation, we concluded that these materials were well pitched. They challenged without dismaying the user. Those in the modules, especially, were clearly written, professionally presented and well organized.

Participants in Core Subject Training

The core subject team initiated the training of selected teachers through Center-based training seminars. The trained teachers then took on a training function themselves, cascading the training to their colleagues through school-based subject-focused workshops. The project provided intensive face-to-face training seminars for some 305 teachers, including 45 subject coordinators and 140 specialist teachers of the three core subjects from the 15 participating schools (see Table 5). The in-school follow-up workshops cascaded the training to many more teachers. Refresher seminars were provided to maintain skills in Cycle 1 and 2 schools.

Additionally, although seminars were not included in the project proposal (AKF-T, 1995), they were organized for Standards I and II teachers, to include them in the program and induct them into the project objectives for the teaching of the core subjects.

Table 5: Number of Participants in Core Subject Training Sessions

Target group	Number trained
Subject coordinators	45
Subject teachers – Standards III to VII	140
Teachers – Standards I and II	120

Note. From SIP project director.

Approach and Content of Training

Project schools nominated up to four subject specialists to attend the seminars. The selection process was mainly the responsibility of the head teacher, who usually consulted senior staff and officials. For Cycle 2 and 3 training, project team members were also involved in the selection of participants. The teachers from each subject area received four week-long seminars during the year in which their schools joined the project. These seminars involved about 100 contact hours of face-to-face training at the Kinondoni TRC and were conducted by the subject specialist of the project team, with the support of other team members.

Decisions about the content of the courses addressed the project objectives directly, were based on a baseline survey of training needs and responded to local perceptions of immediate need. While it is undoubtedly true that the majority of Tanzanian primary school teachers lacked subject content knowledge (more than 60% being Grade B/C teachers who themselves only had 7 years of formal education), the project team had reasonably decided that it could not tackle all such shortcomings. We judged that the declared project focus for training the teachers, Methodology, was appropriate in the circumstances.

Cascade of Training

Every training seminar at the TRC was cascaded to teacher colleagues by those who had participated in the seminars, with each project school organizing an in-school workshop for the other teachers to be trained. A project team member for the subject evaluated the workshops to provide feedback both for the teachers involved in cascading the training and for team colleagues back at the TRC. All these standardized, form-based evaluations were filed systematically at the TRC. These files were always available, comprehensive, and a vital resource for charting the progress of the training in schools. Head teachers confirmed that project team members also made follow-up visits to observe teachers implementing the proposed methodology in the classroom and offered guidance and comment. These visits were also recorded systematically.

In the school-based workshops, people who had attended seminars made presentations to the other teachers in their subject area in the school. Workshops were held around the middle of the day to facilitate teacher attendance since, in schools operating a double-shift system, the changeover of shifts can also involve a changeover of the school's teachers. Attendance varied across the schools in our study. In some, only the subject team attended these workshops, while in others all staff were invited or expected to attend. Attitudes to attendance varied, from enthusiastic to indifferent, and our interviews with teachers elicited the following key points:

- teachers from subjects other than English, mathematics and science said they felt left out of the program, with their subjects and efforts being of lesser importance in the eyes of the school community;
- teachers from Standards I and II expressed similar feelings, although they were incorporated in the training program in the later cycles of the project; and
- core subject teachers who had not been selected to attend the seminars at the Center wished to be similarly valued by being trained directly, not through the cascade.

The project team stressed to head teachers the importance of managing these sources of discontent among their staff and also included training for early years teachers at the TRC.

Training rewards were limited to attendance certificates. We recommended that there be more significant recognition of trainer effort and expertise. This issue is discussed under Training of the Trainers toward the end of this chapter.

Effect on Schools and Classrooms of Teacher Development Training

School-Level Effects

This section starts with general impressions of school change then goes on to directly address the effect on classrooms of teacher development. Everyone interviewed during the project evaluation had heard of the project. Their grasp of its objectives was impressive, although language precision varied across the stakeholder groups. There was far less knowledge of why particular schools had been selected to join the project and the criteria that had been applied in that process. Every individual or group was asked, "Do you have a general impression of change in your school that comes from being a project school?" and, "In order of positive impact, what are the three most important benefits?" Table 6 provides a crude summary of the relative frequency of mention by the main stakeholders of various effects of the project. We pick up particular points from the table in the sections that follow.

Not surprisingly, the head teachers and their assistants had school management at the top of their perception, while teachers and pupils had a classroom focus. The school committee members had a slightly wider vision than other parents did, with management, especially financial, figuring more often in their responses.

Table 6: Stakeholders' Perceptions of the Benefits of the Project, Ranked in Order of Crude Frequency of Mention

Benefit	HTs	Asst. HTs	Teachers	Pupils	Parents	School cttee.	Officials
Planning	1	2	9	-	-	4	5
Management	2	1	8	-	6	3	4
Financial aspects	3	4	10	-	-	2	2
Groupwork/seating	4	3	1	2	3	5	9
Records	5	5	7	-	-	-	6
Pupil activity	6	6	4	3	2	6	7
Teaching aids	7	7	2	1	5	6	8
Improve school	8	8	5	4	1	1	1
Staff training	9	10	3	-	7	7	3
Staff coherence	10	11	11	-	-	-	-
Improve school environment	11	9	6	5	4	8	-

Note. HT = Head teacher

Classroom Effects

Obstacles to Implementation

There were problems associated with or inhibiting the implementation of project objectives in the schools. In some, taking key teachers away to attend seminars was said to leave classes without teachers. In others, this was not allowed to happen, with other teachers covering for absent colleagues. However, it is true to say that the most effective teachers were those selected to attend the seminars, and their classes, where covered, were taught by less accomplished teachers. On their return to school, the selected teachers inevitably felt they had some ground to make up with their classes to get them to a point with which these teachers were happy.

The number and size of classes were frequently mentioned. The methodologies associated with groupwork are difficult to practice when classes are very big, as they are in all of the project schools. (Despite a calculated average teacher-pupil ratio of 1:44, the evaluators did not see a class smaller than 60 pupils and most had over 100!). Grouping of the desks sometimes left little room for a teacher to circulate, and the desired subgroup sizes were almost never achieved. Consequently, project objectives associated with effective groupwork were never fully realized. The solutions, more and bigger classrooms, are beyond the scope of the project, although parents often could not really see why. They seemed to expect that the very presence of a project would draw large financial resources into participating schools.

Teachers' Subject Knowledge and Planning Skills

Evaluation data in this area came almost entirely from classroom observations of teaching in the core subjects. In the lessons observed, the knowledge underpinning the teaching was generally adequate. The teachers were confident both with the subject matter and the methodologies they attempted. Few teachers had their lesson plans in the class with them. The few plans seen were fairly skeletal, inadequate in terms of content, organization and connection with the scheme of work. Teachers are expected to have their lesson plans inspected once a week and this inspection recorded on a chart in the school's administration center. In the schools visited, the recent records were very sparse, with very few lesson plan inspections having been recorded in the previous 3 months.

Teachers' Pedagogy and Attitudes

There was ample evidence of the methodology focus of the project. All classes visited had adopted desk layouts to facilitate groupwork. The desks in most schools were large, heavy structures seating several pupils and, once positioned, could not easily be moved into new configurations. Desks were almost invariably turned to face each other in rows leading away from the front of the class, but there was often insufficient space in the class to separate groups laterally from each other. The corridors between rows were narrow for the most part. The effect was often to create three or four rows with 40-plus pupils in each. Teachers had made valiant attempts to signal the boundaries among groups. True groupwork and effective management of group dynamics, not easy at the best of times, were very difficult in these primary schools, especially when team teaching was not being practiced. We saw plenty of skillful teaching, deserving of admiration and commendation given the difficult physical circumstances. Teachers used real objects, occasionally with sets supplied to groups. High-cost equipment was in evidence, bought through the project equipment grant and used almost exclusively for class demonstrations. Low-cost items were also used, and it was these we saw supplied to groups. Teachers introduced and used these teaching aids well.

Questioning techniques were much in evidence and, again, used with confidence, although to variable effect. This is likely to be as much due to the variation among teachers as to unevenness in the training, as we saw as much within-school variation as between-school variation. The use of test results in formative assessment was not developed, and the project team identified assessment as an area for future development. Teachers were spending much time setting and marking monthly tests. We recommended that the ways in which such test outcomes could be used to inform teaching could serve as a valuable future focus of teacher development. Results of the monthly tests were rarely used as a criterion for improving classroom practice; for example, in grouping pupils, focusing teaching on specific weaknesses and so on.

Teachers were generally confident and cheerful. Their degree of engagement with the project naturally varied. It was noticeable that staff from some schools were less positive than those from other schools. Our probing of the causes of this potential disaffection suggested that the prime reason was poor relationships with the head teacher. Where the head teacher was seen as autocratic and secretive, not communicating well with staff nor involving them in decisions, there appeared to be far less motivation to fully apply the project initiatives.

Stakeholders were asked how the training that teachers received might be recognized. The recognition that comes from being part of the training, and hence the personal improvement derived from it, was seen as a key reward. There were few mentions of monetary reward, even by the teachers themselves. Whether this omission was for the benefit of the evaluators is hard to say, but there was a remarkable similarity in responses. However, one suggestion, made by more than one group of teachers, was that they would be willing to attend weekend- or vacation-time training, or both, if they were to receive slightly more generous travel allowances. We think that such out-of-school-hours training is worth serious thought, since it would also mean that teachers and their head teachers would be less anxious about the extra work caused by teachers leaving classes to attend training seminars.

Learner Interaction and Engagement

Pupil engagement in lessons was high. All the pupils in classes and in the focus groups said they liked their schools and usually explained this by saying they liked their teachers. They linked their satisfaction to the methods teachers employed.

Some of the older pupils said that, at first, they had found it difficult to get used to turning their heads sideway to see the board or the teacher (a consequence of grouping desks to achieve groupwork objectives), but that the chance to work together was a great improvement over previous solo efforts. No pairwork was observed, but a variety of group sizes were in use. In the lessons we observed, groups appeared to be used for quite short periods of time, the teacher spending more time using whole-class techniques usually requiring chalk-and-talk methods. Demonstrations using high-cost items sometimes involved the teacher moving from group to group; however, alternate activity was not provided for groups not being addressed directly.

There was no evidence of differentiated work for groups. All the groups we saw were of mixed ability and both pupils and teachers explained that this allowed weaker pupils to be coached and taught by their quicker learning peers. Pupils were very articulate about this strategy and saw not only its benefit for the slower ones among them, but also the real gains for those who had the explanatory role. However, this peer teaching: fast learners within groups helping slower classmates, was not seen in the classes we visited. It was also true that teachers gave little time for individuals in groups to make contributions to group outcomes. Groupwork and its proper use was a key and continuing training goal of the project team, who rightly expected that it would take time for groupwork's use to improve.

Almost invariably, pupils gave one of the core subjects as their favorite school subject. Asked why, they said that these subjects were where the interesting teaching aids were in use. They also said that these were the difficult subjects and they were doing better in them now, and that their teachers seemed more alert and interested in them. It was difficult to judge whether the evaluators were being fed answers the pupils thought they would like to hear since, like similar pupils the world over, they were very alert to signals associated with the interview process.

Use of Teaching Aids and Equipment

Mention of teaching aids has already been made, under the description of groupwork. Teaching aids are one of the aspects of change in schools associated with this project, and it is noteworthy that pupils mentioned them most frequently as an indicator of that change. Teaching aids in use were most frequently real objects, though stimulus-question cards were given to groups as well. Teachers seemed uncertain about why they had adopted the strategy, and the content of questions seemed to bear little relation to the content focus of the lesson that ensued. Furthermore, although such teaching aids would have taken time to construct and were written on relatively expensive cards, one teacher failed to collect them and some pupils proceeded to write rough notes on them. Otherwise, the materials chosen to enhance learning were used in an appropriate manner in the classrooms observed.

Classroom Management Skills

Pupil seating, task allocation and behavior were managed appropriately. All teachers scanned their classes and were observed to chide pupils for off-task behavior. Sometimes teacher reaction time was slow, but with up to 160 in a class this is not difficult to understand. On no occasion was inappropriate or excessive reaction to misbehavior observed.

Classroom Learning Environment

Classrooms varied. In some, the roofing sheets had rusted and there were holes that would allow ingress of water. All had concrete floors, though some floors were badly cratered. Windows usually had heavy metal mesh over them, though the windows could not be described as secure. Several classrooms had doors missing and almost all were far too small to seat all the pupils comfortably. Only in two project schools were pupils seated on the floor, the rest having sufficient desks. Walls were almost always very dirty and had not been painted since the schools were built. They were, however, in a good enough state for teachers to mount display materials. In only one school was there a significant use of displays. In this same school the majority of classrooms had either a collection of materials for measurement or a 'shop' area. Other display materials were suspended from string lines across the classroom. There was no children's work on the walls in any school visited, although one had plants in decorated tins on the windowsills. The pupils had planted and were looking after the plants.

Enhanced Pupil Learning in the Core Subjects

Almost every stakeholder group spoke of improved learning by children as a product of the project. However, evidence of this improvement was much harder to find, and the general understanding of both what would constitute evidence and the significance of that evidence was very rare. A variety of statistical data were collected by the project and the schools to provide an indicator of project success in enhancing pupil learning overall and in the core subjects. The press, parents and pupils judged school success by the number of graduates who took up the small number of secondary school places in Tanzania. The overall percentage of those sitting the PSLE at the end of Standard VII who proceed to secondary school is around 15% (URT, 1998), and thus schools exceeding that proportion are rated highly. The project had these data for its schools, but they showed fluctuations from year to year independent of the project intervention. This is hardly surprising, since all such figures must be considered proxy data as none specifically measured the effect of the intervention. The number of variables influencing the Primary School Leaving Examination (PSLE) scores and hence the variance in the number of graduates selected for secondary school is large indeed. No significance can be attached to this statistic as a measure of the effect of the project on pupil learning.

Another statistic, the success rate in District standardized exams, which pupils take at the end of Standard IV, was also available. The Standard IV exam performance data are, if anything, an even less robust indicator than the PSLE figures, based as the former are on tests that take less pupil time and are rarely trialed. Test results are not calculated separately for the core subjects, except for mathematics. Standard

IV performance in mathematics across the district was very low, with many pupils scoring 0, making this a poor indicator from which to extrapolate performance to measure project effect.

Year-end tests are set and marked. Files of test results were stored by project schools, but little use was made of these results. No subject averages were calculated for each class to measure progress in the core subjects, nor did teachers use pupil averages, other than to place pupils in rank order. Teachers did not appear to understand the way test results could be used to group children or to pitch lessons according to the results. Monthly tests are scheduled, but without systematic recording of test results. Teachers did not spend class time going through monthly or year-end tests so that pupils could understand where they made mistakes.

Monthly test data could be used to measure crude impact of the project interventions, but they would not be robust enough to confirm anything other than rough hunches. Teachers had neither the skills to see the significance of such data, nor the time to laboriously calculate subject averages for 100-plus pupils at regular, frequent intervals. The tests were not standardized anyway and would not withstand reliable comparative analysis. The project was addressing two issues at the time of our evaluation: the improvement of teachers' assessment strategies and techniques, and the provision of harder evidence of the effect on pupil learning of project interventions.

Sustainability of the Program for Teacher Development Work Within the Schools

There was a firm commitment among Ministry, City Council and District officials to continue to promote project objectives and to continue with the initiative beyond the present contract. Teachers who had been trained directly through the seminars held at the Kinondoni Teacher Center (KTC) were in no doubt that they had been effectively trained. They had been awarded certificates recognizing their growing expertise, and they were well disposed towards the project. These teachers constituted the core around which the future of the project could be built within schools. Head teachers and their assistants were fully involved, and they too said they would take forward the lessons from the management and the core subject training in supporting the teachers in their schools. School committees were becoming more and more knowledgeable about the details of the project and how it affected the teachers in their school. They will need refresher seminars, as the composition of the committees is not constant.

Sustainability of a program is a complex issue and depends on many variables. A major issue in a school improvement effort like the DPSP is the constant need for fresh inputs, continued support from a project team with the full support of key stakeholders. It is unlikely that most members of the school community, teachers especially, can maintain impetus without such inputs. This 'energizing' can come from a variety of sources, and not necessarily the explicit continuation of the DPSP. However, in its absence, the Ministry, Council and District will need to commit resources to sustaining training if the project is to continue to develop in the project schools. At the time of our evaluation, it was difficult to know whether such resources were or could be made available.

The sustainability of the incentive grant initiative after the withdrawal of external funds was likewise open to question. Although we have described the amounts involved as modest, two of the seven schools we investigated were unable to raise any funds through community events and thus did not qualify for a matching grant under the funding formula of the incentive grant. The experience of these schools suggests that, even where local and central government take the place of the project in offering incentives for continued improvement, there will be winners and losers in the grant allocation.

Teacher Resource Center Capacity Building

The training seminars developed under this project were given in the Kinondoni Teacher Resource Center. Considerable effort is being expended in Tanzania on the development of teacher resource centers as a vehicle to increase learning in classrooms. The various initiatives are being coordinated through the Ministry's TRC Coordination Unit. We caution that careful research and thought are needed in this effort. We draw attention to the findings of the recent Department for International Development (DFID)-funded study *The Effectiveness of the Teacher Resource Center Strategy* (Knamiller et al., 2000), which concluded that there was little evidence that TRCs themselves have a direct impact on classrooms. The major exception to this finding was based, coincidentally, on detailed observation of the AKF/AKES-sponsored Mombasa School Improvement Project, which is teacher center based and forms the subject of Chapter 6 of this book. The classroom-level effects of that project were greatest when in-service activities at local primary Teacher Centers were combined with in-class follow-up by teacher development personnel associated with the centers. The work of Joyce and Showers (1980; 1988) is key to understanding the impact of TRC-mediated training on the work of professional teachers. Our advice to the DPSP project was to consider whether the TRC is the best vehicle for the changes it wishes to bring about. Target group needs analysis might show that it is not necessary to have a TRC for the training; any classroom might do, especially if it was close to where teachers lived or worked and it attracted staunch community support. It is in this light that TRC capacity building should be viewed.

It is our observation that teachers only appear to use TRCs to further their own professional development when under pressure from government to do so. For instance, there is current pressure in Tanzania for all Grade B/C teachers to upgrade to Grade A, and the relevant training is delivered through the local TRC network. It is against this background, and the continuing insistence of the donor community that

TRCs represent the best means for teacher improvement, that the project will orient its future training.

Kinondoni Teacher Center: Developing Capacity

There is no doubt that TRC in-service training capacity was considerably enhanced by the DPSP intervention. The key trainers who worked as project team members gained skills, confidence, competence and exposure. As well, their local reputations were enhanced. Management and organizational strategies were strengthened in similar ways. The project provided the vehicle for this capacity building. Materials too were added. Their relevance was considerable since they served as the core training materials created by the project team. However, despite the above, the records showed that use of the TRC had declined over the preceding 2 to 3 years.

Parents in Tanzania are expected to make a financial contribution directly to schools, the Universal Primary Education (UPE) levy, which goes toward the cost of their children's schooling. A proportion of this levy is allocated to TRC support designed to ensure sustainability of the TRC as a means of providing training for teachers. It is paid to the TRC by its local cluster of head teachers. This levy is a recent innovation, so the effectiveness of the transfer of these funds was unevaluated. Time would be needed to develop the levy's ability to fund training courses at the TRC before the impact of the project on the desired functional capability of Kinondoni TRC could properly be judged.

Extent of Dissemination of Project Expertise to Kinondoni Teacher Center and other Teacher Resource Centers

The Ministry's TRC Coordination Unit in the Department of Teacher Education had yet to become functional at the time of our study. The Kinondoni Teacher Center (KTC) will come within the Unit's focus, and its effectiveness will be intimately connected to the future actions of the Ministry's Coordination Unit. The District Education Office (DEO) expressed its commitment to the continuation of the project's training objectives through the KTC and, given DEO staff's presence on the KTC team, their skills are likely to be transferred to the KTC.

Other TRCs were being established in the district. A TRC was established in Kunduchi (a peri-urban area of the district) through an initiative called District-Based Support to Primary Education (DBSPE). It was said to be functioning well and has a high level of input from that district initiative. The DBSPE Coordinator at MOEC headquarters communicates well with the DPSP at a variety of levels, and lessons learned from the DPSP, both those directly relating to TRC management and use in other areas, had been effectively disseminated in his project with some vigor. There was evidence of similar project input to other initiatives. The Save the Children Fund Mtwara Project had benefited from the DPSP's influence and advice on TRC management (SCFR, 1998). The Morogoro Kilosa TRC had similarly received considerable input from the project in developing guidelines for TRC managers and their operations (Kilosa Education Department, 1999). The DPSP team's work in disseminating their TRC experience was noteworthy and admirable.

Dar-es-Salaam Primary Schools Project and the Wider Context

In relation to education as a whole, it was hoped that the DPSP would contribute to the development of a school improvement model that would help revitalize the primary education sector. Central to this hope was the premise that both schools and Teacher Centers would need to develop their capacities and abilities to become focal points for academic, financial and organizational change. A strong feature of the DPSP was its communication with those who may be interested in aspects of its work. A key relationship was that between the project and the Kinondoni District Education Office. Ministry, City Council and District officials described relationships as very good, with mutual respect existing between District officers and the project team. There was frequent and regular contact with the project at all levels, but especially through the project director, the district education officer, and the District Education Office. The DEO was always fully briefed and very enthusiastic in her support of the program. The project in turn showed itself to be very responsive to the DEO's needs and imperatives. District representation on the project's steering group encouraged and enabled input to project policy development. The District Academic Officer described how the District had encouraged project school head teachers to work with other head teachers to disseminate their improved (project) management practices and provide support. In turn, project team members described how the District had responded to their feedback about ineffective project school staff.

Involvement of the Ward Education Coordinators (WECs) is one of the keys to sustainability of the initiative. Each coordinator oversees between one and eight schools in their ward in the District, which may include one or two project schools. The project's seminar program included WECs in its training. WECs were not only informed about the project objectives, but, as active participants in the management training, they were able to support the intervention in project schools and disseminate good practice to others in their wards.

Formal mechanisms for informing and being informed by government and sectoral initiatives such as the national Education Sector Development Program (ESDP) were established. Representatives from the MOEC's Primary Education Directorate, the Zonal School Inspectorate, and the Dar-es-Salaam City Education authorities all served on the project steering committee. These bodies were kept informed through meetings and verified minutes. Less formal structures were also established. One of the project team members was a member of the Zonal Inspectorate, linking that body and the project. The Project Director set up meetings (or was included in them) with the MOEC and the ESDP, continuing the exchange of information and allowing the dissemination of project experiences to encourage continuing reform. Other programs were kept very involved too. Discussions the evaluation team had with DBSPE, Save the Children Fund (SCF), VSO and AKES-T representatives revealed a common theme. The project had taken great care in keeping others informed and in providing impressive detail of the lessons learned from the project intervention. Project materials had been widely distributed both for use in other initiatives and for critical feedback and comment. We concluded that the potential to communicate and receive information had been realized. However, closer cooperation, such as coordinated efforts to ensure efficient use of overlapping expertise, had not yet been achieved. This was understandable, given the range of initiatives and the scope for intervention in primary education improvement in Tanzania. The Ministry was addressing the issue of coordination through its TRC Coordination Unit. The project was a key informant for this Unit and was discharging this information-giving role. The overall ability of the system to sustain initiatives remains unproven.

Lessons from the Dar-es-Salaam Primary Schools Project Experience

In this section, we pick up a number of issues arising from the evaluation exercise and explore them in more depth, our object being to examine some of the lessons learned from the project and from the evaluation process itself. Inevitably, the lessons are derived from looking back, and the dilemma is whether to look at them with a view to improving an existing program or with an undefined future in mind. It is our intention to look forward as much as possible.

Baseline Data and Performance Indicators

Comments on the collection and use of baseline data have highlighted the issue of quantitative measures of pupil performance. The project invested much effort in developing a good range of indicators to be used in a quasi-quantitative way. School improvement is largely a qualitative process, but the project team developed scales of performance on a range of indicators, some of which were used in awarding the innovative incentive grants. However, it remains that the ultimate beneficiaries of school improvement interventions such as the DPSP should be pupils. The question most often asked is, "How much better are the children learning as a result of the intervention, and what measures are there for evaluating that learning?" The TOR for the project's evaluation study focused the question still further by expecting a cost-benefit analysis, especially with reference to the incentive grant. There were good records of the cost of the grant, indeed of the whole DPSP program, but one benefit, improved learning, had not been quantified. The project team had looked hard at ways to derive a measure from existing data: Standard VII results, number of graduates proceeding to secondary school, and so on, but recognized that such data were too deeply flawed to use in calibrating learning derived from the project. Our discussion, which now follows, is focused on what might be involved in carrying out an accurate measurement of pupil learning so that the AKF and other developers of school improvement projects could more closely measure the impact of their interventions on children in the classroom.

Pupil Performance

If the impact of an intervention similar to the DPSP is to be measured, a key indicator must be shifts in pupil performance resulting from the intervention. It is necessary to measure pupils' performance at the program entry point, perhaps again at staged intervals, and then at the exit. The change must be large enough to measure and not obscured by errors inherent in such measurement. This calls for a pre-test/post-test strategy. It has to involve high quality tests, measure progress on a limited range of educational objectives, and ensure secure administration of the tests. The tests have to have known parameters, drawn probably from a bank of trialed items and administered in such a way as to minimize the errors due to random item variance and random pupil variance.

Such an undertaking is not to be taken lightly. It would be expensive, time consuming and require the input of experts in the field of performance testing. It could involve a sampling method whereby a sample of pupils attempt all the test questions, but where aggregation of results allows for a high degree of reliability of outcomes generalizable to the target population (Johnson, 1988). The performance of the individual on the test is not usually important. S/he represents the population that is the subject of the intervention. For instance, in Kinondoni District there are some 110,000 Standard VII pupils (30,000 in the project schools); a random selection from both project and non-project (control group) schools would be chosen to sit the test. Different random samples of pupils could attempt subsequent administrations of the test. Their performance would be generalizable to the population of the Standard VII pupils in the schools. It would also be important to have a large number of test items from which to select, but they would be broken into test packages administered to subpopulations of the target pupils. In that way, many hours of test time could be provided, but at a cost.

The major errors in performance testing are associated with the following parameters.

- Question selection – Test item writers write hard and easy questions, but random selection from a tried-and-tested bank of questions representing a domain (the curriculum area under test: arithmetic, for instance, or addition) ensures a range of questions in the test. The goal is to reduce the error associated with question selection to below one percentage point (since the item selection error will be compounded by other sources of error). This means that errors are multiplied, so an error of less than one percent does not distort real performance estimates. Research suggests that there is a need to select about 66 questions from the bank to keep the variance below 1% (Johnson, 1988).
- Pupil selection – The same research has shown that there are errors associated with selecting a random sample of pupils to sit the test (for example,

> picking a disproportionate number of those good at arithmetic) will distort the performance, and that errors associated with pupil selection are reduced to below one percentage point by having more than 450 in the test group.

Bringing the error below a single percentage point on any variable allows for compounded errors to remain below one percentage point. This permits confidence in what may only be a shift of 2 to 3 percentage points in test performance over the period of a school improvement intervention. Since the error from the pre-test does not overlap with that of the post-test, the shift can confidently be said to be a result of the intervention, when comparing groups of schools. Multi-level modeling can be used to estimate the contribution of other variables. However, it may be that comparisons among project schools can not be done with the same degree of confidence, as the number of pupils in a target-year group may not be sufficient. A different test strategy would be needed if a measure to compare schools was required.

The major investment in a testing regime is test-item writing. Test materials need to be developed and they need to be validated as relevant to the testing context, in this case, Tanzania. This is a long process, but should not be avoided. Acceptable test items need to be described and then 'banked' for random selection on the various test administrations. Test development as described above is expensive, but rigorous. Other projects might want to bite this bullet, along with the AKF; such a test development process would be unique in African school improvement work. It might have commercial potential; it could be sold to other NGO and government programs. A number of test agencies have experience. The Educational Testing Service (ETS) in the United States, the Assessment of Performance Unit (APU) in the UK, and the people at the University of British Columbia, Canada who handled the recent Third International Math and Science Survey (TIMSS), for example, are all capable of supplying expertise.

Once such test data on the intervention are available, a project such as DPSP could engage in all kinds of hard-data cost-benefit analysis related to impact on learning.

Use of Data from a Control Group of Schools

No control group of schools was identified in the project against which to measure the effect of the DPSP intervention. We recommended that thought given to monitoring progress of the project schools compared with progress of a similar group of non-project schools where there was no intervention. We have just commented on the use of control group data in quantitative comparisons. It also holds true that use with control schools of the performance indicators deployed in the DPSP schools would have added to the force with which the project and its evaluators could have commented on project impact.

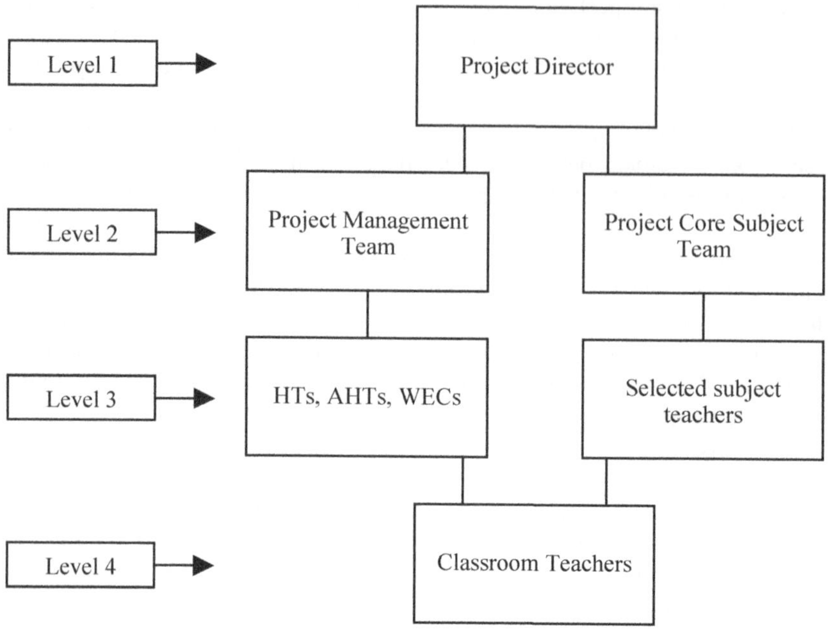

Note. HT = head teacher, AHT = assistant head teacher, WEC = Ward Education Coordinator

Figure 2: Tiers in the Training Model of the Dar-es-Salaam Primary Schools Project

Training of Trainers

Training is needed for those who themselves train trainers. Figure 2 illustrates the tiers of training undertaken by this project. The approaches of the project meant that those at Levels 1, 2 and 3 all had a training function to perform, since training was eventually cascaded to the classroom teachers (Level 4) through in-school workshops. The project team members (Levels 1 and 2 in Figure 2) were trainers of trainers. The skills needed to train trainers and the approaches used were not the same as when the training was directed at the ultimate target group, classroom teachers. There was no formal, specialized training of the trainers, although there were several inputs that inducted team members into their role as trainer-trainers. Team members' needs were not ignored, but perhaps not fully met. This aspect could be significantly developed with a purposefully designed training-of-the-trainers course.

Some of the trainer-training input might come from the implementation of an award-bearing program for team members similar to award programs associated with the Kenyan SIPs in Kisumu and Mombasa, described in other chapters. Key training personnel in those projects have been provided with a purpose-built graduate degree program by the AKF and AKES in Kenya, largely funded by the two agencies. No doubt this degree program is expensive and may be judged as inappropriate for DPSP personnel. However, the impact of studying for the degree on the confidence, status and self-esteem of team members has been documented by the Mombasa SIP project director (Bendera, 2000). He noted the growth in expertise of his project team resulting from the process of studying for the degree and commented on their increased potential usefulness as trainer-trainers in the further development of the Mombasa project.

These points link to the issue of recognition of expertise. It was clear that, while there was a valued measure of recognition in the DPSP through the award of certificates for those directly trained by the project (Level 3 in Figure 2), team members (Levels 1 and 2) had not themselves received any formal acknowledgement of their skills and expertise. The motivational effect of recognition, acknowledged for those attending the project's seminars, applies no less to team members and is a key issue for future sustainability. Rewarded colleagues are much more likely to see as valuable what others have valued explicitly.

Publication of Project Materials

The project's printed training materials were good in content and organization and deserved to be used more widely. The management materials were being organized into four booklets, each focused on a module of management. Two of these had already been printed at the time of evaluation and the other two were in various stages of readiness. One volume of the subject-based printed materials had been produced, but there was potential for more. Publication and dissemination could be accelerated and strengthened, with a process of contribution from and review by peers to strengthen the materials. There are a number of potential users in the Tanzanian context, for example, DBSPE, and other interested experts from the University of Dar-es-Salaam, teacher training colleges and other institutions. The latter groups could be called on to assist in refereeing and reviewing and then might use the materials in similar work. The financial possibilities of sustaining the writing and producing of a series of booklets should not be overlooked. This might defray the cost of the time that has been devoted to production to date and the time yet to be incurred. More time would need to be devoted to the writing process, to convert the collection of separate sheets used for individual training sessions into coherent thematic booklets. The development of two of the three finished booklets has required full-time input from a team member. A similar time investment might well be needed to complete the production of other manuals.

Appropriate pricing might require advice from a publisher, who would take into account production costs, potential market and levels of financial return desired from such an enterprise. Discussions with a publisher might indicate that the most commercially viable solution would be a standard book contract. Publication and dissemination would add value to the project materials, beyond the purely financial. They would increase the profile and visibility of the project and its experts.

Such considerations are not trivial when assessing the long-term sustainability of project interventions.

Developing Training and In-School Follow-Up

Curriculum management depends on knowing whether syllabus objectives are being met, with children learning as required. That knowledge comes partly from effective implementation of good tests and from use of test results. Teachers can find out whether they are pitching lessons appropriately and moving at the right pace; managers can make decisions about staffing inputs and set targets to maintain improvements. The three areas of the DPSP that would benefit from further development of training are curriculum management, assessment and the development of more child-initiated classroom interactions.

Curriculum Management

The project's management training was concentrated on senior managers in schools and on support personnel: school committees, WECs and so on. It was the responsibility of the project management experts to conduct and follow up this training in schools. However, one of the things that head teachers and their assistants, trained through the management component of the project, were responsible for is the checking of lesson plans. The teachers writing the lesson plans were trained by the core subject team or by cascade. The evaluation revealed that, while school managers checked for the presence or absence of lesson plans, they did not actually review the plans' content for effectiveness in achieving lesson objectives, leaving this to subject leaders. This in-school division of responsibilities may have been an artifact deriving from the different training inputs and accentuated by the different focus of in-school follow-up by the two parts of the project team. This highlights a potential problem that could be solved by a blending of the follow-up carried out in school by team members. We would argue that management of the curriculum through effective assessment and teaching approaches falls on school managers *and* on the classroom teachers.

Assessment

All the schools we visited carry out Ministry directives to conduct monthly tests. There are end-of-term tests and end-of-year tests. Those at the end of Standards IV and VII have greater significance, the latter especially since the few secondary school places are decided by success in these examinations. Massive effort is expended on setting, administering and marking tests, yet test results are not used to guide teaching, pupil placement or pupil grouping. Teachers mark hundreds of test questions each year, recording faithfully individual pupil scores, but none we spoke to did anything with these pages of numbers. School managers checked that tests were administered and marked, with records of Standards IV and VII tests stored especially safely. However, they made no comment to the teachers about how to use the results, or what calculations (means, distributions and so on) might be useful to management. The project's core subject team identified assessment as a key training input for further intervention. Assessment and using assessment outcomes are key areas to address so that teachers know how to produce useful test results and managers use outputs effectively.

Test outputs were used neither to look at individual or whole-class difficulties with the subject matter (diagnostic or formative assessment, or both), nor to compute class performance at the end of a teaching sequence (summative assessment) in project schools. Schools said they used the test results, but closer questioning showed limited or non-existent understanding of what to do with the results of the tests on which they spent so much time. No teachers used calculators for relevant calculations; the effort of summing strings of numbers for a class of 160 pupils and then computing means without such aids would daunt even the most dedicated teacher. The project team will have to face this issue when they tackle assessment with project schools.

Development of Child-Initiated Interactions in the Classroom

The project is to be applauded for the significant shifts in the nature of classroom transactions in the project schools. Compared with our experience of other Tanzanian primary classrooms, there was movement away from lessons reliant entirely on teacher talk. All members of the project's school communities spoke of groupwork and of pupils being asked to participate more actively. Difficulties have been noted that inhibited the implementation of such changes: large class sizes, limits in seating arrangements, and so on. The process of persuading teachers and parents that there are significant benefits to be derived from persevering with new ways of working in the classroom had only just started. It would require much more training in the future if these changes in the classroom are to be successfully implemented. Despite our systematic probing of the issue, it was not clear to us that teachers, head teachers and team members had a shared understanding of what is meant by a child-centered approach to teaching and learning. They used the words, but showed little understanding of the necessary shift in locus of communication in the classroom. This is not surprising, as the phrase has become part of the jargon of reform in recent years and is often not defined explicitly through action. This is another area for future effort, as the team seeks to deepen understanding of the implications of the project's interventions.

In our evaluation, we argued that there was a need for more imaginative use of human resources in the schools, of breaking up large classes, and if using more teachers in a single class at any one time. Again, this is an area where the management and core subject teams could work together to convince head teachers that smaller groups can be achieved. Head teachers need to know that this is essential if the groupwork and interactive teaching approaches are to work effectively, a curriculum management issue leading to improved pupil and teacher motivation. The teachers and the pupils interviewed were all convinced that the objective is desirable. The teachers were more skeptical that they could achieve it under present constraints. Head teachers recognized the physical limitations of having too few classrooms, but had yet to work out imaginative ways of deploying staff to realize the best use of human resources.

Future Directions for the Project

The AKF is considering ways in which the DPSP can be developed into other initiatives in Tanzania. It has a number of possibilities under discussion, including

- expansion of the project within Kinondoni District;
- application of the DPSP model to the rural schools in Kinondoni District; and
- extending the model to rural schools in another district in Tanzania.

Expansion of the Existing Program in Kinondoni District

Expansion would involve a significant change in the scale of the project, from working in 15 schools with a population of around 30,000 pupils, to working in 73 schools serving nearly 111,000 pupils. Expansion could proceed through continued stepwise growth, with new schools being included in each cycle, or through a more rapid scaling up by, say, doubling the number of schools each year. Whichever method of expansion is adopted, there would be a demand for considerably more trainers and the introduction of additional tiers of personnel into the model. However, the training materials have been largely developed and, while there will always be a limited demand for new development of training materials, the main demand on the training team will be in revision and production of materials, freeing team time for other training functions.

Dissemination of the project objectives to the increased number of stakeholders in the schools' communities calls for a real focus on 'marketing' the project's image and constraints to ensure better penetration and comprehension than perhaps was achieved in the initial phase. The use of a category of personnel to maintain a dialogue with local communities, the community liaison officer, modeled perhaps on a similar position in the Mombasa SIP (AKF, 1999), would be an advantage in an expanded operation.

We have commented already about the constraint on the project imposed by the lack of sufficient classrooms in the district. We suggested that closer integration of project operations with other initiatives in the district might be a profitable way to proceed. To meet some of the infrastructure challenges faced by schools, the AKF could work more closely with the DBSPE to complement their work, by offering management and core-subject training expertise to enrich the DBSPE activity and make use of the improved buildings that are a part of the DBSPE program. Translating these ideas into action would be a real challenge to the SIP project if it decided to expand.

Application of the Urban DPSP Model to Rural Areas of Kinondoni
This option would take the view that the DPSP has been a pilot for operations in a densely populated urban environment with particular constraints and conditions. It would also suggest that the DPSP's future lies in conducting another pilot, perhaps on a similar scale with similar funding, this time investigating a different set of constraints in a more sparsely populated environment with schools more widespread and fewer pupils per classroom. It is probable that these new conditions would require additional trainers to reach the schools effectively. It may be that there is less need for close and active collaboration with another provider (such as the DBSPE) than just described. The advantages of undertaking both an urban and a rural SIP in Kinondoni are that the current project team would be able to exploit and develop existing relationships. The project is known beyond the existing project schools, and the 'new' schools would require little convincing of the advantages of participation.

Extending the Operation to a New Rural Context: Morogoro Rural District
The DBSPE is already active in Morogoro Rural District, some distance away from Dar-es-Salaam and Kinondoni. The DBSPE plans to create three TRC clusters, thereby injecting aid into an otherwise assistance-deprived area. The Aga Khan Foundation has suggested that it might set up a new school improvement project in another area of Morogoro Rural District to ensure a balance in aid provision within the district. Becoming operational in Morogoro Rural District would require a different way of working and would throw up new problems for project design and implementation. It would require collaboration with other parties and would be complex in many ways. The AKF would also have to design an exit strategy for its present project schools and the Kinondoni District, as part of its adoption of this third option.

Concluding Comments
Whatever the decision about how to proceed in the future, it must be said that, overall, the Dar-es-Salaam Primary Schools project has been successful in achieving its objectives. It might also be said, however, that any injection of resources above the extremely low levels in most Tanzanian primary schools would result in enthusiasm for the intervention and produce similar impressions of success. It might also be argued that the present project was very small, insignificant even, in comparison with the overall picture of primary school output in Tanzania. An evaluation informant said,

> … the AKF is very good at carrying out pilots, but has never been faced with implementing programs on a larger scale; indeed it has chosen not to risk such an undertaking.

It may be that this is a widely held perception. It would provide an explanation as to why two larger interventions: the DBSPE in Kinondoni and elsewhere, and the SCF

Mtwara Project, have proceeded without AKF involvement, despite significant input of DPSP experience and materials in the planning stages of those other initiatives.

Such an explanation is too simple. There has been scope in Tanzania for many organizations to work together and exploit each other's strengths, and yet few do so. Indeed, the collaboration on this project between the AKF and the British Voluntary Service Overseas suggests that both the necessary mechanisms and the will to collaborate exist in AKF-T and in at least some other NGOs. The training approaches of the DPSP would dovetail well with the training inputs and the building programs of larger scale initiatives such as the DBSPE. The challenge is for the AKF and the DBSPE to integrate at the variety of levels necessary to ensure that both organizations retain their character and expertise while exploiting the synergy of collaboration. The dialog between AKF and other interested parties should be continued, and actions identified and implemented, both to increase sustainability and to show that collaborations can produce action rather than rhetoric.

8

Curriculum Materials and In-Service Training for English-Medium Instruction in Zanzibar

Ann Brumfit and H. R. H. Hikmany

This chapter describes a specific curricular innovation project. The objective of this project, the Secondary English Language Orientation Project (SELOP), was to provide a practical response to the impact of a change of instructional medium from Kiswahili to English at the point of primary-secondary transition in Zanzibar, Tanzania. One purpose of this chapter is to investigate ways in which use of new teaching materials may affect the attitudes, perceptions, and practices of teachers and learners in the classroom. Unlike most of the school improvement projects reported in this volume, characterized as 'broad repertoire' approaches (see Chapter 1), the SELOP embodied a 'packaged teaching and learning' approach to pedagogic change. Some further issues, particularly prospects for project sustainability, will also be discussed.

Introduction: Background to the SELOP

The Educational Context

A distinctive aspect of Tanzania's education policy, dating from the key 1968 policy document *Education for Self Reliance*, was the establishment of Kiswahili as medium of instruction throughout primary school. This policy decision was based not only on pedagogical considerations but also on the intent to make primary education complete in itself. The 1968 overhaul of the education system was designed to turn an elitist system, in which primary education was seen to function de facto as preparation for the 'narrow pyramid' of secondary school entrants, into a system that more closely reflected the needs of the vast majority of primary school leavers. Primary education was thus reconceived as integral to a society (a) with an agriculture-based economy, and (b) committed to principles of egalitarian development. Kiswahili, Tanzania's natural lingua franca and national language, was the obvious choice of instructional medium, in preparation for the new drive toward Universal Primary Education (UPE).

Within the United Republic of Tanzania, the Islands of Zanzibar have a separate Ministry of Education, and, while the Islands share their language policy with the mainland, their practice of providing 2 years' secondary education for all primary school leavers is distinct from the mainland practice of selecting only a limited number of secondary school entrants; indeed, the provision of 10 years' basic education is possibly unique within Africa. This distinction has implications for educational language planning, in that the change of instructional medium takes place within, rather than after, the period of compulsory education, and therefore affects a higher proportion of every cohort and involves a much wider ability spectrum.

Provision of this extended period of education, interrupted as it is by a sudden change of learning medium, has put increasing strain upon resources. Concern has often been expressed about a process of perceived 'spiral decline' in standards, from tertiary level downwards. One way in which this decline in standards manifested was in pupils' "handicapped mastery of the English language," an effect exacerbated as soon as English became the medium of instruction. This handicap entails a complex two-way relationship of cause and effect: of language upon learning and vice versa.

One clear cause of concern was that the transition to the heavy demands of an English-medium secondary curriculum was too abrupt. With the exception of private, English-medium, primary schools, founded in response to a rising tide of public demand during the late 1990s, it is still true to say that the pattern of primary education in Tanzania as a whole (on the mainland as well as on the Islands) shows English being taught for a limited number of hours a week, and then only as a subject. Although there have been recent moves to extend the teaching of English (as a subject) downward into primary Standard I, those completing primary school are not prepared for the sudden transition to learning through English. Concern about this is repeatedly expressed on the mainland, but the structure of the Zanzibari education system has rendered the problem more acute.

The Sociolinguistic Context

The Islands of Zanzibar (Unguja and Pemba) share Kiswahili as their national language and lingua franca with mainland Tanzania. Indeed, Kiswahili is the first language of almost all the Islanders; the Zanzibar dialect is highly valued as a national model, and Unguja is home to the Institute for Kiswahili and Foreign Languages.

Domains for use of Kiswahili are uncontested. However, as on mainland Tanzania (and indeed elsewhere in the world), a widespread and insistent demand for English has been intensifying in recent years. But, whereas on the mainland it is arguably less likely that large numbers of primary school leavers will in fact have a real need for English, especially in the rural areas (although this is a controversial and locally contested point), the small size of Zanzibar and recent 'liberalization' of tourism and trade mean that English is likely to impinge upon a much larger proportion of the population (90% was an estimate received from several independent sources). Furthermore, Zanzibar, with its fluctuating relations with the Tanzanian capital Dar-es-Salaam, is more reluctant than the mainland to risk isolating itself from a world language. This groundswell of demand for English in Zanzibar has led to support for the implementation of SELOP at all levels, from the Ministry to the most remote of project schools. Perhaps the most striking feature of the project is the unanimous support, and the degree of enthusiasm and commitment, that it has provoked: 'unique' in the words of British Overseas Development Administration (ODA) personnel in Dar-es-Salaam.

Responses to Needs in English Language

Discussion in Dar-es-Salaam of the possibility of changing to Kiswahili-medium secondary education has been long standing and remains unresolved (Rubagumya, 1989). In response to continuing problems caused by the language transition, various initiatives have been taken. The first of these was the publication in 1967 of *Learning through Language* by Tanzania House, an intensive reading course consisting of a collection of multiple-choice comprehension exercises in English, for use in the early weeks of secondary school Form I. *Learning through Language* is significant in that it set two precedents: Not only did it attempt to combine the use of cross-curricular subject matter with systematic treatment of reading skills, it also introduced group discussion as an essential component of the methodology. From a perspective 30 years on, the material looks overacademic, with multiple-choice variants providing focus for the group discussion often depending upon inappropriately sophisticated nuances of language. In addition, the repetitive classroom procedure, innovative in its time, now looks unadventurous. Unfortunately, no evaluation was undertaken when the reading course was first developed to assess the effect of these materials upon pupil learning, less still upon teachers' professional practice. Although the materials formed part of the English language program, and would have been taught by English language teachers, these teachers themselves usually taught a 'second subject.' It would have been interesting to investigate whether similar groupwork approaches made any impact at the time on the ways in which other subjects were taught.

Over the years, the training and resource demands entailed by the introduction of UPE, together with the increasing development of Kiswahili as a viable and valued medium in most domains of life, including many domains formerly occupied by English, have meant a fall in the standards of English in schools. As a result, *Learning through Language* would now be far beyond the capability of most Form I entrants. Continuing concern over declining standards of English gave rise to the

Criper-Dodd report of 1984, and the resultant secondary-level English Language Teaching Support Project (ELTSP), set up by ODA. The ELTSP sought to raise standards through a process of 'input flooding' consistent with the theories of Stephen Krashen (e.g., Krashen, 1981), which were particularly influential in the language-teaching field at the time. A large-scale reading program was established, supported by a team of expatriate English Language Teaching Officers dispersed across the country. In the early stages, effort was directed toward setting up and administering class libraries and sets of class readers; during the later stages of this project the focus was redirected toward a more general program of teacher training.

Simultaneously with the 4-year secondary school reading program, two much smaller 'orientation' projects were designed as part of the syllabus for the first 4 to 6 weeks of the secondary Form I course: Baseline (funded by Volunteer Services Overseas and the British Council), and the Form I English Language Orientation Course (FIELOC, originated at Aga Khan Mzizima Secondary School, Dar-es-Salaam). Both may be seen as direct descendants of *Learning through Language* and were designed as short orientation courses to take place at the beginning of secondary education. Whereas the Baseline materials emphasized oral and aural skills, and were linked only loosely to the curricular content of Form I, FIELOC took the form of subject-specific course materials in modified English, with separate booklets provided as introductions to Form I curricular content. The aim of both projects was to cushion the impact of the sudden changeover to English. While the Baseline philosophy emphasized general activity-based processes of language learning, with a particular emphasis on oral pairwork, and was only loosely connected with curricular content, FIELOC was a more traditional 'product-based' model, with a stress on specific subject matter, an emphasis on reading skills, and more traditional assumptions about the nature of learning.

In parallel with ELTSP on the mainland, the Zanzibar English Language Improvement Program (ZELIP) was established and administered, from 1989 to 1994, by the British Council on behalf of the British government's ODA. A similar, secondary-level, extensive reading program was set up. In 1992, however, a more radical approach was decided upon by Zanzibar, when the Ministry announced its decision to include within the 10-year period of compulsory free education an 'Orientation Year,' between primary and secondary education, producing a new schooling pattern of 7-1-2 (Primary-Orientation Year-Lower Secondary). An important catalyst toward this decision was Ministry familiarity with one of the two mainland 'orientation' initiatives.

The SELOP materials resulting from this decision consist of a series of curriculum-based books in simple English, designed to promote a process-based, interactive style of learning through English. Alongside the introduction of the materials, provision was made for in-service training for secondary school teachers. Both the curriculum and training are described below.

The SELOP Program

Origins

The Secondary English Language Orientation Project (SELOP) was established in 1994 and implemented by the Zanzibar Ministry of Education, under a grant from the Aga Khan Foundation-United Kingdom, with cofunding from the UK Overseas Development Administration, now the Department for International Development (DFID), the Canadian International Development Agency, and the Commission of the European Communities. The following account is drawn from evaluation reports commissioned by AKF.

Previous contact with Aga Khan Mzizima Secondary School in Dar-es-Salaam had meant that the Mzizima FIELOC materials were already familiar in Zanzibar, particularly in the large and prestigious Haile Selassie Secondary School, whose headmaster had a personal and professional interest in strategies to bridge the language gap.

As mentioned in Chapter 3, predecessors of the FIELOC materials first appeared in Mzizima in 1986 in the form of worksheets for strengthening the English-language proficiency necessary for Form I. By 1988, FIELOC had been further developed and the material published in booklets (under Mzizima SIP, Phase 1) as an orientation course in English language to cover the initial 8 weeks of Form I. Interest in FIELOC from Zanzibar dates from this time, as a Zanzibar representative was invited to observe the FIELOC course at Mzizima in 1988, and the pre-course training in 1989. The materials were subsequently made available for consideration by the Zanzibar Curriculum Development Department. The Mzizima materials, however, had been developed within and for an atypically well-resourced urban setting in Dar-es-Salaam, for a highly selective school with an English-medium primary school among its feeder schools and a relatively educated and 'elite' catchment, a school that achieved Form-IV-leaving results well above the national average. Mzizima's FIELOC materials were never widely adopted elsewhere on the mainland, where offers of use and associated 'outreach' training met with very limited uptake. Conceptually and linguistically, the level of the materials was widely agreed to be 'too hard' for 'typical' schools: In the words of the outside evaluator, "beyond the scope of the majority of students operating in the average or below-average secondary school" (Williams, 1989, p. 77). This was particularly true for Zanzibari schools where, as already mentioned, mixed-ability intake continued up to Year 10. The 1989 evaluation also reported negative comments from students on the (teacher-centered) methodology. But, perhaps most decisively, the FIELOC materials found themselves compared directly with the simpler Baseline materials, which were being developed at the same time as a separate and independent initiative by a small group of expatriate Voluntary Services Overseas (VSO) teachers. Baseline, a "step-by-step approach to basic lessons in functional English," was not only simpler but also methodologically more innovative (more open ended, more learner centered, more process oriented). As a course, it was also much cheaper to implement, being designed as a simple series of three teacher's books with lesson plans outlining straightforward activities to be mediated entirely by the teacher. Possibly as a direct reflection of concurrent English as a Second Language (ESL) practice in British primary and middle schools (from which VSO teachers were largely drawn), Baseline was importantly underpinned by the idea that the material should be taught, and thus 'owned,' by teachers of all subjects, not just by the teachers of English. An important purpose was to provide a training resource for 'other-subject' teachers, and a means of raising awareness of language needs across the curriculum. This idea met with some understandable resistance from 'other-subject' teachers, and when Baseline was expanded and established as a 6-week orientation course within the ODA's English Language Teaching Support Project (ELTSP), the notion that language should be the responsibility of all teachers across the curriculum was, and remained, a necessary focus for pre-Baseline training. For similar reasons, when Zanzibar's SELOP program was eventually developed (1994-1997), Baseline was chosen as an introductory, 4-week, cross-curricular preparation for the new Zanzibar orientation course materials.

In contrast, in spite of project support under Mzizima SIP, Phase 2, the influence of FIELOC remained negligible on the mainland, and the Mzizima material was never widely disseminated or adopted beyond a small group of schools in the southern highlands with which Mzizima had informal links.

Following the early interest of Zanzibar's Curriculum Development Department, the FIELOC materials exerted more influence in Zanzibar, serving as an early catalyst for a similar model of curriculum development. ELTSP did not extend to Zanzibar (the ZELIP program was under separate management), and the FIELOC materials were unchallenged. As on the mainland, the materials were considered too difficult as they stood, but the concept of subject-specific booklets was retained and teacher-led subject teams were established in 1989 in order to adapt the FIELOC units. Progress was slow, as planning was retrospective, and each version, once drafted, was submitted to separate teams of 'language editors' and 'content editors' for rewriting. Principles were not articulated but, as the materials evolved, there emerged a recognizable emphasis on the acquisition of terminology as an end in itself, at the expense, often, of meaningful concept development. Meanwhile, as these materials were reaching final draft form, and were being trialed in 10 schools (during 1993), Ministry thinking had moved toward establishing an extended, year long orientation program. A clear need emerged for extended and redesigned materials. A critical evaluation report led to the work of the 'subject teams' being discarded, but an invaluable legacy from these 3 years' work (1989-1992) was the strong sense of ownership, indeed confidence, generated by these first locally produced materials.

As a separate, autonomous ministry, independent of Dar-es-Salaam, the Zanzibar Ministry of Education was in a position to develop its own strategy. This was initially driven forward by a small group of change agents, enthusiastic individuals notable among whom was the then headmaster of Haile Selassie Secondary School, Haroun A. Suleiman (subsequently Project Director and now Zanzibar Minister of Education). The SELOP, possibly unique in terms of curriculum development, thus came to be launched in 1994. It is worth noting that, since inception, the project has been characterized by a continued sense of distinctively local ownership, invested with a symbolic 'Zanzibari' identity that has permeated public awareness far beyond the confines of the schools it has served. Thus, although in relation to other case studies in this volume, the SELOP may appear as an example of 'top-down' centralized Ministry planning, it may also (if seen against the wider Tanzanian context) be considered an example of an influential, small-scale local initiative.

Set up in January 1994, with materials introduced into 20 pilot schools at the end of that year, the program was subsequently expanded annually so that, by 1997, all 95 schools with an Orientation-level intake (11,000 Orientation Year students) were included. Estimations are that, between 1998 and 2003, 60,000 students and 400 teachers will have had direct practical experience of the Orientation Year.

Project Design

Management Structures

As context for the following discussion of project processes and project outcomes, a brief outline of SELOP management structures is provided. Note that project implementation has been affected by geographical constraints, that is, the necessity to liaise across the two separate islands. Indeed, 'isolation,' or lack of optimal communication, has been a recurrent problem, with liaison problems among agencies (such as the examining bodies) on Zanzibar and in Dar-es-Salaam, among colleagues and counterparts across the islands, and among separate agencies within Unguja. In addition, communication (and continuity) has been complicated by the replacement of personnel; within the 3 years of the project there were repostings within the Ministry and numerous redeployments within and among the schools. The SELOP itself underwent staff changes, with new appointments during the course of the project to the posts of project coordinator and materials coordinator.

While the project was overseen by a steering committee and a project director, responsibility for everyday project management lay with a British expatriate project coordinator (previously with ELTSP in Dar-es-Salaam), together with two part-time Coordinator Counterparts, representing Unguja and Pemba respectively. A small task force of five, representing both islands, was established to help plan and coordinate all aspects of project activities.

Writing of the Orientation Secondary Course (OSC)

Initially, materials writing continued, as before, to be the responsibility of teacher teams, with separate writing teams on the two islands. Drafts were submitted for editing to the expatriate VSO Materials Coordinator. This arrangement presented problems of communication and coordination, so, in 1995, the writing process was centralized and taken over by the materials coordinator and a newly appointed counterpart from Pemba, working together from the SELOP headquarters in Haile Selassie Secondary School in the town of Zanzibar. Duplication of materials for trialing was also organized from Haile Selassie School, and became an increasingly onerous task, as increasing numbers of schools joined the project (in 1995, 45 schools; in 1996, 64 schools; in 1997, 95 schools). Selected teachers on both islands were responsible for responding to feedback questionnaires. Final editing of the materials after trialing took place at Haile Selassie School. Materials were then checked by an independent reader, printed by a local printer, and distributed by the Ministry. The ongoing supply of supplementary reading materials (class readers and class libraries) inherited from the ZELIP program came under the Ministry-funded management of a reading program coordinator. The supply of Baseline materials, adopted to cover the first 4 weeks of the Orientation Year, was organized separately.

Teacher Development

In-service support A centralized, 'cascade' model of annual pre-OSC training has been complemented by systems of local advisory support. The project coordinator increasingly delegated to a team of 10 central trainers the responsibility for planning regular cycles of 'Central Workshops' (held annually to train new teachers for the next Orientation Course) and for training the trainers. Locally, these 10 District Orientation Secondary Course Advisers (DOSCAs), seconded from their own schools for 1 day a week, have worked in an advisory capacity in each of the 10 districts, providing classroom-based support and disseminating good practice. The DOSCAs' role was defined as supportive and advisory, and Ministry recognition of this positive role was instrumental in the remodeling of the Inspectorate at the end of the 1990s. DOSCAs' school visits are, in principle (but rather rarely in practice), supplemented by teachers' peer observation visits, both within and among schools.

Head teachers have, from the start, been closely involved with the project. They attend the initial training sessions and are briefed on staffing, timetabling and record-keeping. They are encouraged to observe OSC classes and help to set up school-to-school observation visits. A high level of commitment was noted among head teachers consulted in our evaluation. The Inspectorate staff, on the other hand, while also supportive, were less closely identified with the project. Contact arose from attendance at training workshops and observation of OSC classes, but there were no formalized feedback systems linked to project records, or project planning.

Pre-service support During the lifetime of the SELOP, it had not been possible to direct pre-service teacher training toward the content and methods of the OSC. This was partly due to inflexibility arising out of centralized control (in Dar-es-Salaam) of the Grade A teacher-training curriculum. Not only are no English-language courses provided for teacher trainees in subjects other than English (although they will be expected to teach their subjects through English), but there is also no interdepartmental contact to enable discussion of language issues (whether English or Kiswahili) that affect the wider curriculum. The OSC initiative remained outside official, institutionalized, pre-service structures.

The Significance of the SELOP: Project Processes and Project Outcomes

The 'SELOP approach' was defined by stakeholders as "a participatory approach," as "talking, discussing and working together," and as "active learning." At least two assumptions are implicit in these definitions. The first assumption is that curriculum development, supported by appropriate teaching and learning materials, is capable of influencing classroom practice. The second assumption is that 'active,' 'participatory' approaches will lead to more effective learning. To what extent does the SELOP experience suggest that these assumptions may be upheld? These issues will be considered in the context of the separate strands of SELOP development.

The OSC Curriculum

The syllabus was essentially a list of topics and items drawn from the syllabus documents for Standard VII and Form I, together with language items recycled from the primary syllabus. It did not specify classroom methodology.

Endorsed in 1994 by the Executive Committee of the Ministry of Education, the new syllabus subsequently underwent various modifications. The number of core subjects was reduced in 1995 from six to four (English, mathematics, science and social science). Similarly, in response to classroom pacing demands, the number of units for each subject was reduced in 1996 from eight to six. Free of external benchmarks such as national examinations, project staff could accommodate the changes described above within the project design, but contemporary evaluation reports warned against allowing undue slippage of syllabus coverage.

A tendency had evolved for teachers (whose constant cry is still "time is not enough") to set the pace of instruction, and for their feedback to act, de facto, as an automatic restraint. It was well known that stretches of teaching time were lost every year for a variety of reasons, such as late school openings, 'disruptions,' and teacher absenteeism, but it would be possible to envisage a downward spiral of mutual cause and effect. There had in the past, been a danger that OSC end-of-course objectives, lacking externally defined standards, might be perceived as self-validating, defined by what teachers felt students (and they) could comfortably cope with. 'Evolutionary' retroactive alterations to the OSC syllabus may exemplify this. In addition, there may have been a danger, inherent in its history, of the OSC being perceived (perhaps subconsciously) as an add-on, a luxury, with learning due to be seen as a bonus rather than as part of a longer term progression toward the levels of English knowledge required ultimately for promotion from Form I to Form II.

The prime emphasis of the OSC syllabus, however, was upon language, and language skills were a focus for each unit. Content was treated as a 'vehicle' for language learning. Some explicit links were made between content and language learning. Certain topics clearly predetermined particular language items (especially vocabulary); other relationships emerged later, at the writing stage, out of activities selected for the different units. But, whereas it was possible to see a link between 'Describing' (English) and 'Living and Non-Living Things' (general science), 'Instructing' (English) was only tenuously related (if at all) to 'Matter' (general science) or 'Geometry' (mathematics). It was not, therefore, possible for the syllabus writers to sustain one specific language item across a series of lessons in the different subjects, in order to establish the item firmly and highlight its varying uses. Instead, the writing team kept a retrospective inventory of structures, lexicons and topics, with an indication of whether each entry is a 'focused' item in any particular lesson, or whether it is just 'used.' Although this retrospective inventory was an excellent example of the 'task-based' approach in action, with language items arising as needed out of tasks, we felt that the initial design of the syllabus might have closed off potential opportunities for making meaningful language links.

As for sequencing, the syllabus proved a satisfactory guide for the OSC writing teams, although, horizontally, some opportunities seemed to have been missed for integrating all four subjects conceptually within a shared topic area. With some exceptions (e.g., the order of the mathematics topics), the vertical sequences for each of the four core subjects appeared arbitrary, without consistently clear thematic links or obvious progression in conceptual difficulty. Without such links, fragmentation of learning is an obvious danger. But, officially described as "a revision year," the Orientation Year was treated as a year of content consolidation. It was free of pressure from national tests and deliberately provided space for recycling of content, allowing for contrast and variety. The telling point: Teachers were adamant that motivation and interest not be adversely affected by the repetition of subject matter previously covered; indeed, many maintained that similar topics, now taught in English, were unlikely even to be recognized. Although this may suggest that many of the concepts encountered in primary school had never been properly assimilated, it must be remembered that few OSC pupils ever had books to work from while in Standards I to VII. It is easy to underestimate the challenge presented by visually unfamiliar material, together with all the conventions associated with the textbook page, let alone the presentation of material in a new language. Nevertheless, if there is any truth in these teachers' assertions, there are clearly important implications here for any education system that introduces a sudden cross-curricular change of instructional medium. Attention needs to be paid to ways in which conceptual links can be maintained (rather than fragmented, see comments above) across the whole process of instructional medium change; this is clearly a large issue that extends far beyond the context of writing materials for Zanzibar. It is pertinent to note in this connection that, according to the 1996 Zanzibar Education Masterplan (ZEMAP), bilingual approaches were to be investigated as a possible extension to the official language policy. This strategy might be supplemented by a more gradual phasing-in of English-medium instruction subject by subject. Both approaches would be consistent with policies being adopted elsewhere, such as Kachru's 1986 'inner circle' educational systems, in which an 'outside' language is used as medium of instruction.

The OSC Materials

The OSC was introduced by a 4-week, initial, full-time period of Baseline, after which the Zanzibar OSC books were used in each of the four core subjects (English, science, social science and mathematics). Timetabling for the English component of the course allowed extra time for continued use of the ZELIP program of extensive and intensive reading (using class library books and class readers), as introduced earlier by ODA (see Responses to Needs in English Language, above).

The ODA reading programs (ELTSP and ZELIP) were not a focus of attention in our evaluation. Extensive reading was retained as part of the SELOP program and embedded within it as a supplementary activity; it was not separately evaluated for SELOP.[16] Observation and discussion suggest that reading is a potentially valuable component, but that drawbacks such as an undeveloped methodology and ongoing problems of book supply reduce reading's impact.

As for the purpose-written OSC materials, the 1997 SELOP interim evaluation report commended this final generation of materials for their clarity, simplicity and general appropriateness. These new materials were intended (in the words of the 1994 Ministry OSC Syllabus) to "revise and strengthen the basic concepts and operations introduced in the primary school," with language "integrated within the subject content of OSC." The material embodied a range of imaginative ideas for presenting cross-curricular concepts either visually (e.g., 'Waridi the Bird' from social science) or by means of simple and accessible language. Above all, the materials were commended for having achieved such an enthusiastic reception in the schools. It might be argued that any materials are better than none, a 'no-lose' situation, but, in contrast to the early FIELOC-based versions, content, language level and methodology had been more successfully honed and adapted to fit teachers and learners, and good lessons successfully embodied the OSC 'interactive' philosophy. The introduction of the OSC materials, supported by a program of training (see SELOP Teacher Training, below) had produced a teaching style that included generally well-managed pair and individual work, enabling teachers to stand back and yield the floor. Teachers using the OSC materials described their classes as "participatory," with "active learning," and teachers of other classes were encouraged to observe their OSC colleagues in order to try out similar techniques.

While 'active learning' rhetoric had clearly been accepted by teachers and pupils, to what extent is it possible to evaluate genuine learning outcomes? The value and efficacy of devolved, activity-based methods in an educational system that traditionally favors a more teacher-centered approach is no longer taken as axiomatic (Roberts, 1998; Chapter 1 of this book). While such a complex issue will hardly be resolved here, the following issues seem worth raising. Two in particular emerge: first, questions of definition, and second, in the context of materials-led innovation, questions of design.

Questions of definition The value of active learning was emphasized during OSC training workshops, and when asked how they would describe the 'OSC approach,' teachers selected active learning as its most distinctive feature. What teachers meant here was that students were, in every lesson, asked to work in pairs. The claim to active learning, in this sense, was well upheld.

Pairwork in its obvious sense (i.e., pupils working together instead of listening to the teacher) had had a visible impact on Zanzibar's OSC classrooms. The OSC materials prescribed pairwork as part of every lesson, and in-service workshops provided support in the form of suggested class management techniques and discussion of rationale. As a result, OSC classes were distinctive in that teachers no longer depended solely upon a 'transmission' mode of imparting knowledge, and pairwork took up a significant proportion of each lesson. Some school visit reports by the DOSCAs attempted to quantify this proportion, with figures recorded as high as 90% per lesson. Against a long traditional background of 'talk and chalk,' this shift of emphasis was in itself a considerable achievement, recognized as such not only by the teachers but also by students (who describe the OSC as different because "we can talk in English"). This opportunity to 'talk in English' was one important reason given by students for their enjoyment of the OSC; in short, pairwork was motivating and made OSC lessons fun. What more justification could be needed?

It is here, importantly, that questions of definition arise. While it is easy to assume that pairs of students apparently working together on the same exercise must be more 'actively' learning than individuals working alone (or, worse, listening to the teacher!), it is important to take into account the fact that the nature, level and quality of the tasks vary widely. At one end of the spectrum, two students may be sitting together but working individually (no interaction); further, they may be working on cognitively undemanding tasks such as copying (little 'activity'). Alternatively, the task may be cognitively more demanding (matching, for example, or gap filling), but one student may be 'actively' working out answers while the other looks on. Further, the 'cognitive level' of tasks will clearly vary: from a reliance on the more or less mechanical (e.g., slot filling to establish the fixed phrases targeted in many of the units, see discussion below) to the cognitively demanding, where choices, or 'active' decision-making, invite, a genuine understanding of the content or language system, or both.

The issue here is partly one of simple classroom management. Clear teacher instructions (leading to a clear understanding by pupils of what pairwork is, and is for) should discourage 'individual work seated in pairs.' The issue, however, is also one of materials design.

Questions of design A random selection of lessons observed yielded a number of occasions when pairwork activities seemed to miss opportunities for more 'active' learning. Students were, for example, asked (in the social science lesson Looking at Things) to draw side views and plan views of common objects and then, as described below, to rehearse in writing the related question-and-answer phrases: "What's this ... ?" "It's a ... ." Perhaps an alternate and more stimulating strategy would have been to exchange books midway and turn this exercise into a simple information-gap activity. Similarly, students were sometimes asked to repeat or rephrase 'given' information from the unit, instead of using the sample sentence pattern to tell a pairwork partner about 'real' events. There seemed to be missed opportunities for personalization as an additional 'student-generated' form of information gap. Personal investment in real meaning is often rare in traditional African classrooms (see the literature). It has been shown to be one way of pushing learners toward formulating more interesting (and accurate) output.

Along the same lines, it was felt that some OSC tasks relied too heavily on copying. The proportion of 'active' time thinking, as opposed to copying, sometimes seemed disproportionate. There were justifications for copying as a class activity. For example, in the absence of OSC books to take home, copied material has value as a substitute for textbooks; also, the process of copying may build learners' confidence. The copying process may also directly contribute to learning, but it is dependent upon the learner's level of 'attention,' a factor difficult to guarantee or, indeed, measure. There were few if any devices, such as requirements for underlining after copying, of salient features significant to the unit under study, or the use of simple substitution tables. Such techniques might have been particularly valuable for drawing attention to language regularities such as inflection or word order.

Further assumptions about language learning processes may be raised here. The syllabus for English was announced as functionally based, with units labeled as follows: Describing, Comparing, Instructing, Reporting, Planning, and Expressing. Although each unit was supported by a list of language exponents, the approach, in common with many of its functional forebears, relied largely upon the presentation and practice of unanalyzed language 'chunks.' As already mentioned, horizontal links between subject areas were not always fully exploited, so that, in any one week, learners were exposed to a great many fixed phrases or unrelated chunks of language. Presentation frequently consisted of a short reading passage, often in the form of a dialog to be read aloud in pairs; practice usually took the form of simple activities (written or oral exchange pairs) such as selective gap-filling or matching tasks, or the asking and answering of simple questions. The point here is that, although the tightly guided nature of the activities meant that the students could 'do' them, there were no devices to encourage students, even implicitly, to notice how these various pieces of language might fit together as a system. An enormous memory load is involved in the learning of vast numbers of apparently unrelated expressions, a type of formulaic learning reminiscent of South African classroom 'safetalk' (Hornberger & Chick, 1998), and which is unlikely to aid conceptualization. This problem was exacerbated by the facts that students were not allowed to take textbooks home, there was negligible opportunity in the environment for independent language acquisition, and the only means of classroom consolidation was a series of sentences in an exercise book, possibly inaccurately copied and probably lacking context.

Roberts (1998, p. 110) suggests two paradigms: (a) person centered and (b) model based, which may offer 'necessary compass points' in our thinking about knowledge and learning. These, more generally reflecting cultural and political differences, are closely associated with differences of approach to language teaching. As suggested in the discussion above, the use of OSC had not moved the learning/teaching model directly or wholly into a person-centered, or experiential, paradigm. While teachers were correct to perceive their innovations as being, with OSC, more 'person centered' and less directive than formerly, there were still recognizably model-based elements, entailing a focus on ends, or product, rather than process. It is important to guard against simplistic equations here: Person-centered methods may, for example, fail to value teachers' hard-won expertise (Pennycook, 1994) and may conflict with approaches presupposed elsewhere in the curriculum. It is possible to place insufficient value on outcomes associated with a model-based paradigm; for example, learners' capacity to internalize language patterns through memorization.

Conclusions

In relation to the design and implementation of the OSC materials, three conclusions may be drawn:

1. There was evidence that introduction of new materials is capable of playing a key role in altering classroom interaction patterns. There was also converse evidence that very few, if any, non-OSC teachers invited to watch OSC classes but lacking access to equivalent materials made parallel adaptations in their own teaching. (Although an additional explanation might be their lack of OSC in-service training.)
2. The OSC teachers had indeed transformed interaction patterns within their own classrooms, but it was not possible to sustain a straightforward equation between the existence of particular classroom interaction patterns and active 'learning.' Particularly where language learning is concerned, research increasingly suggests that peer interaction needs to be supplemented by high-quality input from other sources. In contexts where there is little exposure to the target language outside the classroom, this input may be supplied by the teacher (implying at least some teacher fronting in language classrooms), but 'good input' will depend upon teacher proficiency and teacher confidence in English, both problematic in the Zanzibar context. The need for teacher English-language improvement was recognized by SELOP but not addressed. Future planning would benefit from incentives to teachers to upgrade their English-language skills.
3. Materials naturally embody certain assumptions about the nature of learning, including assumptions that may take inadequate or inappropriate account of culturally or context-specific features. Some suggestions relating to the OSC have been made above, but these clearly need to be substantiated by research. At the same time, it would be wrong not to take serious account of the voices of teachers and pupils, in whose words the program was clearly endorsed.

> All the time we speak English … You can see the difference … The program feels positive … Teachers can speak now in English … Students can talk to each other [no criticism intended] … though their language is broken, something is better than nothing.

SELOP Teacher Training

Background As it was not possible to visit an OSC workshop, the following comments are based upon discussion and upon documents: workshop planning notes, timetables, handouts, reports, memorandums, DOSCAs' school visit record sheets, head teachers' records and statements of intent.

A cumulative program of teacher development ran throughout the 3-year life span of the SELOP, as new schools were brought into the program annually and new teachers were deployed within schools to teach the Orientation Year. The challenge was considerable; 1997 figures for Unguja showed that 39% of the teachers at post had had no initial training. In the town of Zanzibar, it was reported by head teachers that staff were leaving the profession, following the growth of opportunities elsewhere for those with knowledge of English. As in many other places worldwide, English-language programs find themselves de facto 'draining' the school system and instead training for the economy at large. As a consequence of this drain of trained teachers, an untrained, part-time, teaching work force was increasing, a further challenge to the OSC training system. In addition, the shortage of teachers across the Islands (an estimated shortfall of 10% at secondary level) led to large classes, and the introduction of new interactive techniques presented teachers (and trainers) with new implementation concerns. Communication between Pemba and Ongaju and provision of funding for 'equal' levels of training were further issues.

On the positive side, head teachers commented on the commitment of most OSC teachers; staffing the OSC was, in general, no problem and a (relative) continuity in OSC classes was ensured. It was clear that the professional support available was valued by most OSC teachers (with a minority preferring to avoid OSC, seeing it as constantly under the microscope of the classroom observation program or else as involving too much preparation). Besides the attraction of the materials themselves, the school visits and workshops were an incentive to teach OSC, although it must be said that the associated per-diem allowances also, understandably, contributed to teachers' motivation.

Central workshops The main emphasis of the training component of the SELOP was upon classroom methodology, based upon use of the OSC materials, including the reading program. There was comparatively little emphasis upon a parallel need: teachers' own language improvement (see Human Resource Outcomes and Capacity Building, below). The main thrust was toward in-service work with serving teachers (see prior comments on lack of pre-service 'permeation' by OSC).

The mode of training delivery was described as experiential and participatory, to parallel the recommended OSC methodology; tasks and demonstration lessons preceded trainer input. Stimulus for a workshop conducted just prior to a SIP evaluation visit, for example, took the form of a fictional case study, Kichocheni School, in which workshop participants were invited to evaluate the fictional teachers. Worksheets on large classes invited teachers to widen their horizons by making comparisons with teaching conditions in Nigeria and Indonesia.

Instruments to measure the impact of training workshops were not formally established. At the end of each workshop teachers were invited to indicate what they 'liked' and what they 'learned'; unfortunately, no summary statistics are available. New recruits at Nkrumah Teacher Training College who had previously taught OSC classes and attended OSC workshops were described as being more confident than their peers, and trainers reported 'increased participation' in workshop activities by the more experienced OSC teachers.

Role of Head Teachers While only some head teachers taught OSC, all were invited to the OSC workshops. The SELOP experience demonstrated that the influence of the head teacher was likely to be the single most important factor contributing to the success of the project in any one school, as has been widely observed in connection with projects elsewhere. Head teachers are in a strong position to facilitate innovation, such as teacher-to-teacher observation. Their management skills are important in such matters as timetabling, troubleshooting, and record-keeping. Their support and encouragement is important for maintaining teacher morale. Head teachers were invited to draw up 'statements of intent' for their schools. These plans included a variety of interventions; for example, undertakings to maintain teacher continuity in OSC classes, the holding of regular monthly meetings for OSC teachers, and ways in which the use of English can be encouraged outside the English classroom. These statements served as useful means of policy formulation. The level of support from head teachers was informally estimated at about 80% by the DOSCAs. As for head teachers' response to the DOSCAs, relations were good, with head teachers describing the DOSCAs as 'very helpful.'

In-school observation If new approaches are to be assimilated, classroom-based training is generally considered essential. Classroom-based training for the SELOP was the responsibility of the itinerant team of ten DOSCAs, who (in the words of the Ministry) "play a very big role in the Districts now." Commitment among the DOSCAs was evident: "We are trying to work hard in a difficult situation," but staff turnover was, as in the schools, an ongoing problem. The concept of DOSCAs: advisory teachers whose role was to support rather than inspect, had made a significant impact in Zanzibar and was acknowledged by the Ministry to have been an important influence behind the new Advisorate model that was to be integrated within the new decentralized structure of Teacher Centers. DOSCAs' visits covered administrative matters (checking on staffing issues), curricular matters (coverage of the OSC syllabus), and classroom observation.

Comments on the school-visit record forms revealed the SELOP's criteria for good teaching. The following positive descriptions of teachers' instructional behaviors are constructed from typical entries by DOSCA's in the record forms:

- teacher language – straightforward, simple, clear, with vocabulary at the level of the learners;
- questions – understandable, short, nicely distributed, open, in the form of active learning;
- classroom management – wonderful participation, active pupils (most frequently recorded entry), students being given chances to work things out on their own, encouragement of weak pupils, enjoyment of pupils;
- teacher approach – enthusiastic, relaxed, knows what he [sic] is doing;
- feedback to students – encouraging, use by teacher of words such as 'good,' 'nice,' and 'well done,' teacher elicits correction of errors from class;
- teacher English proficiency – satisfactory, simple, correct, includes self-correction.

While comments such as those summarized above served to reinforce desired teacher behaviors, the DOSCAs' school visit records also highlighted needs for teacher change:

- classroom management – "How do you help students who do not participate?" and "How do you praise students who give right answers?";
- quality of teacher language – "May sometimes slide to big words," "instructions wrongly given, need for more guidance in use of leading-in questions";
- teacher linguistic competence – "Not good enough in terms of structure," "How do you help yourself in speaking English grammatically?"

Many of these points were featured for discussion in the 'case-study' training worksheets mentioned above (worksheets also included an emphasis on the dangers of time wasting and lack of good classroom organization, as well as warnings against a 'passive' role for the teacher during pairwork). Together, these sources provided a starting point against which project objectives could be measured. Unfortunately, no quantitative data on teacher performance were gathered by the project, but participant comments revealed generally agreed-upon perceptions of the 'good teacher,' and a degree of consensus on the nature of the ideal classroom, coinciding generally with project perceptions. In line with the DOSCAs' support role, positive comments by participating teachers tended to outweigh negative, and about current 'classroom realities' we could only add a few first-hand observations. These may help to fill out the picture, although they cannot be seen as representative.

If one generalization can be drawn from classroom visits conducted during the evaluation, it is that teachers' classroom behavior was constrained by their lack of confidence in the English language. All classes were making use of pairwork, many classes showed impressive levels of pupil participation, and, in most classes, the stages of OSC lessons were being followed as intended. However, classes were also generally characterized by an indecisive quality that hampered the achievement of objectives. The start of lessons and the transition stages between 'steps' were unclearly marked. Feedback was frequently ambiguous; students' responses would often go unacknowledged (right or wrong?), and, although some teachers were clearly attempting to follow up incorrect answers by further elicitation, the direction of their strategy was rarely clear from the back of the class. Further, when acceptable answers were produced, it was remarkable how few teachers acknowledged them with any form of positive feedback, let alone enthusiasm or praise. In our view, a system of signaling clearly whether or not a response is acceptable is essential in the classroom (particularly, one might argue, in a language-oriented class, where language form provides an important focus), but this lack of signaling has been widely noted elsewhere in African classrooms (e.g., Snyder & Ramatsui, 1990 on junior secondary schools in Botswana) and is evidently widespread in African classroom culture. The OSC trainers explained the lack of signaling in terms of the hierarchical nature of African society, in which it is not customary to thank or praise 'downward.' The workshop sessions devoted to this very question seemed to have had little impact on the teachers visited, and one suspects that another plausible explanation would be the teachers' lack of confidence about what was in fact acceptable, or not, in the target language.

A similar lack of linguistic confidence was also manifested in teachers' failure to 'get behind' the materials. For example, persistent 'common errors' at the pairwork stage were rarely picked up and generalized. It was rare to find any teacher insisting upon a decisive rendering of any text-based task as a means of clearly establishing correct answers. Teaching styles generally lacked decisiveness. It was unfortunately not possible to visit classes in Kiswahili in order to make comparisons, but, even taking into account recognized cultural factors, we concluded that lack of linguistic confidence was a major determinant of teaching style.

Teacher-to-teacher visits Classroom visits are a client-centered and cost-effective means of stimulating, exchanging and testing new ideas. The most developed category of classroom visits were those among OSC teachers with the specific purpose of sharing ideas about the teaching the OSC material. While these were unstructured observations and no written records of these visits were kept, teachers confirmed their value. A second category of classroom visits was those (more rarely made) by non-OSC teachers to OSC classrooms (see below). A third category of school-to-school visits was set up by the SELOP and appreciated by teachers while it lasted; however, these visits lapsed for administrative reasons before the end of the project.

Conclusion The project did not build in instruments for systemically measuring change in teacher behavior, so no quantitative data were available. Impressionistically, development at trainer and teacher levels was judged a successful outcome of the project. The final evaluation revealed a high level of commitment permeating the project taskforce, its central training team, and the DOSCAs, who all exercised a range of administrative and pedagogic skills allowing them to develop workshop structures, procedures and teaching aids (e.g., tasksheets). We suggested that these skills should be capable of sustaining the program well into new decentralized era of District-level in-service provision. The impact of the DOSCA system as a model for Ministry planning has already been mentioned. However, although OSC, with its privileged status and new textbooks, was generally acknowledged as 'special,' and although many head teachers saw OSC teachers as potential change agents, the impact of the OSC on teachers outside the project seemed to have been minimal. Head teachers and 'outside' teachers explained that it is not easy to introduce 'participatory' methodology without the appropriate materials. While simple-language materials were essential to the language-learning aims of OSC, the 'OSC model,' based on a series of subject-specific, prescriptive, printed texts, may have implied to 'outside' teachers a restricted notion of 'participatory methodology.'

SELOP: Student Outcomes and Reflections on Future Directions

Project activities consisted of (a) the production of the OSC materials, and (b) the associated teacher development program. These have been discussed above. Student outcomes associated with these activities will be discussed below.

Student Outcomes

Students' attitudes were not systematically investigated. The data were impressionistic and anecdotal and suggested that OSC lessons were generally found by students to be enjoyable and motivating. How much this was an effect inherent in the materials and associated methodology and how much it was due to the perceived 'specialness' of the new program (as mediated by newly enthused teachers) was impossible to say.

The program design included two types of test: program-specific achievement tests and the national (Zanzibar) Form II examination. Valid measurement of student performance was difficult in that, until the final year of the project, the OSC draft materials were constantly evolving and being rewritten in response to feedback. New tests were also developed in the course of the program. Longitudinal comparisons were therefore not possible. Test results for one cohort that might (with all the usual caveats) be attributable to one particular draft of the program materials were not comparable with test results for the following cohort, who might have been tested on an adapted or rewritten version of the materials. Initially, the tests used were those administered at the beginning and end of a year of the ZELIP reading program (in common with the mainland ELTSP). These were the Edinburgh series of simple cloze tests, approved as a reliable and valid measure of learning on the basis of an item analysis carried out at the University of Warwick, in connection with an evaluation of the parallel, mainland ELTSP reading program (Rea-Dickins, 1996). Under the SELOP, these tests were rewritten (as pieces of connected prose rather than as a series of 'isolates') and supplemented with four others based on the four OSC subject areas. The rewritten tests were later collapsed into a single, one-hour, integrated test. This was designed as a skills-based test and was used as a post-OSC achievement test, administered only at the end of each academic year. Thus, no baseline figures were provided. The tests did, however, yield some demographic data, which showed that (predictably) urban students outperformed rural ones, and that (less predictably) rural students on Pemba Island achieved higher scores than those on Unguja Island. Overall scores were extremely low (on the new cloze test0, from 28.8% down to 3.8%), suggesting that the tests were too difficult. These 'local' tests were supplemented in 1997 by an analysis of the national (Zanzibar) Form II examination results. This analysis showed significantly higher scores for the 20 project schools, but, as very little of the OSC course was in fact available at that time to the relevant cohort (in their Orientation Year during l994), the findings might possibly be explained by teacher and student enthusiasm arising from the initial novelty of the project (the Hawthorne effect). One final aspect of testing should be mentioned: Some schools set their own midyear tests. This subverted the purposes of the SELOP in that the content-focused questions encouraged pre-test cramming.

Resource Outcomes

The SELOP represented a considerable achievement. The OSC materials were the tangible outcome: 10,000 copies of each coursebook were printed, 6,000 to be shared by 12,000 pupils and 4,000 to sustain the program over the following years. Continued supply of the extensive reading books was covered under Zanzibar's educational plan (ZEMAP, 1996), as was provision of training for both these elements of the Orientation Year program.

As for the processes of materials production, the balance between local and outside expertise is an important issue. It was impossible to judge the value of time spent and experience gained, pre-SELOP, on locally produced but then discarded drafts. SELOP's alternative, with its centralized approach and provision of outside-author input, presented its own problems. Financial and other considerations led donors to employ VSO middle school teachers who were young, enthusiastic, and had a feel for 'participatory' classroom interaction and related classroom activities, but who had little experience of Zanzibari classrooms. In 1996, classroom links were strengthened through the recruitment of a materials writer from Pemba, and classroom piloting mechanisms were streamlined through the appointment of smaller teacher teams of eight on both islands, specifically tasked with supplying feedback. However, feedback tended to be restricted to two areas: (a) reproduction and clarity of illustrations, and (b) timing or pacing issues. The pacing issue was important. As discussed above, the natural tendency for teachers was to 'drag' rather than 'stretch' content coverage, but the writing team increasingly recognized that teacher confidence was closely linked to clearly defined and clearly limited coverage within each unit. Supplementary components led to anxieties that seemed associated with 'test dread,' on the assumption that, ultimately, the unit objective was to cover content rather than to impart generic skills (see comments, below, on test development). The feedback teams rarely commented upon substantive problems of content or language. The importance of feedback instrument design is emphasized here. Teachers might have been more willing to make quick annotations or suggest substitutions on draft copies than to complete more time-consuming feedback forms that called for explanation and its associated metalanguage.

The VSO teachers were untrained in the technical skills of textbook design and textbook writing. The first coordinator appointed (in 1994-1995) was also untrained in ELT. This presumption may smack of gross "native-speakerism" (Phillipson, 1992), but would convey an unduly negative picture of the writing process, which did, as emphasized above, result in imaginative and largely appropriate materials. It also meant, however, that, in the later stages of the project, a large proportion of the (ELT-trained British-expatriate) coordinator's time was spent on final editing and preparation of camera-ready text. Nor should the additional logistical problems associated with a small-scale, local, publishing enterprise be overlooked: arrangements for illustration, for printing, for distribution, not to mention supply problems (in obtaining writer reference materials for writers, for example, or upgrades for word-processing packages, or paper).

Human Resource Outcomes and Capacity Building

As outlined above, the SELOP was instrumental in developing expertise in a range of areas: classroom teaching; teacher training; school administration; materials design, writing and editing. One important area remained less developed: teachers'

English-language proficiency. At the time of the SELOP evaluation report, the Ministry was considering introducing incentives (certification leading to salary increases, possibly with release time) as a reward for individual skills upgrading. Tanzanian teachers are very poorly paid and most are forced to take additional jobs; their career prospects are extremely limited. The privileging of OSC teachers through provision of extra training and new books seemed unlikely to remain a permanent source of motivation. Salary increases could help counter the generally low levels of teacher morale. In addition, new Teacher Centers, developed subsequent to our evaluation, seemed ideally placed to support language improvement. It is likely that such support, however, would lead to non-teaching careers for a number of teachers, and this would need to be taken into account by planners. A second area neglected by the SELOP was the development of testing and monitoring skills. The backwash effects of inappropriate tests have been mentioned above. Finally, although the SELOP impinged (if rather indirectly) upon parents and the wider community, and although the project took into account, where appropriate of school committees, there was less emphasis in this project than in others reported in this book on community responses, community relations, and community project input.

The Future

The OSC became firmly embedded in the school system. The 1996 draft *Education Masterplan* (ZEMAP) included a firm commitment to "consolidating the achievements of SELOP and ZELIP" and to "sustaining the two projects' objectives beyond the year 2001 through planned in-service training of teachers, with supervision and support to schools." We concurred that further drafts, to develop simple language-focused materials for selected subjects at the lower secondary level, would build upon the SELOP's foundations and would, along similar lines, provide an integral training for teachers of other subjects, enhancing both their language skills and (through the teaching materials) their methodology. This strategy would reflect Ministry policies "to improve the quality of education ... through provision [sic] of adequate and appropriate instructional materials" (ZEMAP, 1996). It would also address the commonly voiced concern that, following the Orientation Year, pupils 'lose the language' in an environment where English-language input is less carefully structured and where teachers concentrate on 'giving out the facts without the medium.'

At the time of our evaluation, the long-term future of the OSC remains uncertain. Some officials at the Ministry recommended that an intervention such as the OSC should eventually eliminate the need for itself and, thus, lead to its own withering away. Depending upon the success of the Ministry's present policy of strengthening primary-level English, it is possible that Zanzibar's compulsory basic education phase may eventually be reduced from 10 to 9 years, resulting in a 7-2-2-2 model, with no space reserved for the Orientation Year. Others suggested that the OSC, or a similar bridging course, should eventually be delivered in primary Standard VII (a suggestion with rather radical policy implications, with Kiswahili-medium instruction at present a cornerstone of Tanzanian primary-level language policy). Meanwhile, reliance on strong forms of immersion was being questioned, and it seemed possible that alternatives could emerge. One possibility (apparently not seriously considered when the OSC was planned but advocated within the Zanzibar Ministry of Education's *Masterplan*) might be the integration of bilingual approaches within and across the education system, possibly by means of a more gradual, phased introduction of English-medium instruction (as, for example, in Brunei). This could possibly be implemented by establishing an agreed-upon specific role for Kiswahili within the English-medium classroom. However, although it is possible that, within Zanzibar's 'small system,' the value of such an approach would be agreed upon, it is unclear how far Zanzibar could move away from mainland policy on this important and very controversial issue.

9

The Aga Khan Foundation Experience Compared with Emerging Alternatives to Formal Schooling

Joseph P. Farrell

My task in this chapter is to situate the various case studies contained in this book within the context of what we are learning on a broad scale about educational change, or lack thereof, throughout the 'developing' world.[17] The case studies represent a rich and instructive addition to the growing body of knowledge about how we may improve education, and more fundamentally, *learning*, in economically poor nations. In Section A, I provide my understanding (I emphasize that this is *my understanding*, as there is considerable disagreement in the policy-related and theoretical literature about what we can make of the accumulated experience) of the growing body of international, comparative experience. In Section B, I apply that interpretation to the particular set of cases in this book.

Section A: The Comparative Experience, an Interpretation

Starting about 30 years ago, and gaining momentum over the past decade, there has been a quiet revolution in schooling in the 'developing' world, that is in many cases radically transforming the 'forms' of formal schooling as we have come to know them, particularly at the primary level. These mostly successful radical alternatives are little known and seldom remarked upon by educators and scholars of education in the 'developed' world, and are often not known, or if known not well understood, in their 'home' nations. A small published literature about them is beginning to grow, but it consists mostly of individual case studies with few references to other 'cases,' or comparative analyses of small sets of such cases, generally from the same geocultural region (e.g., Reimers, 2000, and this book). But this literature is still scanty, fugitive, and hard to find, and much of the information consists of 'lore' passed informally among practitioners and scholars. Over the past few years, I have been working, with a team of graduate students,[18] and local and international colleagues, to systematically gather information about these alternative models or programs, and to begin to try to make sense of them, to 'analyze' them. We now have over 100 'cases' in our database and are in the midst (early stages yet) of that broad comparative analysis. This analysis provides the backdrop for the African case studies in this book. But, in order to make some meaning of these attempts at radical change of the 'forms' of formal schooling, we first need to consider how those forms themselves arose, how they have changed (or not), and what they now are.

The 'Forms' of Formal Schooling: The 'Bad News'

The History

We need to remember in our discussions that 'schooling' as we have come to know and understand it, is only one of a vast array of social institutions that humans have invented through the millennia to provide opportunities for young people to learn. It is a human invention of relatively recent origin, at least on a mass scale. The broad-scale provision of education as an instrument of statecraft and state development, through schooling, was effectively 'invented' in Prussia after, and as a result of, the Napoleonic invasion of that nation; it spread quickly throughout Europe and other relatively wealthy nations of the time, and more gradually across the world through borrowing or colonial imposition. But, in the broad sweep of history, it is a quite new social institution. In its fundamental forms (hence the term 'formal' education) it was set by the experience, attitudes and understandings of the mid-to-late-19th-century elites of the then newly industrialized nations. Those basic 'forms' have spread around the world (Fuller & Rubinson, 1992; Fuller, 1991) and have hardly changed, even in the wealthiest nations, in the past century (Tyack & Cuban, 1995; Ravitch & Vinovskis, 1995).

As I will argue in more detail below, those fundamental 'forms' of primary education are also, given what we now know about how children best develop and learn, antilearning and antidevelopment, and the most promising experiences of primary education, in terms of *learning*, are those that alter the basic patterns in very fundamental ways. 'Education' also packages, authorizes and legitimizes only certain forms, categories and styles of 'learning' (this is what curriculum setting is about). And those decisions about what is 'real' or 'important' learning are typically made by people who are very far removed (geographically, socially, and experientially) from the lived reality of many, if not most, children. We also know that there is a great divide between what 'education' teaches, and what children actually learn from the experience of schooling (Farrell, 1997, pp. 300-303). Beyond this, we also know that children learn much of great importance to them outside of the walls of the school. In this sense, a 'hidden cost' of participating in 'formal schooling' is sometimes the loss of opportunity for other forms of learning (Robertson, 1984).

We now know from the anthropological and learning theory literature that, as Hall has said (Hall, 1986), the 'learning drive' is almost as powerful as the sex drive in motivating human behavior. Children, and adults for that matter, do not have to be coerced or forced, or motivated into learning, except in highly artificial settings like 'school.' Learning is what they do naturally if given the opportunity. That is not to say that all learning is easy. It often involves very hard work. But, if you have ever watched an individual child or a group of children actively engaged in trying to learn a new skill or master a 'body of knowledge' that is really important to them, you will know how much time, energy and effort they will put into trying to 'get it right.' As adults who are working on something that really interests and engages them know, when you are in that condition, 'hard' becomes 'easy' and the distinction between 'work' and 'play' disappears.

The Possibility of Change

In a recently published article, I ask the basic question here: "Why is educational reform so difficult?" (Farrell, 2000). On this question, we have a vast amount of information and experience, from nations rich and poor, that has been accumulated over at least the last 50 years, in some nations over the past century. Several major 'state of the art' papers have been published in the past decade or so attempting to summarize and make sense of what we might learn from this experience (e.g., Caillods, 1989; Fagerlind & Sjosted, 1990; Shaeffer, 1986; Davies, 1996; Farrell, 1997; Fullan, 2000; Levin, in press). Beyond that published literature, there is a great amount of more fugitive material, and much 'lore' (the experience-based knowledge and 'tales' of those working actively 'on the ground' in trying to get educational change working). All are valid forms of 'evidence.' Unfortunately, that general experience is not particularly hopeful. In 1997, I summarized this body of literature as follows (Farrell, 1997, p. 298).

> One general lesson is that planning educational change is a far more difficult and risk-prone venture than had been imagined in the 1950s and 1960s. There are many more examples of failure, or of minimal success, than of relatively complete success. Much more is known about what does not work, or does not usually work, than about what does work. ... Moreover, when planned educational reform attempts have been successful, the process has usually taken a long time, frequently far longer than originally anticipated. In recent decades there have been a few examples where an unusual combination of favorable conditions and politically skilled planners has permitted

a great deal of educational change in a relatively brief period, but these have been rare and idiosyncratic.

All of the broad surveys noted above come to roughly the same conclusion. For example, a very recent review of educational reform efforts in Latin America observes, " ... there is a high degree of failure of reform plans. We don't know what factors favor the implementation of reform plans, nor what conditions or institutional capacity are needed for the reform plans to affect wide sectors of society" (Alvarez & Ruiz-Caseres, 1997). Centrally driven, top-down, elite-defined and command-oriented educational change programs rarely work, and, if they do, they typically create resistance and resentment among teachers, which makes it even more difficult to 'push through' the next wave of change. This has been the case for a very long time, even in the richest nations. Two recent major books from the United States, that both analyze the history of failed educational reform in that nation over the last century, come to the same conclusion. The vast majority of the various and numerous educational reform movements and campaigns in that nation either (a) were never effectively enacted, (b) if enacted, were never implemented, except in some cases in small pockets of schools, (c) if implemented more widely, were not implemented well, and (d) if (in very rare cases) implemented widely and well, had little or no discernible impact on the individual learning of most of the 'target' students, and no long term effect on their life chances later, either as individuals or as members of marginalized groups (Ravitch & Vinovskis, 1995; Tyack & Cuban, 1995). Over a century in that richest of nations, it was all essentially wasted time, energy, and money. (And because the US is a very rich nation, one is talking about, by other nations' standards, *huge* amounts of wasted time, energy and money). Recent reviews of the massive, and very expensive and disruptive, educational reforms in England come to the same general conclusion (Peters, 1998). What went so terribly wrong?

The world literature I have noted above provides a variety of possible explanations, rooted in a wide array of differing 'theoretical' views. Among the many different axes of disagreement, two basic, and starkly different, points of view on this question are captured in the two books on US educational reform noted above. Both trace out an essentially identical historical account of the mostly failed educational reforms in that nation. (Some differences in details of the account can be found, but they are not ultimately important to the overall story.) But their diagnoses of the nature of the problem, and hence their proposed prescriptions for its cure, are fundamentally opposed, and reflect almost perfectly a basic disagreement in the worldwide debate on this question. In the Ravitch and Vinovskis view of things, the problem is with *teachers*: they resist and reject the reform proposals and projects advanced from the 'centers of knowledge': the universities, the research centers, the foundations, the national governments, the international agencies. The teachers resist, and subvert (or in the most benign interpretation, do not understand), the change proposals of what Tyack and Cuban call the 'policy elite.' In these authors' view of things, the problem is precisely the opposite. It is the 'policy elite' who consistently get things wrong, oversell reform propositions, exaggerate possible outcomes, and ignore, or try to coerce into compliance, teachers. In their view, effective change will only occur when the 'policy elites' 'enlist and honor teachers.' All of the international evidence of which I am aware suggests to me that, on this fundamental argument, Tyack and Cuban are right. "Change where it counts the most – in the daily interaction of teachers and students – is the hardest to achieve and the most important. ... To bring about improvement at the heart of education ... will result in the future more from internal changes created by the knowledge and expertise of teachers than from the decisions of external policy-makers.' They, thereafter, exhort those policy elites to 'enlist and honor teachers as the key people in reforming schooling.' (Tyack & Cuban, 1995, pp. 10, 134). Perhaps their most important 'lesson', for our purposes, because it matches well with successful educational change programs in developing nations (that I will outline below) is that, if change is to happen at all, it must be at the level of the classroom, not at the level of national policy.

What are Those Unchangeable Forms of Formal Education?

This evidence and interpretation is all very well, but at the end it misses (or at best glances off) a more fundamental problem. As I have suggested above, the basic 'model' of schooling that we have all come to accept as 'normal' is fundamentally antilearning. If that claim is true, then no amount of internally driven or externally supported change or improvement in one or another aspect of schooling can really change the learning opportunities available. That standard model, derived as I noted above from late-19th-century understandings among the then elites, generally comprises, around the world, the following basic elements.

- One hundred to several hundred children or youth assembled (sometimes compulsorily for at least a period of time) in a building called a school
- from approximately the age of 6 or 7 up to somewhere between the ages of 11 and 16
- for 3 to 6 hours per day, where
- they are divided into groups of 20 to 60
- to work with a single adult (a 'certified' teacher) in a single room
- for (especially at the upper 'grades') discrete periods of 40 to 60 minutes, each devoted to a separate 'subject'
- to be 'studied' and 'learned' in a group of young people of roughly the same chronological age
- with supporting learning materials; for example, books, chalkboards, notebooks, workbooks and worksheets (and in technical areas such things as laboratories, workbenches, practice sites), all of which is organized by
- a standard curriculum, set by an authority level much above the individual school (normally the central or provincial or state government) that all are expected to 'cover' in an 'age-graded' fashion.

'Adults', assumed to be more knowledgeable, 'teach' and students 'receive instruction' from them in a broader system, in that the 'students' are expected or required to 'repeat back' to the 'adults' what they have been taught, if they are to go any 'higher' in the educational system. Teachers and a central examination system(s), or both, evaluate students' ability to repeat back to them what the students have been 'taught' and provide formal recognized certificates for 'passing' particular 'grades' or 'levels.' Most or all of the financial support comes from national or regional governments, or other authority centers (e.g., churches) well above the local community level.

There are a variety of explanations (or 'theories') regarding why and how this particular way of trying to deliver opportunities to learn to young people on a large scale has become well-nigh universal (see Fuller, 1991 for a useful summary). However, a careful examination of the cross-national literature from anthropology (Hall, 1986) and learning psychology[19] regarding how young people and adults actually best learn suggests that this 'standard model' of schooling and education is inherently ineffective and inefficient. People of whatever age simply do not learn well under these arrangements. These traditional, but now nearly universal, patterns of 'education' are an artifact of the misconceptions of a different place and a different time. But, now that we have them, we cannot easily get rid of them. In their analysis of the problems with US educational reform, Tyack and Cuban suggest that a major part of the explanation for the failure of reform is that there is a standard 'grammar' of schooling, a set of expected patterns we have historically constructed regarding what a 'real school' is. Anything that deviates substantially from that 'real school' or 'grammar-of-schooling' image will, by their analysis, be resisted by teachers, parents and students.

There are a number of things central governments *can* do with respect to education: build schools, train, employ and pay teachers for them, supply furniture, textbooks and other learning materials, alter curriculum documents and so on. But, even in nations that are wealthy enough to do all of these things reasonably well, there are still two central problems: (a) these mechanisms do not alter the traditional model; indeed, they reinforce some of its key elements, and (b) neither these, nor other policy instruments available to central governments can alter the 'core' of the process, the teaching and learning that occurs in thousands of classrooms. Change at that level is *not* what central governments or outside agencies do well. The international record of educational reform attempts over the past several decades shows quite clearly that change at this fundamental level rarely, if ever, occurs as a result of centrally driven, 'top down', 'decree-and-regulation-driven' change models. Central governments can 'provide space for' change at this level; for example, by providing resources for teachers who want to experiment with new ways of doing their work, or by loosening controls and regulations and examination systems that often tie teachers to traditional practices, but they cannot effectively *command* it.

But there are, nonetheless, many quite good schools out there, places where, somehow, individual teachers or small groups of teachers have found ways to break the traditional model and create a place where learning thrives, often in some remote village or in a corner of a desperately poor urban slum. Stumbling upon one of these schools is like discovering a beautiful flower thriving in the midst of desolation. The problem has been for many years that these remain isolated examples. As a recent review by the United States Agency for International Development (USAID) of its efforts to support educational reform in Africa indicates, "There is an ultimate irony to education. Good schools and good teaching can be found in any education system, sometimes under very adverse conditions. The problem is that they cannot be found everywhere. The challenge confronted in supporting education reform is exactly that: how to help good practice occur on a larger scale." (Crouch et al., 1997, p. iii) Or, as Fullan recently noted (with respect specifically to School-Based Management [SBM] but the point is generic), "One final, and absolutely critical caution. There is a fatal flaw in the research. Even the best research on SBM identified factors and conditions associated with success, *it does not tell us how to establish those conditions when they do not exist* [italics in original]. Put another way, research portrays successful cases once they are 'up and running' (and in a very few cases how they evolved); it provides little insight on how to get there." (Fullan & Watson, 1999, p. 11) Fortunately, we are beginning to accumulate examples from around the world of how that challenge can be met. This is the good news.

An Emerging Alternative: The 'Good News'

As noted in the introduction to this chapter, there are a growing number of examples in developing nations of programs that have fundamentally altered the teaching and learning process in primary schools that serve very poor and marginalized children (sometimes rural, sometimes urban), that have dramatically increased the percentage of poor children who enroll in school and stay in school, and that have produced learning results (as 'measured,' for example, on standard national tests) among these children that are as good as (in many cases better than) those of children born into more favorable circumstances. That is, by fundamentally changing how teaching and learning occurs in the classroom, they are successfully providing opportunities to learn well for children who have previously been thought to be the hardest to reach and the hardest to teach. Some of these programs are still at a very early stage of development and experimentation (perhaps involving from 20 to a few hundred schools). Others have grown to large systems involving thousands or tens of thousands of schools.

It should be clearly understood that these 'revolutionary' new ways of providing learning opportunities to the most marginalized children are almost all very new. Even the 'oldest' of the programs noted below have only 20 to 30 years of history (depending upon what one considers to be the 'starting point'). And more are 'springing up' or 'being invented' every year. Because of this newness, there is not an organized and clear research and evaluation literature about these programs that is easily available to an international audience. That is our own research agenda. But, even at this early stage, there are a few things that are becoming quite clear in terms of 'lessons we are learning,' and other matters that are still 'best guesses.' And there are some questions that do not at this time have any clear answer at all. This general situation indicates that we are on the beginning of a steep learning curve, which suggests that we have to take a very experimental and flexible approach to this level of educational change; nationally set regulations and commands, universal supervisors' checklists of what constitutes 'good teaching', or national and international 'test scores' will not serve the process well. But, we should not let the lack of complete knowledge keep us from taking advantage of what we *do* know.

The International Experience Base

As noted, there are many alternative primary schooling programs around the developing world. Below, some of these programs from our database are listed. This is not meant to be a complete or exhaustive list; rather, a few cases or examples are cited that (a) indicate that these kinds of changes are being successfully introduced in widely differing cultural communities, and (b) are reasonably well documented. There is not space here to provide a description of each of these examples, but documents that provide such descriptions are found in the References section at the end of this book. What I do provide, following the list of examples, is a set of characteristics that many, if not most, of these alternative programs share, and a list of what appear to be major 'lessons learned' thus far.

Some selected examples

- The *Escuela Nueva (New School) program* in Colombia is, in many ways, the 'grandmother' of many of these alternative programs. It is also the most thoroughly documented and evaluated. It has been implemented in over half of the rural schools in Colombia (over 20,000 schools) and has been adapted for use on a small-scale or large-scale basis in at least 10 other nations in Latin America. (Colbert & Arboleda, 1990; Schiefelbein, 1991; Psacharopolous, 1993; Arboleda, 1994; Siabato, 1997; McEwan, 1998; Miwa, 2000; Pitt, 2001).
- The *Non-Formal Primary Education program* of the Bangladesh Rural Advancement Commission (BRAC). This is another of the 'grandparents' of this alternative school movement. It has been successfully implemented in thousands of villages in Bangladesh. It started in 1985 with 22 schools. By 1999, it had expanded to 35,000 schools in one-quarter of the villages in the nation and had recently been adapted for urban children, particularly children working in the garment industry. Its pedagogical model has spread to other nations in the region. It originated and was developed by an NGO (BRAC) but key elements of its model are now being adopted by government schools (Ahmed, 1993; Sarkar, 1994; Scott, 1996; Sweetser, 1999; Haiplik, 2001).
- The *Escuela Nueva Unitaria program* in Guatemala. This program started with 100 government schools in 1992, under the initial leadership of one of

the 'founders' of the similarly named Colombian program. By 1997, it had spread to over 1300 schools, privately funded by charitable foundations (Kraft, 1997).

- The *Fe y Alegria (Faith and Happiness) schools* in Latin America. This program, aimed at children 'where the pavement ends,' started in Venezuela in the late 1950s, with a very few schools, and has slowly spread to well over 500 schools in 12 Latin American nations (Swope & Latorre, 1998).
- The *'multi-grade program'* in Guinea, that started in 1991 in 18 schools and has now spread to almost 1,300 schools serving over 40,000 students (Bah-Layla, 1998).
- The *'multi-grade program'* in Zambia, that started in 1985 with four schools and has now spread to about 150 schools (Kelly, 1998).
- The *Convergent Pedagogy program* in Mali, which grew out of experiments that started in the late 1980s and has now grown to close to 700 schools (République de Mali, 1998).
- The *Community Schools initiative* in Zambia, that now involves more than 200 schools serving about 25,000 children (Kelly, 1998).
- The *Community Schools program* of UNICEF - Egypt. This program, aimed particularly at girls in remote hamlets in Upper Egypt, started as a small pilot in 1992. It has now expanded to about 200 schools. Of particular interest is that a process has recently begun to diffuse key aspects of its pedagogical approach into the government's One Room School program, which aims to establish 8,000 small schools in educationally underserviced (for girls) areas of rural Egypt within a few years (Zaalouk, 1995; Hartwell, 1995; Farrell & Connelly, 1998).
- The *MECE RURAL (Improvement of Educational Quality and Equality in Rural Schools) program* in Chile. This program is now present in over 3,000 rural schools and has been adapted for use in several other nations in the region (San Miguel, 1996).
- A *network of 'education for production' programs* in Latin America that are providing basic education and productive self-employment skills for youth in poor urban neighborhoods, and, in some cases, in rural communities (Corvalan-Vasquez, 1988).

As noted, these are just a few of the representative and best-documented examples of such programs. They are not all identical. Each is, as it should be, adapted to the local conditions and culture and to the history of educational thought and practice in the nation and culture in which it works. But, even in these early days of our analysis of this set of cases, these examples do tend to share some important common features. Typically, they have some or all of the following characteristics.

Some common features

- Child-centered, rather than teacher-driven, pedagogy.

- Active, rather than passive, learning.
- Multi-graded classrooms with continuous-progress learning.
- Combinations of fully trained teachers, partially trained teachers and community resource people (parents and other community members) are heavily involved in the learning of the children and in the management of the school.
- Peer tutoring: Older or faster-learning children, or both, assist and 'teach' younger or slower-learning children.
- Carefully developed self-guided learning materials, that children, alone or in small groups, can work through themselves, at their own pace, with help from other students and the teacher(s) as necessary. Children are responsible for their own learning.
- Teacher- and student-constructed learning materials.
- Active student involvement in the governance and management of the school.
- Use of radio, correspondence lesson materials, in some cases television, and, in a few cases, computers.
- Ongoing and regular in-service training and peer mentoring for teachers.
- Ongoing monitoring, evaluation and feedback systems, allowing the 'system' to learn from its own experience, with constant modification of, and experimentation with, the methodology.
- Free flow of children and adults between the school and the community.
- Community involvement that includes attention to the nutrition and health needs of young children long before they reach school age.
- Locally adapted changes in the cycle of the school day or the school year.
- A focus for the school that is much less on 'teaching' and much more on 'learning.'

While these school change programs do not all share each and every one of the features noted above, they do appear to have a few fundamental elements in common, that provide us, even at this very early stage, with some basic and important 'lessons learned.'

Some Lessons Learned

It is instructive to compare the list of characteristics just above with the list of features of the 'traditional' school model provided earlier and with the 'image' we all carry around in our minds of what a 'standard' school looks like and what happens there. The most important thing that these alternative programs teach us is that the traditional model can be broken, on a large scale and generally at a per-pupil recurrent cost that is no greater than, and often less than, that of the traditional model. These programs demonstrate that child-centered, active pedagogy, with heavy involvement of parents and the community in the learning of the children 'works.' It *can* be done, and, where done, it generally produces remarkable learning gains among even the poorest and most 'disadvantaged' children. It is extremely important to note that these change programs do not simply alter one feature of the 'standard traditional school' (for example, changing one part of the curriculum) or strengthen one or a few parts of the standard schooling model (for example, adding more textbooks or improving teacher training) or provide a small amount of extra local money to the school. Rather, they represent a thorough reorganization and fundamental revisioning of the standard model of schooling, such that the 'learning program' for the youngsters, although occurring in, or based in, a building called a 'school', is far different from what we have come to expect to be happening in a 'school' and far more effective than what we have typically seen in even very 'good' schools, even for very well-off children.

These programs also demonstrate that, contrary to a very popular belief around the world, teachers are not obstacles to fundamental school change: When it happens, they are the promoters and agents of such change, even when they are working in very difficult situations, are not formally very well trained, and are very poorly paid. They, like the equally disadvantaged children in their charge, can accomplish remarkable feats of learning and change, in quite a short time, under these alternative primary education programs. There is an important parallel here between children and their teachers. Just as the success for children seems based fundamentally on a focus on 'learning' rather than 'teaching,' so too do the changes in teachers. These successful change programs typically spread (or are 'taken to scale') not by a centrally planned and commanded 'reform plan,' with goals and objectives set from afar and agents or supervisors from the national center, or perhaps a regional university, going out to 'teach' the teachers about the latest new educational scheme. Rather, they spread by an *innovation-diffusion* process: Teachers learning from other teachers, sharing their knowledge and skill with other teachers, and, together, exploring how their shared and growing knowledge and experience can help them all, together, experiment with ways to improve their ability to serve the children in their care. It should be noted, however, that administrative *leadership* does play a critical role in these programs. The vast and recent literature on school improvement or school effectiveness, mostly from and about 'developed' nations, places much emphasis on the importance of *school-based* leadership. This does not generally apply directly to these cases, as, typically, the schools involved are too small for the roles of 'principal' or 'head teacher' to have any real meaning. But there *are* people who carry out the pedagogical leadership function, who provide initial formation and in-service mentoring of teachers, who arrange regular meetings of teachers from neighboring schools to discuss problems and share ideas, who travel from school to school helping to solve problems and spreading innovative ideas and practices. At more senior levels, there *are* people who design and produce learning materials, secure and manage funds, and carry on all the other standard tasks of management and administration that a 'system' requires. And, at the 'core' there is typically a single individual or a small team who provide 'ideational' and 'inspirational' leadership. The important difference from the traditional model is that, in these cases, the leadership and management is carried out in a fashion and spirit that enable and provide space for teacher learning and teacher-driven change, rather than attempting to control and regulate teaching, which, in turn, encourage and develops teachers' capacity to be creators of their own knowledge rather than simply consumers and deliverers of somebody else's knowledge, and thus enables teachers to be agents of change.

What of the role of government itself, and 'policy'? Some of these programs have grown under government sponsorship, others have grown outside of government sponsorship or control, and still others have been, or are, working well under various forms of combined sponsorship or ownership. This is a critical issue, from the experience at hand. Government agencies and bureaucracies have a predictable and natural tendency to want to command, decree, regulate, control, supervise, organize, and generally keep administratively tidy all things that fall within their official jurisdiction. That is part of what has led to the standard model of schooling: Whatever else, it is predictable, controllable to a degree (as long as one does not ask what children are actually learning), and potentially (if not actually) administratively tidy. In contrast, the programs discussed here are somewhat anarchic, unpredictable, uncontrollable, and constantly changing as they learn from their own experience. This is typically very unsettling to the hired hands of a bureaucracy intent upon control, regulation, and standardization. The first role of government in these cases is to get out of the way, to loosen control and regulation. The second role is to provide space for, and indeed encourage, a healthy degree of anarchy, frightening as that may be to those who comfortably inhabit offices in the national capital or its regional outposts (and often to officials of international donor agencies who also require predictability and accountability before they will release funds). If you are seriously interested in promoting learning rather than teaching, the last thing you want is predictability. All of this requires a massive attitude and behavioral change among those in positions of political and bureaucratic power. What is perhaps most remarkable about these cases of successful major educational change is that, somehow, they have managed to find or develop that bureaucratic attitude change, to 'get past' or 'get around' the power of the traditional bureaucracy. Investigating how exactly this has been accomplished is a major task before us now.

There are a few other 'lessons learned' worth noting more briefly from this experience.

No need to coerce children Children do not have to be 'forced' or 'coerced' into learning. It is what they do naturally, indeed what they are genetically compelled to do, if given the opportunity. In the late 1980s, I was in a small village in rural Colombia visiting yet another of the early examples of *Escuela Nueva*. I was chatting with a group of the parents of the children in this 'new school.' I asked them what they saw as the main difference between the experience of their older children who had attended and mostly 'failed' in the traditional school that had more or less functioned in the village, and the experience of their younger children in the 'new school.' After some mumbling among themselves, one mother replied, obviously speaking for the group [rough translation from the Spanish dialect here], "Look, with my older kids, they hated going to school. I always had to force them out of the house in the morning. Both of them repeated first grade twice. They weren't learning anything, and they were not happy, so finally I said OK, you can go with your father out in the fields and work, at least you'll be doing something useful there. My younger kids, in the new school now? The problem is the opposite. I can't keep them away from school, even when they're sick. And they don't come home in the afternoon; they're in the school working on things. I have to send my oldest down there to tell them that they have to come home to eat dinner and do their chores, and they complain to me, 'Mamita, it's so much fun at school, and we're doing such interesting things, how come we have to come home and do these boring chores?'" Traditional schooling restricts and tries to channel the learning potential of children; these new alternative programs unleash it.

Multi-grading is superior Multi-grading is not simply a second-best expedient for use when there are not enough children in a school catchment area to support age-graded schooling. It is, in and of itself, pedagogically superior to age-graded schooling; it matches much more closely what we now know about how children actually develop. It permits cross-age peer tutoring and continuous-progress learning. This is actually a very old idea (indeed, almost all 'traditional' societies, from the anthropological data, depend upon older children teaching younger children). And some Europeans caught on to this idea long ago. On the grounds of Oxford University there is an historical plaque that reads as follows:

> Andrew Bell. 1753-1832. It was while serving in Madras in India that he developed a form of schooling where the older pupils taught the younger. When he returned he introduced his 'Madras' or monitorial system as an economical form of mass education. The idea spread, Madras schools appearing in Canada and Australia.

But these schools were soon washed away by the new, industrial society creation of the age-graded school. Almost two centuries later, we are learning that the 'traditional' societies had it right all along, and when they go back to their own original model of learning, the children actually learn well.

ECE is as important as primary school Early Childhood Education, or more properly put, attention to the nurture, health and learning needs of children before they reach formal school age, is as important as, probably more important than, the primary school itself in improving ultimate learning outcomes. Children start learning at birth, and learn continuously thereafter. Nothing fundamentally changes in that pattern of continuous development at the age we have arbitrarily set for starting formal school attendance. And children's development thereafter does not come in the system-convenient 9-to-10-month bursts that the age-graded school year assumes. Continuing education, or lifelong learning, is not something that starts *after* one has completed formal schooling. It starts at birth and ends at death, and formal schooling can, by all evidence, be more a hindrance than a help to its continuance in the crucial years between early childhood and adulthood. But, as some of these programs demonstrate, that dysfunctional pattern can be broken.

Grow programs slowly These sorts of successful learning enhancement programs must be grown and nurtured carefully and slowly. Trying to go too fast can destroy potentially promising innovations.[20] The useful motto is Things Take Time (TTT). Unfortunately, this does not match well with the frequent need of governments to have 'demonstrable results' in the short term, nor with the 'short term project funding' rituals of most donor agencies. If one is, as a government or a donor agency, to be in the *learning enhancement* enterprise, one must be in it for the long haul or not at all. There are no large-scale quick fixes. What we do have is an increasing number of slowly and carefully developed long-term successes. Those are what we must learn from.

There are two other points worth noting that arise from at least some of these successful cases. They are, in my view, less well established but sufficiently important to warrant further thought.

National high-stakes examination systems (e.g., primary-school-leaving examinations) do not *necessarily* have the frequently observed 'backwash' effect that mitigates against child-centered teaching. Some of the successful alternative programs described above have grown and thrived within educational systems where such examination systems exist, and they have demonstrated that children do well on these traditional examinations. Indeed, in such situations it appears to be imperative to be able to demonstrate, to gain parental support and broader political legitimacy, that the children are able to 'pass' the examinations. This appears to be relatively easy to do, given that the children being served are typically from groups for which the traditional model of schooling has failed utterly.

Active child-centered learning requires specially designed learning materials. This is not simply a matter of adding teacher- or student-constructed learning materials to the classroom, although that is important. Even more important is developing new forms of 'textbooks' or 'workbooks' that assume or require that children work independently or in small groups. Traditional textbooks (if they are available at all, not a common condition in much of Africa) are designed with traditional schools and classrooms in mind (which is hardly surprising) and do not serve child-centered learning well. Specially designed texts and workbooks not only support and guide the children's learning; in conjunction with specially tailored teachers' guides, they become key instruments for in-service education and self-instruction for the teachers themselves. This approach was pioneered by Escuela Nueva, but has now spread. As Welford and Khatete (1998) have noted in a recent, and generally negative, review of the impact of Teacher Centers in Kenya,

Obviously both textbooks and teachers are important to improving the quality of education. What we are suggesting is by no means a new idea – that is, developing textbooks and workbooks with which children can engage with a degree of independence. The role of the teacher is not to teach the content, but to support the children in their use of texts (and other learning materials). Our suggestion acknowledges that teachers most commonly teach from the textbook and use it as their curriculum guide. Why not, then, focus on producing textbooks that engage children actively? In-service training would then be straightforward and principally for orientating teachers to the new texts. Such training is immediately relevant and does not require a great deal of restructuring of lessons in the way that six pedagogical principles, 12 teaching skills, and how to do hands-on investigations in science and mathematics does. If teacher reflection is a goal, the text and its accompanying teacher notes provide a concrete focus. An additional attribute of such learning materials is that children can get on with learning when their teacher is absent.

Summary of Section A

I return to a central point: As educators, our ultimate objective is learning and its enhancement. 'Schooling,' as we have come to know it, is but one tool available to use in pursuit of that end, and it has turned out to be a not very useful tool, particularly for poor and marginalized children. Moreover, efforts to improve that tool have turned out generally, no matter how wealthy or poor the nation, to be very expensive failures. And, in the few cases where a group or a nation does manage to 'get it right,' what they generally end up with is a slightly improved version of the standard model with only marginal learning gains, bought at great cost. The standard-change-model, top-down, centrally driven, regulation-ridden grand 'reform' schemes have proven to be about as intransigently ineffective as the schooling system they attempt to 'improve.' In contrast, we now have several large-scale successful examples of programs in very poor nations that are succeeding in significantly improving learning, especially among very disadvantaged children, by adopting a radically different form of pedagogy, and spreading it via an innovation-diffusion process that, inter alia, demonstrates that teachers can and do quickly and easily learn a wholly new understanding of their role and a very different, better way of doing their work in the classroom. It is the contrast between these two sets of experiences that provides a fundamental template against which we can consider the cases presented in this book.

Section B: The Cases at Hand

It should first be noted clearly that the two contrasting 'lists' of characteristics developed in Section A are of different orders of 'truth.' The first list, the 'forms of formal schooling,' rests upon a huge volume of research and descriptive reporting. I can quite safely claim that almost all formal schools throughout the world exemplify all, or almost all, of the characteristics on that list, or vary from them only in small degree. Among the millions of traditional standard schools in the world, there are of course exceptions, but they are small in number and generally modest in their degree of variation from that standard. That, in a sense, defines the educational reform problem. The second list of characteristics, of an emerging alternative model, is a different sort. It is tentative, based, as noted, on a limited amount of research and reporting using information that is often scanty and fugitive. It might best be thought of as a kind of Weberian 'ideal type.' There are programs that include most, if not all, of those characteristics. The Escuela Nueva program in Colombia and its adaptations in Nicaragua and Guatemala, the Community Schools program in Egypt, and some others in Africa and Asia seem to have most of these characteristics, but none have all of them; for example, the Escuela Nueva program does not use distance education as a core part of its work, but some other programs do. Moreover, the distinctions between the two 'models' should be seen not as dichotomous 'either/or' differences, but rather as continuums of difference. Thus, some programs may have more or fewer of the characteristics of the 'ideal type.' And, on any given continuum, programs can be placed toward one 'end' or the other. This way of viewing things can allow us to properly understand the Aga Khan Foundation school improvement programs profiled in this book, what they have accomplished and the difficulties they have faced, within the overall context of educational change in developing nations. Comparing these cases to the 'ideal' template, it is clear that they have some, but far from all, of those characteristics.

Where the Aga Khan Foundation SIPs Match

A core goal of these projects is a *more active and child- or learner-centered approach to classroom work.* As Siraj-Blatchford, Odada and Omagor note in the evaluation of the Kisumu project (Siraj-Blatchford, Odada & Omagor, 1997, p. 17),

> The intent ... is to increase the extent of activity-based learning through greater student participation in activities and group discussions. Children are encouraged to question more and teachers are encouraged to have pupils work in small groups and to make and use materials to support learning. These instructional strategies are referred to by AKF as a child-centered approach to teaching and learning and are central to the school improvement program ....

It should be noted that this is a rather limited version of 'child-centered or activity-based' learning, but it is clear from the case report that even moving that far from the traditional model has been quite difficult. But there has been some progress. From the Zanzibar case (Chapter 8) we learn that

> Pair work in its loosest sense (i.e., pupils working together instead of listening to the teacher) has indeed had a very visible impact on Zanzibar's classrooms

> … Against a long traditional background of talk-and-chalk, this shift of emphasis is in itself a considerable achievement ….

The Mzizima case report (Chapter 3) notes a ‘remarkable’ consensus on growth of teacher confidence in the classroom and observational evidence that teachers are using more interactive teaching strategies. But students reported considerable variations in the degree to which teachers employed more student-centered and interactive methods (no surprise there), and it is noted that the “discussions were mostly teacher-pupil exchanges. Whole- or small-group discussion among students was rare …. While teacher awareness of alternative teaching strategies had increased, there was less progress in implementing these strategies.” The results of the Mombasa case evaluation (Chapter 6)

> … strongly supported the overall finding that teachers and pupils in project schools demonstrated more the child-centered teaching and learning behaviors than teachers in nonproject schools … however … the fidelity (closeness to ‘ideal’ child-centered practice) of teacher implementation remained highly variable in project schools. A majority of project school teachers were still judged highly teacher centered … The proportion of project school teachers who did not engage pupils in activity-based learning tasks was also high (59%).

The report from Kampala (Chapter 5) shows roughly the same pattern. Thus, on this dimension, active child-centered learning is a *basic goal* of these projects (albeit a fairly limited version) and some progress is being made in that direction.

Ongoing and regular in-service training and peer mentoring for teachers is a key feature of all of these programs. Indeed, variation among them on this feature appears to be key. As Anderson notes in Chapter 1, “Teacher development activities are a cornerstone of all of the projects.” Over time, the various projects have ‘covered almost all of the bases’ among standard methods to accomplish this. We see Teacher Centers, school-based workshops, in-class coaching by consultants or master teachers or peers, collegial working groups within schools, teachers visiting each others’ classrooms, and so on. Yet, from the reports noted in the previous paragraph, none of this has seemed to have any great effect, certainly not an effect proportional to the costs of these programs. As Greenland notes in Chapter 2, the shift from teacher-centered to child-centered practice “implies a much bigger change in teachers’ understanding than program resources can easily afford.”

Specially designed learning materials are a key feature of several of these programs. Indeed, such materials, for English language instruction across the curriculum, are the core focus of the Zanzibar project (which is not a whole-school improvement project). While it does not appear from the case report that these materials are designed for self-guided learning, they do have built into them ‘activity-based’ learning. The early-stage results seem highly positive, at least for these teachers and their students. As the report notes, "The positive reception that these materials have received cannot be emphasized too strongly" (Brumfit & Hikmany, 1997, p. 10). However, it is also noted that, at the time of the evaluation, the teachers did not yet understand the methods embedded in the materials well enough to transfer the activity-based approaches to areas of instruction beyond the English language orientation program. Learning materials provision is also a key feature of the Kisumu, Kampala and Mombasa projects, which *are* whole-school improvement efforts. These projects have emphasized and supported the development of teacher- and student-made teaching and learning aids (e.g., manipulatives, visual aids) to support more activity-based learning. The materials appear to be popular and frequently used by teachers. They are, however, 'add ons' or 'supplements' to an otherwise traditional curriculum and set of teaching methods and learning materials. These projects have not gone so far as to work with teachers to develop 'programs' of study and core learning materials that would support activity-based and self-guided learning across the curriculum.

A key and notable feature of the overall Aga Khan Foundation program is an *ongoing monitoring-evaluation-feedback system.* Regular, careful and well-designed evaluations are an integral part of the entire history of this program, and, as one reads through the set of reports chronologically, it is very clear that the 'system' is indeed learning from its own experience (as well, of course, from the experience of others). The resources devoted to ongoing evaluation appear strongly to be 'money well spent.' Both Anderson and Greenland, in Chapters 1 and 2, use the phrase 'work in progress.' This is apt, and the key operative word is 'progress.' While this is true for the AKF school improvement program as a whole, the effort has been much less successful at developing within any of the individual projects an ongoing capacity at the local school or local system level to learn from experience and adjust accordingly. There are some examples in the various case reports of learning from experience, but they do not seem, in any of the cases, to have become ongoing and systemic. At least part of the 'problem' here (perhaps most of it but I cannot say for sure with the information at hand) is that the only sort of 'evaluation' that is deemed appropriate and useful is the highly complex and technically sophisticated sort (using baseline data, technically excellent test scripts, careful tracing of individual students from 'pre' to 'post' intervention states, etc.) that is needed for publishable articles in 'respectable' academic journals but that is generally impossible to achieve in schools like these, in places like these. This appears to lead to a kind of 'technicist paralysis' in that the 'best' does indeed become the worst enemy of the 'good.' The experience of the 'good news' cases discussed in Section A of this chapter indicates that one can, on a regular basis, learn a great deal about what is working and not working, and how to improve, without benefit of the armamentarium of technically precise and sophisticated analytical techniques. What is required is careful observation, close listening, and hard thinking, on a regular basis within the system.

Leadership is discussed above with reference to the alternative model, but it is not included as a category in the 'template' because, as noted, it takes on such different forms and meanings depending on the context. It is important to note that, given that they are working within, rather than trying to fundamentally change, the traditional schooling model, these AKF programs have placed much emphasis on head teacher management development. Given the context, this seems a quite appropriate emphasis, although the 'results' are difficult to ascertain from the case reports, and 'management' is only one part of a potential head teacher leadership role.

Where They Do Not Match

While these cases share some key features of that 'ideal' template discussed in Section A, to one degree or another, obviously there are elements that they do not share. These schools are not multi-graded, which means, among other things, that there is little, if any, room for continuous-progress learning or for cross-age peer tutoring. There are not varying combinations of teachers, parents and community members in the classrooms. Learning materials are provided in several cases (in some cases constructed by teachers or students) but they are not designed for self-guided learning, and they do not permeate the curriculum and its teaching. Students are not actively involved in the governance and management of the school. There is some increasing community involvement, but it focuses on management and funding, not on the teaching and learning core of the school. Nor does it focus on early childhood needs. The standard school day and year cycles are maintained.

The presence or absence of any of these individual elements may or may not have a great deal to do with how the schools in these projects produce learning improvements. Among the successful programs described in Section A, many have gone to some form of multi-grading, but not all. Some actively involve parents and community members in the classroom, but not all. Some use specially designed, self-guided learning materials, but not all. And so through the template list. Again, I emphasize that this template is not at all a kind of supervisor's checklist of required elements.

The question then becomes, "How do we account for what appear to be rather limited 'results' from all the time, effort, and resources invested in these AKF programs?"

The 'Results': A Rough and Ready Cost-Benefit Analysis

Reading through all of the case reports and the overview chapters (1 and 2), it is hard to avoid a sense of communicated disappointment or apology. I have just emphasized the importance of the word 'progress' in the phrase 'work in progress.' But, generally speaking, when we use that phrase (certainly when I do) we mean to imply something like, "We are working hard at this, and there are some positive results, but we sure haven't got it right yet." Throughout the case reports, there is a sense of some progress but those projects have very far to go. There is also analysis of obstacles and patterns that might account for the slow progress or lack of it. The underlying causal model for these efforts is as follows: Change a key set of 'inputs' to a 'whole school,' principally through in-service work with teachers (in some cases, head teachers and administrators) and, in some cases, the provision of new or additional learning materials. This will move teacher behavior toward a more child-centered and active-learning style, which will, in turn, produce improvements in the learning of the children.

We need to keep constantly in mind that improved learning *is* the ultimate objective. Only one of these cases includes a careful study of this objective: Capper, Ogula and Nderitu's (1997) work in Kisumu. Why did this evaluation effort not focus clearly on this ultimate goal? From the information available, the answer appears rather complicated. It does not seem to stem from a lack of knowledge of, or recognition of, that ultimate goal. Rather, it seems to be a problem of the 'technicist paralysis' noted in Section A. The designers of the evaluation specifications and those they hired to carry them out set (explicit or implicit) standards for themselves regarding what would constitute an academically respectable 'learning gain' or 'value added' approach to achievement evaluation; these standards were manifestly impossible to meet in the real circumstances of these projects. Recognizing this reality, they 'abandoned the chase.' I do not mean this as a criticism. I recognize the dilemma they faced. I have 'been there.' I have often received rather scathing 'peer review' critiques from 'methodological-purist' academic colleagues who could not understand that their 'standards' were impossible to meet in the real conditions in which we were working. Again, the 'best' can easily become the worst enemy of the 'good.' Ultimately, the question becomes, "Who is being served: the people on the ground for whom some information (rough and ready, warts and all, as it may be) is better than none at all, or those who do not want to risk committing an academic 'sin'?" Be that as it may, in the Kisumu evaluation (Chapter 4) some learning gains are noticed, but, as always, with caution. "In short, then, there appears to be little achievement advantage to being in a [Kisumu] SIP school until a school has been in the project for at least 5 or 6 years, and then only on performance in English." Moving backwards in my presumed causal model, we find that this lack of advantage should not be surprising, since the changes in actual teaching and learning behavior were modest at best. Going back a step further, to the 'inputs' themselves, we note that the cost, where such data are provided, is extremely high, especially in terms of the capacity of public finances (or family finances) to *sustain* them once the project funding is withdrawn, and for the national educational system as a whole.

I will consider sustainability case by case because it is crucial. I should first note that the original case reports, to which I have had access, contain a variety of attempts at cost analysis, and some attempts at unit-cost analysis or cost-benefit analysis. Undertaking such analyses, for projects such as these, is notoriously difficult for a variety of well-known reasons. Having myself often faced these difficulties, I fault neither the project staff nor the evaluators. Some seem to have 'abandoned the chase' (as above) and others have reported what information they could obtain, with appropriate cautions and qualifications. The Mzizima case is effectively sui generis [of its own kind]. All it really demonstrates is that, if you spend very large sums of money (roughly US$ 100,000 per year by my calculations) over almost a decade, you can produce some possibly important modest changes in teacher behavior that might improve student learning (although this was not systematically assessed).

Some valuable lessons for future projects in the overall program may have been learned, but at considerable cost.

We are not provided with detailed cost data for the Kisumu project, but the report notes that the costs are "fairly high and probably not viable costs for a government agency with limited resources" (Capper, Nderitu & Ogula, 1997, p. 85). For the Dar-es-Salaam project, costs are not given in the chapter, but the original report notes total costs from 1996 to 1999 of US$ 595,300. It also notes that 45% of the expenditure was for an expatriate project director whose role in what success was achieved is identified as crucial. That expenditure is obviously not sustainable.

No cost data are provided in the Mombasa case chapter (Chapter 6). There were exchange visits to Kampala for several personnel, and expatriate professors from Leeds University in England were brought to Mombasa to deliver, through modules, a Master of Education degree program. All of these activities must have been quite expensive in relation to possible student learning improvements. The original project report (Anderson & Nderitu, 1999) notes that, by 1998, the unit cost per year per teacher was US$ 608 and the cost per affected student was US$ 19.49. Those cost levels seem neither sustainable for Mombasa nor generalizable to the entire system. It is, however, noteworthy that this project was not dependent upon an expensive expatriate project director, except for a 6-month 'fill in' between the terms of the first and the second local project directors.

For the Kampala project, we are provided with some general cost figures (Siraj-Blatchford, Odada & Omagor, 1997). They are not broken down by category, but it is noted that there were many overseas study visits, and that several expatriate trainers were brought in. The importance to the overall project of an expatriate project director (from the UK) is noted. All of these aspects must have been quite expensive. Over the 3 years reported, the total expenditure was US$ 560,000. The average cost per teacher involved was US$ 1,485; the average cost per student was US$ 46. Even if one uses only Year 3 figures, after the project was taken to scale, and discounts all previous costs as early-stage 'development' expenses, the costs per teacher are US$ 460 and per student US$ 12.60. These would be absolutely minimal costs. I cannot imagine how such costs could be sustained across the entire Ugandan educational system.

The Zanzibar project had overall costs that are comparatively quite low (but this is, or was, a limited program in a relatively small place). Most of the costs for developing and producing the learning materials were borne by the Ministry of Education. The major donor contributions were for equipment and fuel. A problem is that the equipment, and the materials themselves, will eventually wear out, and, if replacement costs are not provided for in routine local budgets (unlikely, given the state of the local public finances), the results of the original donor investment will suffer the same fate as many externally supported, learning materials provision programs over the last decades: Once the originally supplied material or equipment wears out or breaks down, the impact of the entire original effort is lost (Farrell & Heyneman,

1989). So we have some very mixed results here, not much different from the 'bad news' discussion in Section A. How might we account for this situation?

Possible Explanations

Some of the problems can be attributed to the predictable difficulties of being on the beginning of a learning curve. Overall, these AKF African programs are relatively new. The earliest of these cases, Aga Khan Mzizima Secondary School, started in 1985, but this school is, as noted above, sui generis and thus not directly relevant to the other cases, although important lessons have clearly been learned from the Mzizima experience. The other five cases date from the 1990s. Given the international experience of the successful school improvement programs, it is far too early to judge these African cases one way or the other. What is key here is that the long-term AKF African program is learning from its own experience. Many 'classical' mistakes have been made (e.g., heavy reliance on expensive expatriates, lack of consideration of ongoing cost implications of program development decisions, unquestioned acceptance of the traditional school model as 'given'), but learning how to do school improvement better is clearly taking place. I understand that several of the projects that earlier depended upon expatriate leadership have since gone to local leadership.[21] I note as well the efforts, mostly in Mombasa but also in Kampala and Kisumu and in a different fashion in Zanzibar, to develop locally sustainable support systems for ongoing teacher and school development and improvement, and the apparently reasonably successful integration of these support systems into the routine public school management systems in Mombasa, Kisumu and Zanzibar. And, if the overall program continues to keep this 'self-learning' as a key aspect of its agenda, that learning will continue and will be an important part of the overall learning agenda I outlined at the start of this chapter. However, there does seem to me to be a deeper conceptual problem here, the consideration of which may hopefully contribute to the pursuit of that learning agenda.

A Basic Problem with the Whole-School Improvement Model

A major issue in all of the cases in this book, in the overall AKF program, and in the conceptualization of the whole-school improvement model itself (on which the projects are based), is that the idea of 'whole-school' is quite limited in its conception of 'school.' I would characterize it instead as fundamentally a 'partial school' model. This returns us to the 'matching' and 'not-matching' discussion in Section A. The whole-school improvement concept has been an important and necessary reaction to the evident failure of the top-down, centralized, 'one-size-fits-all' educational reform efforts. Concentrating on individual schools, or local systems of schools, and trying to take into account all of the key stakeholders involved in those local systems, is a salutary counterpoint to the 'reform movement' strategy. The problem is that the conception of 'school' as the teachers (all of them in the whole-school model), the administrators, those who assist them at immediately higher levels (e.g., supervisors, inspectors, local change agents) and learning stakeholders (students, parents and local community members) are only part of what a school 'is.' A 'school' on any local street is also part of, and a product of, a set of assumed and implemented structures and patterns that are built into the model historically and hardly ever questioned. They are literally 'taken for granted.' This is the 'point' of the traditional school model outlined in Section A. When was the last time you heard any advocate for educational change support strongly a move to multi-grading, even though evidence has powerfully shown it to be the better way to organize opportunities for our young people to learn? A fairly standard, but quite predictable, response for many would be. "How could you even imagine such a thing? It is all of a piece." The local school is not simply a product of the particular adults currently within it (who we may or not be able to change through the outside interventions discussed in this book), the 'students' who happen to be there by accidents of local geography and demography, and the architectural decisions made recently or years ago by government officials or outside agency officials. It also embodies a set of historically embedded structures and patterns of behavior of all involved that form a pattern that is extremely difficult to break. There is a major dilemma here for the 'whole-school' improvement movement (which is not far removed from the 'school effectiveness' movement, but that is another story for another time): Most of what a 'whole-school' *is* is not included in the model and most of it is set (culturally and historically) outside the control of the well-meaning 'managers of school change.' Put another way: You cannot really change the behavior of the actors within the system if you do not change the nature of the system itself. Chapter 2 is replete with references to this problem: "A challenge to capacity-building programs is that implementing any element of the school improvement formula involves actions that lie beyond the competence of the primary stakeholders ... to resolve." And "AKF has little leverage over the formidable bureaucracies that determine postings and promotions [of teachers and other officials in the educational bureaucracy]." And all of the AKF African case reports refer directly or indirectly to this same fundamental problem. It is not simply the actions, or lack thereof, of bureaucratic systems and their denizens or the lack of money; it is a matter, as Tyack and Cuban say, of a fundamental and deeply embedded, cross-culturally set, way of understanding what a 'school' is assumed to 'look like' and 'do;' or what these authors call the 'grammar' of schooling.

But how, then, does one change that? How does one do that necessary fundamental thing? How does one go beyond changing one or a few elements of the traditional model of schooling, that has proven so costly and ineffective as a school change and system change strategy, so as to create a very different and more learning-effective approach? The answer to this question is not clear, but it seems to me to be the key question. We now have some examples of this change taking place, as outlined in Section A. The idea of 'school' in the whole-school improvement model needs to be dramatically expanded. We need to learn much more about how and why and under what conditions these young people in some very poor nations manage to learn as much and as well as they do. How and why do children, particularly highly marginalized children, learn so well under these programs? How exactly do the teachers involved manage to learn so much and change so much so quickly? How exactly have these programs spread and be taken to scale so relatively easily? These are the core questions our team is wrestling with. The AKF projects, in both their successes and their failures, contribute to that learning process. Strong and clear answers are not likely to come to us soon. But if we do not continue to follow the example of these AKF programs, to work hard at learning from what we and others do, then we will never know the answers to the questions posed at the beginning of this paragraph. That is the challenge before us, if we can break the taken-for-granted mental bounds of the 'standard forms of formal schooling' model that we all internalized in our youth, and that, I suspect, makes it extremely difficult to imagine any fundamentally different way of providing to young people much better opportunities to learn. But it is now clear that there are examples from which we can learn, if we choose to.

Perhaps a fundamental problem here is that those of us in the 'educational reform or improvement' community are so wrapped up in what we do day to day that we fail to pay attention to the fact that much of what we do, often with enormous effort and good will, never quite works out because we have not taken into account the mostly failed history of intentional educational reform. And we become so immersed in the taken-for-granteds of the assumed immutable 'forms of formal schooling' that we do not even imagine the possibility of them being changed in any fundamental way. There is, I suspect, a fundamental failure of imagination at work. The 'good news', if we care to notice it, is that some folk in some places, often at large scale and generally with very few resources at hand, have *not* failed in their imagination.

10

The Aga Khan Foundation School Improvement Initiative: An International Change Perspective

David Hopkins

The Aga Khan Foundation (AKF) East African school improvement initiative provides a unique insight into a sustained effort to build school improvement capacity and enhance student learning and attainment in a developing educational context. By referring collectively to the six case studies reported in earlier chapters as a 'school improvement initiative,' I am highlighting the systemic quality that characterizes these improvement projects. As Anderson and Greenland point out in Chapters 1 and 2, it would be misleading to regard these programs as a series of individual projects. Rather, they reflect an evolving school improvement program based on coherent design principles that contain certain distinctive and possibly contentious features.

In Chapter 1, Anderson claims that there are "six key principles underlying the designs of each of the projects." He further implies that these principles, taken together, will enhance "the chances for improving the quality of teaching and learning in schools." The six principles are:

- being based in schools;
- the whole school as the unit of change;
- ongoing professional development of teachers;
- attention to school management and other organizational conditions;

- preparation for institutionalization and sustainability; and
- involvement of stakeholders.

These principles, described in more detail in Chapter 1, form part of the conventional wisdom of school improvement. What, of course, is more problematic is how these principles play out in contexts radically different from those in which they were derived.

If we accept, for the time being, Anderson's claim that all the projects are broadly based on these principles, the other much more contentious feature of the Aga Khan Foundation approach to school improvement is the "four such debatable strategies common to some, if not all, of the school improvement projects." These four strategies are:

- investment in district-level teacher development personnel;
- the emphasis on enhancing teacher's general pedagogical expertise;
- a focus on the adoption and use of child-centered activity-oriented methods of teaching and learning; and
- the promotion of data-based decision-making for school development.

These strategies are what makes the Aga Khan Foundation approach to school improvement distinctive and coherent, and the rationale for the strategies' use in the East African context provides much of the substance for the discussion in the opening two chapters of this book. The task of this chapter is to subject these principles and strategies to some mild critique from "an international change perspective."

There are a number of paradoxes in the application of principles and strategies from a developed educational context to a developing one. Perhaps the most stark relates to the contrast between the Aga Khan Foundation approach to school improvement and the contemporary international policy context. The last 10 or 15 years have been a time of great challenge as well as considerable excitement for educational systems around the world. Governments everywhere have been embarking on substantial programs of reform in an attempt to develop more effective school systems and to raise levels of student learning and achievement. A general strategy has been to centralize educational policy while at the same time placing the responsibility for its implementation on individual schools. This tension has made the task of implementing change both complex and challenging. The balancing of centrally directed change and locally developed improvement has proved most difficult to achieve in practice.

The most common policy approach within this framework is widely referred to as 'performance-based' reform. There are a number of properties that this approach to reform:

- centrally determined and explicit goals and standards for student performance;
- curriculum frameworks and materials for use in accomplishing the goals set for students;
- widely disseminated information about school and student performance;

- a system of finance and governance that gives the school responsibility for producing improvements in student performance;
- a series of agencies that judge the extent to which standards have been met and distribute rewards and sanctions for success or failure in meeting specified standards.

There is now growing research evidence of the inability of performance-based policies to affect learning and attainment (Hopkins & Levin, 2000). For example, on the basis of a review of five well-known cases of performance-based reform: in Kentucky, California, New Zealand, Victoria (Australia), and Chicago, Leithwood et al. (1999) reached two striking conclusions.

1. First, that, based on the available evidence, there was no increase in student achievement in any of the jurisdictions except Chicago, and even there, change was "slow in coming."
2. Second, "the disappointing contribution that performance-based reforms have made to improving the core technology of schooling." These reforms did not:
    - adequately acknowledge local context;
    - take the school site seriously;
    - find incentives that work;
    - contribute to significant increases in professional capacity; or
    - address and diagnose opportunity costs.

Although the impact of large-scale reform on student achievement is notoriously fickle, the fact that these reform strategies neglected to focus on teaching and learning and capacity building must have contributed to their inability to impact positively on student achievement.

In support of this argument, it is helpful to look at the case of Chicago, where there were student achievement gains but not until Year 7 of a 10-year initiative. On examining the evidence, it appears that, for the first 6 years of the initiative (1988-1994), the "system operated in a decentralized fashion with little functional contact between schools and the district. In other words, too little structure characterized the operation" (Fullan, 2000). Since 1994, however, the central district was reorganized, with decentralized development being retained within a context of capacity building and external accountability. During this period, five extra district-level functions were developed that may help explain why students in Chicago began to achieve more during the period:

- policy-making increasingly supported decentralization;
- there was a focus on local capacity building;
- a system of rigorous accountability was introduced;
- innovation was stimulated; and
- external support networks were established.

The Chicago case resonates with other analyses of large-scale reform efforts (for a review, see Hopkins, 2001). McLaughlin (1990), for example, concluded that:

> It is exceedingly difficult for policy to change practice, especially across levels of government. Contrary to the one-to-one relationship assumed to exist between policy and practice, the nature, amount, and pace of change at school level is a product of local factors that are largely beyond the control of higher-level policy makers.

When the implementation of central policies was effective, the following implementation strategies were usually in place:

- concrete, teacher-specific, extended training;
- classroom assistance from local education authorities (LEAs) and universities;
- teacher observation of other classrooms, schools or LEAs;
- regular meetings that focused on practical issues related to teaching and learning;
- teacher involvement in school improvement planning;
- local development of classroom materials; and
- head teacher participation in training.

The analyses of 'performance-based' approaches and centralized policy initiatives are entirely consistent with previous research on the implementation of large-scale reform efforts. It is, however, an irony of quite breathtaking proportions that the dramatic increase in educational reform efforts in most Western countries over the past decade is having insufficient impact on levels of student achievement. The paradox here is that the Aga Khan Foundation approach has tended to focus on the 'capacity-building' dimension, without the emphasis on performance management so ubiquitous in Western educational systems. The awarding of incentive grants to schools based on evidence of progress on indicators of school performance in Dar-es-Salaam, however, is a step in this direction.

It is the implications of these issues that I will be addressing in this chapter. In particular, in exploring the appropriateness of the design features of the AKF school improvement program, I will:

1. Briefly review the current knowledge base of school improvement at the school and local levels, particularly that which has emerged from research in developed countries (see Chapter 9 for a review of school improvement in developing countries).
2. Discuss what appears to me to be the key issues raised by the AKF school improvement projects.
3. Reflect on the strengths and weaknesses of the AKF school improvement initiative and explore the policy implications for moving such programs 'to scale.'

The School Improvement Knowledge Base

In this section of the chapter, I will review the knowledge base about school improvement strategies most closely related to those employed in the Aga Khan Foundation programs. There are two dominant approaches in the policy and practice of educational change: *adoptive* and *adaptive* approaches to change (see Hopkins, 2001, Chapter 2).

The *adoptive* approach tends to disregard variables within the individual school environment. This approach is preoccupied with a top-down approach to change; it assumes that change is linear, initiated by an authority figure, and is motivated by external pressure. This top-down, or center-periphery, approach to change was developed to assist the implementation of centralized curriculum innovations in the mid-1960s and later.

The *adaptive* approach to educational change is more sensitive to the context of the individual school and its local situation. It appreciates the environment in which it intervenes, and is concerned with developing a capacity for change within the school rather than adopting a specific approach per se. Strategies for school development planning as a means of managing and adapting centralized change exemplify this approach (Hargreaves & Hopkins, 1991).

As is clear from the preceding chapters, the Aga Khan Foundation projects work within the 'adaptive' approach, and for good reason. Changes in curriculum, teaching methods, grouping practices, and assessment procedures have the greatest potential to enhance the performance of students and so should provide the key focus for school improvement efforts. The projects have emphasized a particular approach to teaching and learning through in-service professional development of teachers, supplemented in some cases by new curriculum materials, as in the Mzizima and Zanzibar school improvement projects. Beside this, they have also addressed local capacity at the district and school levels to initiate, carry out, adapt, and sustain innovations in the classroom more generally.

Unfortunately, the implementation of changes that positively affect the learning of students is very difficult to achieve. This is because the focus of policy implementation is usually on some aspect of curriculum and the production of guidelines or materials. Yet, educational changes that directly impact on the learning of students usually involve teachers in not only adopting new or additional teaching materials but also in:

- acquiring new *knowledge*;
- adopting new *behaviors* (e.g., modifying teaching styles); and
- sometimes, modifying their *beliefs* or *values* (Fullan, 1991).

If classroom practice is to be affected, then teachers' behaviors and practices as well as their beliefs and understandings need to be addressed. As Fullan reminds us, real change, "whether desired or not, represents a serious personal and collective experience characterized by ambivalence and uncertainty" (Fullan, 1991, p. 32) for the individual involved. There are a number of implications for authentic school improvement:

- change takes place over time;
- change initially involves anxiety and uncertainty;
- technical and psychological support is crucial;
- the learning of new skills is incremental and developmental;
- organizational conditions within and in relation to the school make it more or less likely that school improvement will occur; and
- successful change involves pressure and support within a collaborative setting.

It is exactly because change is a process whereby individuals need to "alter their ways of thinking and doing" that most changes fail to progress beyond early implementation. Fullan (1991) graphically refers to this phenomenon as 'the implementation dip.' The 'implementation dip' incorporates the constellation of factors that creates the anxiety and the feelings of incompetence so often associated with relearning and meaningful change. Many research studies have found that, without a period of destabilization, successful, long-lasting change is unlikely to occur (Huberman & Miles, 1984). These perspectives are highly germane to the AKF school improvement initiative, given its focus on child-centered learning and the professional development of teachers. What is also clear from the educational change literature is that conditions need to be created within the school that ensure that individuals are supported through the inevitably difficult and challenging process of altering their ways of thinking and doing.

It is also important to realize that the change process itself is not linear. As Miles (1986) and Fullan (1991) have demonstrated, the change process consists of three overlapping phases: initiation, implementation, and institutionalization. Although these phases often coexist in practice, there are some advantages to describing them separately. It is particularly important to understand what happens during each phase and which behaviors within each phase make for success.

The *initiation* phase is about deciding to embark on innovation, and of developing commitment toward the process. The key activities in the initiation phase are the decision to start and a review of the school's current state with regard to the particular change. Miles (1986) analyzed the various stages of school improvement. This is a list of factors he found that make for successful initiation:

- an innovation tied to a local agenda and a high-profile local need;
- a clear, well-structured approach to change;
- an active advocate or champion who understands the innovation and supports it;
- active initiation to start the innovation (top down is acceptable under certain conditions); and
- a good-quality innovation.

Implementation is the phase that has received the most attention. The key activities occurring during implementation are the carrying out of action plans, the developing and sustaining of commitment, the checking of progress and overcoming problems. The key factors making for success at this stage, according to Miles (1986), are:

- clear responsibility for orchestration and coordination (whether head teacher, coordinator, or external consultant);
- shared control of implementation (top-down control is not acceptable); good cross-hierarchical work and relations; empowerment of both individuals and the school;
- a mixture of pressure, insistence on 'doing it right', and support;
- adequate and sustained staff development and in-service training; and
- rewards for teachers early in the process (empowerment, collegiality, meeting needs, classroom help, load reduction, supply-teacher coverage, expenses, resources).

Institutionalization is the phase when innovation and change stop being regarded as something new and become part of the school's usual way of doing things. The move from implementation to institutionalization often involves the transformation of a pilot project into a school-wide initiative, often without the advantage of previously available funding. Key activities at this stage, according to Miles (1986), are:

- an emphasis on 'embedding' the change within the school's structures, its organization and resources;
- the elimination of competing or contradictory practices;
- strong and purposeful links to other change efforts, the curriculum and classroom teaching;
- widespread use in the school and local area; and
- an adequate bank of local facilitators, (e.g., advisory teachers) for skills training.

Many change efforts fail to progress beyond early implementation because personnel involved do not realize that each of these phases have different characteristics and require different strategies for success. Differentiating between the three phases of initiation, implementation and institutionalization can be helpful, as can be the articulation of appropriate activities at each stage. But there are two further issues to be considered in terms of the Aga Khan Foundation projects. The first is that, nowadays, one is rarely involved with just one innovation. A school can be going through a number of change cycles at any one time. This places great stress on the organizational capacity of the school and the confidence and maturity of those leading the change process.

In the early phases of a school improvement effort, the process of initiation, implementation and institutionalization will be going on at least two levels. The first is the classroom level: putting into practice changes in curriculum focus and teaching strategy. The second is the school level, where the cycle of initiation, implementation and institutionalization is concerned with capacity building: the process of learning how to change, in particular, the way in which in-service activities, planning, and inquiry are organized to support the school improvement initiative. At the start of a school improvement effort, when the schools' organization is not well attuned to change, more effort needs to be allocated initially to building capacity and possibly limiting the amount of classroom change. Once the capacity is in place, then managing multiple cycles of innovation becomes far more possible.

A second issue raised by differentiating among the initiation, implementation and institutionalization phases is the skills required of change agents, both those within and outside the school. Beside the specific activities required during each of the phases, there are also a series of generic skill clusters that characterize the behaviors of effective change agents, such as those supporting the schools in the Aga Khan Foundation projects. The following abilities appear to be the most important:

- generating trust;
- understanding and diagnosing the state of the school's organization;
- planning in the medium term and seeing the bigger picture;
- working productively in groups;
- accessing the required technical resources and advice, be they research, good practices, or specifications of teaching and learning; and
- giving people the confidence to continue.

These skills also need to become the focus of leadership development and in-service training.

The purpose of the preceding discussion has been to provide a perspective on the change strategies most commonly employed by Aga Khan Foundation projects. The AKF initiative is also committed to the 'school as the unit' of change, so it is important to consider what is known from past research on whole-school approaches (see, for example, Joyce, Calhoun & Hopkins, 1999).

Our research on rapidly improving schools suggests that the classroom level provides the explicit focus for development work (Gray et al., 1999). What is striking in these schools is that they are seriously interested in student learning and classroom practice. They are focused on the 'learning level.' In most cases, the focus on classrooms began incrementally. It was first seen in an encouragement of teachers to talk about teaching. It was as if teachers had to be enfranchised to discuss pedagogy with their colleagues. This talk then served as a basis for some form of experimentation with different aspects of classroom practice. Teachers were actively encouraged to 'tinker' with their teaching; it was then that achievement scores began to rise.

'Rapidly improving schools' display a willingness to go beyond merely incremental approaches to change and, instead, to engage in some form of organizational restructuring to support student learning. They are specific about how they wish to improve pupils' learning, and about how to draw on colleagues' experiences to formulate strategies. They have found ways of helping teachers evaluate and learn from their own and other teachers' classroom experiences. The other significant point is that innovations in teaching are conducted within a curriculum context. Teaching and learning strategies such as induction and groupwork strategies, the metacognitive cues and meaning-making activities, are all embedded within a curriculum framework. Incorporating instructional cues and strategies within schemes of work enables consistency to be more easily ensured across the school and allows students to engage more directly with the learning aspect of the teaching strategy.

According to Hopkins and Reynolds (2001), effective school improvement initiatives share the following characteristics. They:

1. Focus on specific outcomes that can be related to student learning, rather than succumbing to external pressure to identify nonspecific goals such as 'improving exam results';
2. Target the learning level and the instructional behaviors of teachers for explicit attention. Specifications of curriculum and teaching are adopted that extend current practice and draw on theory, research into teaching practice, and the teachers' own experiences in formulating strategies, so that the rationale for the required changes is established in the minds of those expected to bring them about.
3. Develop an infrastructure to enable the knowledge base, both 'best practice' and research findings, to be utilized. This involves an internal focus on collaborative patterns of staff development that enable teachers to inquire into practice, and on external strategies for dissemination and networking.
4. Increasingly emphasize the importance of 'capacity building.' This includes not only staff development but also medium-term strategic planning, change strategies that utilize 'pressure and support,' as well as the intelligent use of external support agencies.
5. Provide for monitoring the impact of policy and strategy on teacher practice and student learning early and regularly, rather than relying on post-hoc evaluations. This includes an audit of existing classroom and school processes and outcomes, and comparison of them with desired end states, in particular, the educational experiences of different pupil groups.
6. Stress the importance of ensuring reliability or 'fidelity' in program implementation across all of the schools' organizational structures, a marked contrast to past improvement programs, which did not have to be organizationally 'tight.'
7. Appreciate the importance of changing the beliefs and norms that comprise the organizational culture of the school, in order to embed and sustain this approach to school improvement. There is a careful balance between 'vision building' and the adapting of structures to support that vision.

In this section of the chapter, I have reviewed research evidence on school improvement strategies relevant to the AKF school improvement initiative. The purpose has been to provide a context for a consideration of the AKF's educational initiative in East Africa from a school improvement perspective. In Chapter 9, Farrell comments on the links between the AKF project experiences and school improvement research in the developing world. In this chapter, the intent has been to create a bridge between those experiences and similar research on educational change in developed nations' educational systems. (For a comprehensive and authoritative review of educational change in these settings, see the four-volume *International Handbook of Educational Change* (Hargreaves et al., 1998).

Key Issues Raised by the Aga Khan Foundation School Improvement Projects in East Africa

In this section of the chapter, I address a number of crucial school improvement issues emerging from a cross-case analysis of the individual projects, ones that characterize the Aga Khan Foundation initiative as a whole. For details about the individual projects, readers may refer back to the summaries provided by Anderson and Greenland in Chapters 1 and 2, and to the project case study evaluations in Chapters 3 to 8. The school improvement issues to be discussed here are:

- the commitment to child-centered learning;
- curriculum versus pedagogic development;
- the focus on teacher learning, professional development and leadership training;
- the school as the unit of change and capacity building;
- local support infrastructure; and
- sustainability.

These issues will be discussed one by one.

The Commitment to Child-Centered Learning

The commitment of the Aga Khan Foundation school improvement projects (SIPs) to child-centered learning resonates throughout the preceding chapters. This is a laudable aim from which few would demur. The issue, however, is not one of aspiration; it is, rather, one of definition. There are many concepts in education that generate passionate support but, in practice, lack operational clarity. Child-centered learning is one of these concepts. This is not to say that the SIP originators are not clear about the concept; the difficulties occur as the concept is translated into practice. As Fullan (1995) said in another context, conceptual clarity at the top often leads to confusion at the bottom: "Clarity must be achieved on the receiving end more than on the delivery end."

There appear to be two ways in which this essentially Western educational concept is used. First, child-centered learning most usually refers either to a particular ethos within a school or classroom that prizes the individuality of the learner, that respects the current phase of their learning history, and that creates conditions within the school and classroom where the learner feels accepted and secure. In simple terms, this approach to child-centered learning refers to a caring, accepting and happy school environment.

Another way of viewing child-centered learning is more technical and precise insofar as it relates to cognitive development. This approach works within the broad parameters of developmental psychology and, consequently, views the child as an active learner and the teacher as the one who promotes learning skill and capability. The aim of the teacher working within this approach is to develop powerful learners who can respond successfully to tasks that they are set, as well as to tasks they set themselves. The key point is that, within whatever context learning takes place, it involves an 'active construction of meaning.'

It is clear from the evaluation evidence presented in the Aga Khan Foundation SIP case studies that, in the main, teachers have a view of child-centered learning that is similar to the first meaning described above, but that program developers and evaluators are committed to the second. For example, in the evaluation evidence from the Kampala SIP, the evaluators report that many teachers remain unclear about the educational objectives of the child-centered approach. Similarly, the Kisumu SIP evaluators found that child-centered teaching behaviors, a central aim of the SIP, were not correlated with higher test scores and, as with the Kampala SIP, teachers' understanding of these methods was quite shallow. Unfortunately, child-centered learning of the second type is not simply achieved by more or better training. There appear to be at least two formidable challenges to be met if authentic child-centered learning is to be achieved.

One of the threats to child-centered learning is the narrowing of the definition of effective student learning to the point where it is equated with a range of test scores, rather than something broader. The emphasis on test-score-focused target setting in many educational systems puts pressure on schools and on teachers to raise levels of achievement in the short term, by means of tactical responses such as homework clubs or teaching to the test. This leads to an impoverished interpretation of what constitutes learning. The first challenge, therefore, is to address this normative and politically attractive view of attainment and learning.

It is evident from the case studies that teachers in Uganda, Kenya, and Tanzania are subject to significant expectations and pressure to ensure student success on national examinations. Any changes in teaching and learning strategies that do not yield demonstrable short-term improvements in student outcomes according to local accountability standards are unlikely to be fully implemented and sustained. This provides a difficult challenge for child-centered teaching methods, given the predictable implementation dip experienced by teachers in the process of classroom change. There is little evidence from the case studies that the Aga Khan Foundation projects have either:

- successfully demonstrated the power of constructivist teaching methods to improve the quality of student outcomes on traditional measures of student performance (such as national examinations); or

- attempted to significantly alter local or national expectations and norms that define effective student learning in an East African context.

Even if this was achieved, a second challenge remains. It is to find ways of raising levels of attainment while at the same time helping students become more powerful learners, by expanding and making articulate their repertoire of learning strategies. Promoters of educational change often focus their efforts on helping teachers modify their beliefs and classroom practices, but neglect the fact that many changes teachers aspire to introduce in the classroom also require changes in student beliefs and approaches to learning. The AKF project case studies give little attention to the important role of students in the process of classroom change. It is now becoming clear that it is extremely difficult to expand students' repertoires of learning strategies outside of a curriculum context and a whole-school approach.

These two challenges are obviously interrelated with the other school improvement issues, so I turn the latter before suggesting a potential resolution of these somewhat daunting challenges.

Curriculum Versus Pedagogic Development

The debate about whether curriculum or teaching should be the focus of reform efforts has a long history. It is also related to discussions of whether teachers should, or should not, be involved in curriculum development. Advocates of child-centered learning often also argue persuasively for school-based curriculum development, saying that it provides a more authentic framework for student learning. Unfortunately, the debate has often been polarized both in terms of policy and practice, and a number of threads to the argument need to be untwined.

- First, when one talks about curriculum, is one referring to curriculum guidelines or (more comprehensively) to curriculum materials?
- Second, if the latter, should they be developed by outside agencies or by teachers?
- Third, should the curriculum focus solely on content or include process and instructional strategies?
- Fourth, if both of the above, should teaching or instructional strategies be classroom specific or school wide?
- Fifth, what are the most relevant curriculum designs and instructional strategies for students, in this case, East African schools?

Some of the answers to these questions lie outside the scope of this chapter, but simply asking them begins to set an agenda for educational reform programs in developing educational systems such as those described in this book. What is clear, however, is that curriculum and teaching as focuses for reform should not be treated as 'either-or,' but that creative ways must be found to integrate them.

The experience of successful contemporary school improvement efforts is that learning potential is best realized, and learning capability best enhanced, through the range of teaching and learning models that teachers use with their students. The deliberate use of a range of teaching and learning strategies that are rich in metacognitive content delivers on both learning and achievement. But, as has already been stressed, teaching and learning strategies are not 'free floating' but are embedded in the schemes of work and curriculum content that teachers use to structure the learning in their lessons. These schemes of work also have potential to be shared between schools and to be more widely disseminated. As we will see, this way of working assumes a whole-school dimension, through the staff development infrastructure the school has established, the emphasis on high expectations, the careful attention to consistency of teaching practice, and the discussion of pedagogy that pervades the cultures of these schools.

By way of contrast, much recent centralized reform places great faith in the ability of curriculum guidelines to codify and ensure consistency of practice. National governments now, almost without exception, outline curriculum guidelines but require teachers to develop the detailed schemes of work involved and to select the most appropriate instructional strategies. Although guidelines ensure a degree of uniformity in the curriculum diet of students, they will not necessarily raise standards of performance. There are three key implications for policy here. The first is that educational programs need to be sufficiently comprehensive, so that they link together, in explicit and concrete ways, curriculum content and teaching strategies. The second is that, as Crandall, Eiseman and Louis (1986, p. 29) have noted, "It should not be assumed that involving teachers in developing innovations will invariably lead to better programs ... The crucial variable is whether programs are implementable and attractive to teachers." This leads to the third implication, the need to increase the number of comprehensive, school-wide models for school improvement that have been carefully developed and can be broadly disseminated. This, possibly contentious, point is returned to again in the final section of this chapter.

The learning focus is the key to effective school improvement. This focus can only be achieved by linking together teaching and learning strategies, within a curriculum context. With the exception of the Mzizima SIP and the Zanzibar Secondary English Language Orientation Program, it is not evident from the case studies that the AKF school improvement initiatives have systematically engaged teachers in changing classroom practice within system-wide and school-wide curriculum expectations. But, even this is not sufficient, as the ability to enhance student learning also depends on building the capacity for professional learning throughout the school. This, as is seen in the following two sections, provides the crucial strategic challenge.

The Focus on Teacher Learning, Professional Development and Leadership Training

The realization of child-centered learning and appropriate curriculum support is not achievable without a strong focus on teacher learning and professional development. The commitment to teacher learning and professional development is a strong feature of the case studies in this book. It is clear that deliberate and focused efforts were made in every case to enhance teacher learning. In fact, this became the key strategic drive in a number of the projects, particularly the Kisumu, Kampala and Mombasa SIPs.

It also appears, however, that the model of training is not as powerful as it could have been. The use of master teachers to support classroom teachers in Kisumu, Kampala, Mombasa and Dar-es-Salaam is in line with effective practice elsewhere. What is less evident is the quality of interaction between mentor and teacher. For example, in the Kisumu SIP, staff were not trained to a sufficient level of mastery to do their jobs effectively. They reported that they:

- had not received any guidelines or training in what to do when they visited a school;
- were not taught how to do a demonstration lesson or give feedback to teachers; and
- would have preferred training and coaching that gave them time to reflect in groups, to read and discuss original source documents and to have the expert(s) observe them in classrooms as they worked with teachers.

Subsequent initiatives, notably the Mombasa SIP, addressed this gap by including professional development programs for SIP teacher development officers in the project designs. An even greater challenge, however, is to develop the expertise of teacher leaders at the school level, leaders whose task is to sustain and extend project interventions. As noted in the Kisumu, Kampala, and Mombasa SIPs, insufficient attention was given to the development and support of teacher leaders, and this became a major obstacle to scaling up, deepening, and sustaining classroom change across the schools. Greater success in this regard is reported in the Mzizima, Dar-es-Salaam, and Zanzibar case studies.

The research on staff development is now very clear on what staff development infrastructure is required to support teacher learning. Joyce's and Showers' (1995) work on staff development, in particular their peer coaching strategy, has in recent years transformed our thinking on staff development. Joyce and Showers (1984) summarized the 'best knowledge' we have on staff development:

- the use of the integrated theory-demonstration-practice feedback training program to ensure skill development;
- the use of considerable practice in simulated conditions to ensure fluid control of the new skills;
- the employment of regular on-site coaching to facilitate vertical transfer; and
- the preparation of teachers who can provide one another with the needed coaching.

Joyce (1992) also distinguished between the two key elements of staff development: the workshop and the workplace. Ideally, the workshop is where teachers gain understanding, see demonstrations of the teaching strategy they may wish to acquire, and have the opportunity to practice the strategy in a non-threatening environment. If, however, they wish to transfer the skills that the workshop has introduced back into the workplace (the classroom and school), then merely attending the workshop is insufficient. The research evidence is very clear that skill acquisition and the ability to transfer vertically skills to a range of situations requires 'on-the-job support.' This implies changes to the workplace and the way in which staff development is organized in schools. In particular, this means the opportunity for immediate and sustained practice, collaboration and peer coaching, and the study of development and implementation. These changes in the workplace cannot be achieved without, in most cases, drastic alterations to the ways in which schools are organized. Yet, newly acquired teaching skills will not be transferred from teacher-training workshops to classrooms without them.

What is being proposed here is a wide range of staff development activity that represents a fairly sophisticated infrastructure for sustained professional development. A key element in all of this is the provision of in-classroom support or, in Joyce's and Showers' term, 'peer coaching' (Joyce & Showers, 1995). It is the facilitation of peer coaching that enables teachers to extend their repertoire of teaching skills and to transfer them from one classroom setting to another. When the elements noted below are incorporated into a school improvement plan, peer coaching can virtually ensure 'transfer of training' for everyone:

- Peer coaching teams of two or three are much more effective than larger groups.
- These groups are more effective when the entire staff is engaged in school improvement.
- Peer coaching works better when head teachers and deputy head teachers participate in training and practice.
- The effects are greater when formative study of student learning is embedded in the process.

From the case studies, it is clear that all the Aga Khan Foundation SIPs have, by design, attempted to combine outside-the-classroom in-service training workshops for selected teachers with in-classroom assistance for all teachers in the school. The workshops have been typically delivered through local Teacher Centers. The provision of in-class coaching for teachers, in particular, has made a positive difference in the use of new teaching methods in the cases of the Mombasa and Kisumu SIPs. While, in the Mzizima SIP, school administrators participated fully in school-based teacher development activities, their involvement is less evident in the other projects. A careful reading of the cases also reveals that on-site coaching in most of the projects has relied more on 'expert coaching' from project consultants or previously trained classroom teachers, using a 'cascade strategy' to disseminate new knowledge and skills. As a consequence, with the exception of the Mzizima SIP, much less emphasis has been placed on creating the attitudes, skills, and conditions required for sustained mutual observation and feedback among teachers at the school level, which of course is the essence of 'peer coaching.'

This discussion raises three issues for the Aga Khan Foundation projects. The first is that the training model could have been more powerful in its design and in its implementation. The second is that the training appeared to focus on generic teaching skills rather than the curriculum context. The third is that there was, until quite recently, a lack of focus on head teacher training. I have already addressed the first two concerns; let me say a word about the third.

During the past decade, the dominant debate about educational leadership has been between the 'transactional' and 'transformational' approaches. There seems to be a presumption of the 'transactional' approach in systems where strong central control has been retained while, in systems where decentralization has been most evident, considerable interest in the 'transformational' approach has emerged. It has been widely argued that complex and dynamic changes, such as the *changes in organizational culture* that are required for sustained school improvement, are more likely to occur as a result of transformational leadership. This style of leadership focuses on the people involved and on their relationships and requires an approach that seeks to transform feelings, attitudes and beliefs. Transformational leaders not only manage structure but they purposefully seek to impact upon the culture of the school in order to change it.

I consider the transformational approach a necessary but insufficient condition for school improvement, for the simple reason that it lacks a specific orientation toward student learning. For these reasons, I am attracted to the notion of 'instructional leadership.' Instructional leadership consists of three broad constructs:

- defining the school's values in terms of its student-learning mission;
- managing the school's instructional program; and
- promoting the school as a professional learning community.

In terms of this argument, the key focus of instructional leadership needs to be two key skill clusters. These are (a) strategies for effective teaching and learning, and (b) the conditions that support implementation, in particular staff development, planning, and learning-focused collaborative inquiry.

The constellation of staff development activities described earlier makes the structural link between the collaborative and reflective work of teachers and enhanced levels of student achievement clear and achievable. The staff development focus has the potential to unite the focus on teaching and learning with the focus on capacity building. In highly effective schools, it is this combination that provides the essential infrastructure for school improvement.

The apparent lack of emphasis on instructional leadership in the Aga Khan Foundation projects has meant that the linkage between teacher learning and organizational learning has, for the most part, been missing in the cases reviewed in this book. As noted by Anderson in Chapter 1, attention to head teacher leadership development linked to other elements of SIP activity emerged through the project experiences in Mzizima, Kisumu, and Kampala but did not really become a key focus of project activity until the Mombasa and Dar-es-Salaam SIPs. Head teacher development in these projects has been complicated by the need to address basic management skills (e.g., record-keeping, accounting, inventory, decision-making, goal setting and planning) before more complex leadership roles and skill sets, such as those required of instructional leaders.

The difficulty of preparing instructional leaders in such contexts may provide a partial reason for the lack of 'change capacity' within the school. There is, however, another aspect of capacity building that is reflected in the networks established to support innovative work between schools. I focus on these issues in the following sections.

The School as the Unit of Change and Capacity Building

The position being advocated here is similar to the 'new professionalism' proposed by Hargreaves (1994). His argument was that, paradoxically, many recent educational reform efforts have the unintended consequence of eroding the traditional culture of individualism in teaching, by forcing structural changes that generate new forms of collaboration. The 'new professionalism' implies a change in the values and practices of teachers as structural changes encourage a move from 'individualism to collaboration' and from 'hierarchies to teams.' In making his argument, Hargreaves sets out two propositions about the relationship between teacher and school development. The first is that there is little school development without teacher development. The second is that there is little teacher development without school development. The truth of the first proposition is axiomatic. The second is more controversial.

What is particularly significant about contemporary school improvement strategies is that they not only focus directly on improving student behavior, learning and attainment but also pay attention to teacher and school development. There is now a significant amount of evidence to suggest that a school improvement strategy is more likely to advance achievement for all students if it addresses not only the learning of individual teachers but also other dimensions of the organizational capacity of the school. This is now often referred to as 'school capacity.' Without an emphasis on *capacity*, a school will be unable to 'transform' itself or sustain continuous improvement efforts that result in student attainment.

Newmann, King and Young (2000) define 'school capacity' as the collective competency of the school as an entity to bring about effective change. They suggest four core components of capacity:

- *knowledge, skills and dispositions* of individual staff members;
- *a professional learning community* in which staff work collaboratively to set clear goals for student learning, assess how well students are doing, develop action plans to increase student achievement, while engaging in inquiry and problem-solving;

- *program coherence* – the extent to which the school's programs for student and staff learning are coordinated, focused on clear learning goals and sustained over a period of time; and
- *technical resources* – high-quality curriculum, instructional materials, assessment instruments, technology, workspaces, and so on.

If these ideas are to impact on practice, it is critical to be able to define 'capacity' in operational terms. This has been a feature of a school improvement program that I directed, Improving the Quality of Education for All (IQEA), which is now operating in a number of developed and developing educational systems (see Hopkins, 2001; Hopkins, 2002). The IQEA experience has demonstrated that, without an equal focus on the internal conditions of the school, innovative work quickly becomes marginalized. The 'conditions' have to be worked on at the same time as the curriculum or other priorities the school has set itself. 'Conditions' are the internal features of the school, the 'arrangements' that enable it to get work done. The conditions within the school that are associated with a capacity for sustained improvement are:

- a commitment to *staff development* and the establishment of a professional learning community;
- practical efforts to *involve* staff, students and the community in school policies and decisions;
- a 'transformational' *leadership* approach that is both instructionally focused and dispersed among staff within the school;
- effective *coordination* strategies that ensure whole-school consistency of practice and high expectations;
- serious attention to the potential benefits of *inquiry and reflection* and the gathering of data on both achievement and learning; and
- a commitment to *collaborative planning* that adapts external change to internal purposes.

In terms of the IQEA project at least, it is these conditions that provide a working definition of the development capacity of the school.

In Chapter 1, Anderson identified a set of design principles underlying the Aga Khan Foundation SIP initiatives that are consistent with the arguments for school capacity building presented here. To recapitulate, these arguments include (a) the school as the unit of change, (b) a focus on school-based in-service development of classroom teachers, (c) attention to school management and other supporting organizational conditions, (d) inquiry-based decision-making for school improvement, and (e) preparation for sustainability of change at the school and classroom levels. As noted in the case studies, however, implementation of these principles has been uneven across the projects. The evaluations of both the Kisumu and Mombasa projects, for example, questioned whether project interventions were really engaging whole schools or just providing individual professional assistance to teachers. Project efforts to develop school and school district capacity to inquire into and reflect on student learning have been frustrated in most of the projects by a complicated array of factors. These include inadequate research expertise at the project and school levels, lack of organizational support (time, funds) for collaborative inquiry, and difficulties arising at the interface between SIP activities and national accountability measures. Thus, while the projects have genuinely identified and addressed school capacity building as integral to long-term school improvement, carrying this out in multiple schools over time has turned out to be a most difficult challenge.

Successful school improvement, however, is not just predicated on regarding the school as the unit of change. As is seen in the following section, it is how the school can be supported over time and, by the same token, support other schools.

Local Support Infrastructure

Another aspect of building capacity is reflected in the networks established to support innovative work between schools. As has been seen, this is an aspiration of many of the Aga Khan Foundation projects, but, in terms of the evaluations, as an aspect of capacity it has largely remained latent. There has recently been much international interest in the role of networks in supporting school improvement. There are, however, various interpretations of the 'network' concept. Although networks bring together those with like-minded interests, they are more than just opportunities to share 'good practice.'

The following definition of networks emerged from an analysis of effective networks identified by the OECD:[22]

> Networks are purposeful social entities characterized by a commitment to quality, rigor, and a focus on outcomes. They are also an effective means of supporting innovation in times of change. In education, networks promote the dissemination of good practice, enhance the professional development of teachers, support capacity building in schools, mediate between centralized and decentralized structures, and assist in the process of restructuring and reculturing educational organizations and systems.

The qualities exhibited by such networks are, however, not easily acquired. A number of key conditions need to be in place if networks are to realize their potential as agents of educational innovation. In terms of school improvement, a number of conditions for effective networks can be identified:

- *Consistency of values and focus* – It is important that networks have a common aim and purpose and that the values underpinning the network are well articulated and owned by those involved. This consistency of values and purpose also relates to the need for the focus of the network to be consistent with the overarching policy framework.
- *Clarity of structure* – Effective networks are well organized, with clear operating procedures and mechanisms for ensuring that maximum participation

is achieved within and between schools. These structures promote involvement that is broadly based, preferably with a whole-organization or systemic focus, rather than a narrow, limited or particular focus.
- *Knowledge creation, utilization and transfer* – The key purpose of networks is to create and disseminate knowledge to support educational improvement and innovation. Such knowledge and practice needs to be based on evidence, focus on the core features of schooling, and be subject to robust quality assurance procedures.
- *Rewards related to learning* – those who belong to networks need to feel that their involvement is worthwhile. Rewards for networking are best related to supporting professional development and the encouragement of learning. Effective networks invest in people.
- *Dispersed leadership and empowerment* – Highly effective networks contain skillful people who collaborate and work well together. The skills required by network members are similar to the skills associated with effective teams and include a focus on dispersed leadership and empowerment.
- *Adequate resources* – Networks need to be given adequate resources, particularly in terms of time, finance and human capital. It is not necessarily the quantity of resources that is important; more crucially, there needs to be flexibility in the way in which resources are deployed.

In line with the argument of this chapter, networks in education have a key role to play in supporting innovation and school improvement. Accordingly, networks need to be regarded as support structures for innovative schools, not only in disseminating good practice but also in overcoming the traditional isolation of schools and, to a certain extent, even challenging traditional hierarchical structures. Such isolation may have been appropriate during times of stability, but, during times of change, there is a need to 'tighten the loose coupling', to increase collaboration, and to establish more fluid and responsive structures.

To summarize, networks have the potential to support educational innovation and change by:
- providing a focal point for the dissemination of 'good practice,' the generalizing of innovation, and the creation of 'action-oriented' knowledge about effective educational practices;
- keeping the focus on the core purposes of schooling, in particular, by creating and sustaining a discourse on teaching and learning;
- enhancing the skills of teachers, leaders and other educators in change agency and management of the change process;
- building capacity for continuous improvement at the local level and, in particular, for creating professional learning communities within and between schools;

- ensuring that systems of pressure and support are *integrated*, not segmented (e.g., ensuring that professional learning communities incorporate pressure and support in a seamless way); and
- bridging the centralized/decentralized schism resulting from many contemporary policy initiatives.

Among the Aga Khan Foundation projects, networking for school and teacher development has been addressed largely through the creation or revitalization of Teacher Resource Centers at the district, school cluster, and, in some cases (e.g., in Mzizima and Kampala), at school levels. The most robust effort to facilitate inter-school collaboration among teachers is found in the Mombasa SIP, where 10 Teacher Centers operating from schools within each of the district's 10 zones were refurbished. These became active sites for teacher in-service training, for advising by local teacher development consultants, for developing teaching and learning aids, for providing access to professional literature, and for facilitating subject teacher meetings and meetings of head teachers within each center's service zone. Project evaluations have tended to focus on initial project efforts to develop, resource, staff, and fund the Teacher Centers, efforts which reflect the mobilization phase of change. What is unclear, however, is the degree to which the district- or school-cluster Centers in Mombasa, Kampala, Kisumu, and Dar-es-Salaam have actually led to widespread school-improvement-linked teacher networking across the districts.

As has been argued at length in this chapter, the main reason why reforms have not had the desired impact is that governments' policies on education have not been adequately informed by what is known about teacher and school development. One issue is certain: The future of schooling requires a systemic perspective, which implies a high degree of commitment to school improvement, teacher development, and an unrelenting focus on student achievement and learning. Networks, within and outside schools, provide a means for enacting that commitment.

Sustainability

'Sustainability' is a broad concept that covers a range of issues related to the institutionalization or embedding of change. The term embraces both human and material resources. In this section, the human resource issues are discussed, with economic considerations being dealt with in the concluding section.

There are a number of issues to do with sustaining change that I have already referred to. The first is the struggle many teachers have in coming to terms with the technical and psychological aspects of the change process. This constitutes the 'implementation dip,' and explains the resistance characteristically associated with school improvement efforts that focus on classroom change. It is these aspects of the implementation process that are not taken into account in most policy formulations. The second issue relates to the quality of implementation. Although all policies and programs require some adaptation to local context, this is not a license to dilute an innovation beyond recognition. It is the quality and consistency of implementation that eventually ensures positive impact on student achievement. Third, most effective school improvement strategies require some form of intervention or active implementation. Whether this is internally or externally initiated, it requires some break with existing practice, with all that entails. Policy, therefore, has to be concerned with helping teachers through the change process as well as the quality of implementation itself.

Sustainability also relates to the system level. Earlier, I implied that effective policy implementation is both system wide and system deep. This applies both to coherence in structures and to coherence in values, aspirations and ways of working. The key point, as Fullan (2000) has noted,

> ...is that working with systems means conceptualizing strategies with whole systems in mind. Working with schools means taking into account the total set of changes facing given schools, and figuring out the best relationship with the surrounding infrastructure. ... Large-scale reform then, will require units to make connections and to synergize activities around common priorities.

The system emphasis is not to achieve control (which is impossible) but to harness the interactive capability of systemic forces. The systemic perspective also applies to the school. As has been seen a number of times, changes in teaching practice only occur when there is clarity and coherence in the minds of teachers. It appears that, the more coherent and collaborative the internal conditions of the school, the more knowledgeable teachers in those schools are of national policy initiatives. In such schools, where staff commitment has been very high, the outcomes secured are unusually impressive.

If a systemic perspective is to be genuinely achieved, a high degree of consistency is required across the policy spectrum. First, policy makers need to continually keep the 'big picture' in mind in searching for connections and ways of exploiting potential synergy. Second, the more the school works collaboratively on improvement at the school level, the more it engages critically with external standards and policy. Third, sustained and authentic school improvement requires a high-quality teaching profession that reflects both a system-level commitment to upgrade the profession and the nurturing of professional learning communities within the school. A policy framework for facilitating this integration provides the focus for the following section.

Toward a Policy Framework

In the previous section, I reviewed the AKF school improvement projects from the perspective of the contemporary knowledge of school improvement. In this concluding section of this chapter, I will attempt to draw the argument together by suggesting a strategic policy framework that may assist individual programs in moving to scale. I am not optimistic, however, about the ability of individual NGOs to carry out such a systemic program without explicit support from national authorities. But, before exploring the policy implications, I would like to reflect a little on the AKF school improvement initiative more generally.

What I have written, especially in the previous section, could be taken as being critical of the Aga Khan Foundation approach to school reform. This is not the case. The more I have read about the initiative in general and the individual programs themselves, the more impressed I have been by the commitment and vision that underpinned them. With the benefit of hindsight, there are obviously things that one would do differently. When the initiative was being conceived in the mid-1980s, however, it represented 'cutting edge' thinking, at least among liberal Western educational change workers, and this was both its main strength and key weakness.

The main strength of the initiative is that it focuses on the crucial lacuna in current reform initiatives, the lack of attention to capacity building. The Aga Khan Foundation initiative recognized that most 'performance-based reforms' currently are not sustained over time without an investment in the learning and development of teachers and their schools. In this respect, the AKF initiative was well ahead of its time and provides a powerful example of tackling head on the formidable challenges of systemic change.

The main weakness of the Aga Khan Foundation approach is that it was insufficiently strategic and lacked attentiveness of the curriculum context necessary to achieve the immediate results it desired. This is not to say that the focuses on child-centered learning and capacity building were wrong; rather, that they need to be conceptualized somewhat differently. As has been seen, an approach such as child-centered learning is open to a number of interpretations and, therefore, needs to be implemented in practical and easily understandable ways. This could have been more easily achieved if the child-centered learning approach had, from the start, been embedded in a curriculum context that emphasized a limited number of culturally appropriate teaching and learning strategies. In educational cultures like those in East Africa, there is a desperate need for curriculum and teaching programs that directly effect student learning in achievable and practical ways. More effort in the early stages of the program could have gone into developing such curriculum and teaching programs. This approach would also have fitted more easily within what appears to be the dominant policy context.

Before turning to a discussion of a potential policy framework, I need to make a brief comment on two key aspects of the AKF school improvement initiative that, to this point, I have ignored. They are the role of parent and community involvement in school and teacher development and the issue of financial resources.

One of the main focuses of the AKF school improvement projects has been the attempt to mobilize and train parents to take increased responsibility for the governance of their schools. This was supposed to involve general management of the school and policy directions as well as responsibility for basic funding and funding for school/teacher improvement, in light of the fiscal realities of local and national governments. This is in line with the community ethic of the SIPs as well as the research on variables that impact positively on student learning. According to the meta-analysis by Wang et al. (1993), the three key variables that correlate with higher levels of student achievement are psychological, instructional and home environments. The clear implications are that any strategy to promote student learning needs to give attention to engaging students and parents as active participants, and to expanding the teaching and learning repertoires of teachers and students, respectively. Despite the rhetoric of parental and community involvement, however, few of the SIP projects have paid much attention to date to the actual involvement of parents in their children's learning. Direct interventions to mobilize parents began toward the end of the Kisumu project, and feature most prominently in the work in Mombasa. However, all projects intend to employ community development officers to work on this side of the equation in future.

To this point, I have also ignored the issue of financial resources. This is not because I regard it as unimportant; on the contrary, it is a vital element of sustainability. The economics of educational reform, particularly in such developing educational systems, are complex and beyond the scope of a chapter such as this. Let me, however, briefly outline two of the key issues. First, the cost of the initiatives: Is the cost sustainable without donor support? In many of the case studies, the cost of the in-school training appears to be realistic, and it is to the project developers' credit that efforts to secure funding for continuation beyond the pilot-project phase of the SIPs has been a priority from the start. To say that the cost seems realistic, however, is not to say that it is necessarily sustainable. As illustrated, particularly in the Mombasa and Dar-es-Salaam initiatives, the daunting underlying issue is that local and national governments appear to have little internal funding to commit on an ongoing basis to school-based teacher and school development. Thus, the big question is not simply how much it costs, but where will the money come from initially and on a continuing basis? This difficult problem is linked to the emerging debate about whether basic education should be free or whether, either on grounds of principle or utter necessity, parents and communities should be encouraged or required to pay tuition. This issue is further complicated by the large number of private independent and community schools, in addition to government schools, that have to be taken into account in any system-wide school improvement initiative in East Africa.

Having critiqued 'performance-based' approaches to educational reform, reviewed the implications for policy from the school improvement research, and discussed the elements of a local support structure for schools and the role of networks, it is necessary to pull these themes together into a set of guidelines for policy-makers. There is no doubt that most governments are committed to the aspirations of school improvement. What they often lack is a considered conceptual framework with which to drive forward and deliver their educational agenda. Successful policy initiatives in the area of school improvement reflect a 'goodness of fit' between the aspirations and implications of the policy being implemented, the values of the school, and the beliefs of teachers. Creating the conditions for such synergy is a key challenge for those at all levels of the educational system.[23]

Most governments now identify targets for achievement in key learning areas. If these targets are to be realized, government policies need to incorporate three further elements:

- First, making available to schools and local agencies strategies to assist them in realizing the goals they have identified, remembering, of course, that all schools are at different levels of effectiveness. Instead of every school 're-inventing the wheel,' governments should encourage the development and piloting of curriculum *and* instructional programs that directly address, in implementable ways, the targets that they and the schools are setting.
- Second (if this could be achieved), a range of policy options related to programs that *really* work. Then schools could begin to select from among a range of options strategies that address the particular targets they have set and the learning needs of their students. With a such series of program options available, schools would be in a position to address more directly the crucial issues of staff development, capacity building and consistency of implementation that are so necessary for ensuring student achievement. The school-level context can facilitate, and to a great extent, determine, successful policy implementation.
- Third, governments and NGOs would then be in a position to target funding to schools with the greatest need, in the far more secure knowledge that what they were going to do would achieve the goals set for the system as a whole. In order to challenge schools to do their best, and to support program development and capacity building, a local infrastructure is also needed. The development of this infrastructure also needs to be a part of policy formulation. Further, there will be the need to focus not just on how innovations impact individual schools but on how such innovations can be scaled up, so that they impact all schools and the system as a whole.

The proposals made here require a fundamentally new and radical way of thinking about educational reform. In particular, they require a new mind-set on the part of policy-makers around the world in attempting to link reform strategies to the achievement of young people in our schools. How might governments move such an agenda forward?

- Even the best of current reform initiatives adopt an a-la-carte perspective to the principles previously outlined. Some principles may be included, but rarely all are; the criteria for inclusion often appear to be serendipitous. For example, an initiative may be strong on 'models of teaching' and staff development but weak on context specificity and capacity building. Most reform initiatives fail to understand the dialectic between classroom and whole-school change; research and dissemination also usually gets short shrift. While the principles outlined above may not be the last word on the matter, they do provide an integrated set of research-based criteria against which policies can be formulated and evaluated.

- Reform does require additional resources, but the critical issue is how these resources are deployed. It is vital to link funding to clear plans for improvement that are based on thoughtful use of the best available evidence. It is also important to make sure that the various forms of additional funding are accessible and coherently focused, so that schools and districts are not trying to manage multiple funding requests and multiple programs that lead them in different directions at the same time. Effective reform may also be impossible without resources for other purposes, such as retaining good teachers, maintaining adequate school facilities, or overcoming some of the deep-seated effects of social disadvantage and poverty. However, in all cases, the essential element is a clear link between resources and outcomes.
- One of the reasons that the relationship between national policy and classroom practice is notoriously fickle is the lack of consistency within and between policies. Policy alignment needs, therefore, to be both horizontal and vertical to reach across policies and through the various levels of the system. Horizontal coordination suggests that policies should be aligned to support instructional goals and strategies. Vertical coordination means that classrooms, schools and local authorities receive consistent messages about what is required for success. In this respect, the proliferation of educational agencies and actors in most systems may not be helpful.
- The building of local capacity is as important as a coherent national school improvement policy. Key elements of building local capacity are often in existence but not well connected nor linked to a capacity-building agenda. There also needs to be a closer connection between school improvement work and initial teacher training, so that new teachers come to their work with at least some understanding of key improvement strategies and skills. Building capacity and managing real change require skills that most administrators now only learn through practice and experience, if they learn them at all. These elements, as well as local research and dissemination, could be integrated in ways that are much more powerful than their current separate embodiments.

Overall, it is unlikely that proposals put forward in this chapter would be any more costly than the programs described in the case studies in this book. Nor should they be politically problematic. Governments could continue to focus, if they wish, on matters of achievement, standards and accountability, but they would be more confident that their policies will bring about the conditions they say they desire. Without such an approach to school improvement, the evidence of practice and research clearly suggests that society will continue to set educational goals that are, based on current performance, beyond the capacity of education systems to deliver.

Notes

[1] The Aga Khan Education Services (AKES) governs a network of private nursery, primary, and secondary schools across East Africa. Regionally and within each country AKES has been an integral partner in the planning and management of the East Africa school improvement projects. The earliest projects used AKES school facilities for project offices and training sites. AKES school staff participated in project in-service training activities along with participants from government and other private schools. Selected AKES schools were promoted as training and demonstration sites for teaching methods advocated through the projects. As the projects grew in scope to entire school districts and became more integrated into district education structures and teacher centers, the role of AKES schools as training and demonstration sites diminished. AKES authorities were active participants in project management (as grantees) in all but two of the projects described in the case studies (See Chapters 7 and 8), where AKF worked with local governments and assumed a more direct role in project management. Since AKF has been the primary instigator and funding agent for the projects, AKF is referred to here as the major mobilizing force behind the projects, while the long-term collaboration with AKES authorities and schools is acknowledged.

[2] Base funding for the Mzizima SIP was provided initially by the Aga Khan Development Network-Geneva. Additional funding in the later phases was contributed by the Aga Khan Foundation Canada and by the Canadian International Development Agency (CIDA).

[3] We employed a multi-method approach to the evaluation, including document review (program plans, previous monitoring and evaluation reports, implementation records), stakeholder interviews (teachers, administrators, students, school governors associated with Aga Khan Education Services, project managers from the Aga Khan Foundation, personnel from outreach schools), and on-site observation of teacher development activities and classroom practice during a 2-week site visit in October 1994. The research design used Stufflebeam's context, inputs, process, products (CIPP) model for program evaluation (Stufflebeam, 1983). Our analysis of SIP activities was grounded in a research-based framework for understanding classroom and school improvement with an emphasis on teacher learning (Fullan, Bennett, & Rolheiser-Bennett, 1990).

[4] See Lassibille, Tan and Sumra (2000) for a historical and contemporary analysis of the expansion of private secondary schools in Tanzania, with detailed comparisons to participation, staffing, resources, student intake, student performance, and efficiency in government secondary schools.

[5] The Phase 1 plan included the appointment of a research specialist to assist in evaluating the curriculum resources created through the project, teacher experimentation with new resources and teaching methods, and the project impact on teachers and students. The plans for in-house evaluation of SIP products and activities in Phase 1, however, were not enacted and the evaluation specialist departed for reasons left unclear in SIP reports. Phase 2 and 3 plans to involve local university consultants as research advisors and to engage teachers in action research on the impact of SIP activities likewise failed to achieve the vision of an in-house process for continuous evaluation of SIP outcomes on curriculum, teachers, and student learning. The project funders and managers ended up relying on external evaluation reports commissioned during and at the end of each phase. None of these reports, including the one conducted by the authors, benefited from systematic collection of longitudinal or experimental data on teacher or student outcomes.

[6] The Kenyan education system currently has two categories of schools: public and private. Public schools are those run by parents in the form of a school committee. The parents provide the school facilities while the central government pays teacher and head teacher salaries and provides inspection and supervision services. Within the public-school category, there are both high- and low-cost schools. In Kisumu, high-cost schools are public schools that were once Asian or European schools with high-quality facilities. These schools pay for the services of groundskeepers, cleaners, and clerks, thereby requiring that parents pay a higher fee to maintain the facilities. Often, they also employ an accounting clerk to maintain school accounts, thereby freeing head teachers' time for other responsibilities. Low-cost schools are primarily community schools that levy small fees for services and school development.

[7] The evaluation study on which this article is based was commissioned and funded by the Aga Khan Foundation (AKF). The school improvement project investigated was cofunded by AKF, the British Department for International Development, the Commission of the European Communities, and the Mombasa Municipal School District.

[8] These approximate figures were reported as target population statistics in the original 1994 project proposal.

[9] The Concerns-Based Adoption Model (CBAM) is a theory and methodology for understanding and measuring the developmental progression of teacher attitudes, behaviors, and variations in practice as teachers implement innovations in curriculum and instruction. The model also provides a framework for analyzing the inter-ventions of persons acting in change facilitating roles (Hall & Hord, 1987; Anderson, 1997).

[10] Legally, PTAs and school committees in Kenya are distinct entities, the former constituted primarily for fundraising purposes and the latter for participation in school governance. One thrust of the MSIP intervention and of PRISM training for head teachers was combining these groups into single organizational bodies with both financial and management functions. In this chapter, we have used the term 'school committee' to reference both PTAs and school committees.

[11] In each school, we observed two upper and two lower primary teachers for one period each, ensuring that we saw a mix of English, mathematics, and science lessons in each school. We observed both female and male teachers, where possible. We interviewed the head teacher and two observed teachers in each school about their contact with MSIP and TAC activities, and how that contact affected their teaching and pupil learning. We held focus groups with four to six teachers per school.

[12] Steering Committee made up of District Education Office, Kinondoni (Chair); DPSP Project Director (Secretary); head of Kinondoni TRC; representative of Canadian International Development Agency (CIDA) and Canadian High Commission; NGO Coordinator, British High Commission; Program Officer, Voluntary Services Overseas (VSO); Program Officer, AKF-T; representative of the Zonal Directorate, Kinondoni; representative of the Primary Education Directorate, Ministry of Education and Culture (MOEC); representative of the City Education Office; representative of the Bureau of Educational Research and Evaluation, Uni-ver-sity of Dar-es-Salaam (UDSM); two chairs of school committees of project schools; head teacher of a project school.

[13] Durkin, J. (1999). *Improving your school and planning for its development.* DPSP School Improvement and Management Series: Module 1. Vimpany, P. (1998). *Financial management for primary schools in Tanzania.* DPSP School Improvement and Management Series: Module 3.

[14] DPSP School Improvement and Management Series: Modules 2 and 4, *The management of your school and School curriculum management* are in preparation for publication.

[15] Waraka wa Elimu na. 2 wa Mwaka 1999 - 'Michango ya Kuendeleza Shule. May 6, 1999.

[16] But see the evaluation report for the equivalent mainland ELTSP program (Rea-Dickens, l996).

[17] Parts of this paper draw upon an earlier paper written for the World Bank Institute: *Improving learning: Perspectives for primary education in rural Africa.* June 1999.

[18] The following Graduate Assistants have worked on this project, or are working on it: Anna Kochan, Melissa White, Obed Mfum-Mensah, and Sarah Haines. Core documents produced thus far are Kochan, 2000 and White, 2000.

[19] Classic works include Case, 1985 and Knowles, 1984, but the cognitive psychology and adult education literature is too vast to cite here.

[20] Escuela Nueva in Colombia was almost destroyed in its middle years by the insistence of a major donor agency (the World Bank) that it expand more rapidly than was possible given the administrative capacity of the educational system.

[21] S. E. Anderson; personal communication.

[22] The quotation and discussion in this section are from Hopkins, 2000, p. 1.

[23] For a further elaboration of these issues, see Hopkins & Levin, 2000 and Hopkins, 2001, from which the following discussion is taken.

References

Aga Khan Foundation. (1995). *The Aga Khan Development Network.* Geneva: Author.

Aga Khan University Institute for Educational Development (AKU-IED). (2000). *A strategy for educational development.* Karachi, Pakistan: Author.

Ahmed, M. (1993). *Primary education for all: Learning from the BRAC experience. A case study*. Dhaka, Bangladesh: Association for Basic Education and Literacy (ABEL).

Ali, M. A. (1998). *Supervision for teacher development: A proposal for Pakistan.* Paris: UNESCO & International Institute for Education Planning.

Ali, M. A., Qasim, S. A., Jaffer, R., & Greenland, J. (1993). Teacher-center and school-based models of collegiality and professional development: Case studies of the Teachers' Resource Center and the Aga Khan school system in Karachi, Pakistan. *International Journal of Educational Research, 19,* 735-754.

Alvarez, H. B., & Ruiz-Caseres, M. (1997). *Senderos de cambio: Genesis y ejecución de las reformas educativos en America Latina y el Caribe.* Washington, DC: US Agency for International Development.

Anderson, S. E. (1997). Understanding teacher change: Revisiting the Concerns-Based Adoption Model. *Curriculum Inquiry, 27,* 331-368.

Anderson, S. E. (2000). *A coordinated district consultant and teacher center approach to school-based teacher development: The Mombasa School Improvement Project.* Paper presented at the Comparative and International Education Society Conference, San Antonio, Texas.

Anderson, S. E., & Nderitu, S. (1999). *The School Improvement Project of the Aga Khan Education Service, Kenya at Mombasa* [evaluation report]. Geneva: Aga Khan Foundation.

Anderson, S. E., Rolheiser C., & Bennett, B. (1995). Confronting the challenge of implementing cooperative learning. *Journal of Staff Development, 16,* 32-39.

Anderson, S. E., & Stiegelbauer, S. (1994). Institutionalization and renewal in a restructured secondary school. *School Organization, 14,* 279-293.

Anderson, S. E., & Sumra, S. (1995). *The School Improvement Programme: Aga Khan Mzizima Secondary School, Dar es Salaam* [evaluation report]. Geneva: Aga Khan Foundation.

Andrews, J., Housego, I., & Thomas, D. (1990). Effective in-service programs in developing countries: A study of expert opinion. In V. D. Rust & P. Dalin (Eds.), *Teachers and teaching in the developing world* (pp. 63-82). New York: Garland.

Arboleda, J. (1994). Participation and partnership in the Colombian Escuela Nueva. In Shaeffer, S. (Ed.), *Partnerships and participation in basic education* (Volume 2, Case 5). Paris: International Institute for Educational Planning.

Arthur, J. (1998). Institutional practices and the cultural construction of primary school teaching in Botswana. *Comparative Education, 34,* 313-326.

Avalos, B. (1980). Teacher effectiveness: Research in the Third World – Highlights of a review. *Comparative Education, 16,* 45-54.

Avalos, B. (1985). Training for better teaching in the Third World: Lessons from research. *Teaching and Teacher Education, 1,* 289-299.

Avalos, B. (1990). Teacher effectiveness: More questions about an old theme. In V. D. Rust & P. Dalin (Eds.), *Teachers and teaching in the developing world* (pp.199-217). New York: Garland.

Ayot, H. (1983). Teacher advisory centres in Kenya. In J. Greenland (Ed.), *The in-service training of primary school teachers in English-speaking Africa* (pp. 153-165). London: MacMillan Education.

Bah-Layla, I. (1998, December). *Les classes multigrades en Guinée. Formation et emploi des formateurs. Les Centres ANAFA ou Écoles de seconde chance. Le Programme de petites subventions aux écoles* [country case paper]. Prepared for the Improving Learning: Perspectives for Primary Education in Rural Africa seminar sponsored by UNESCO and the World Bank, Lusaka, Zambia.

Bartlett, K. (2000). *Supporting NGOs working in education: Nurturing ideas and approaches and mainstreaming these through partnership.* Paper presented at the Comparative and International Education Society Conference, San Antonio, Texas.

Bendera, S. R. (2000). *An attempt to evaluate the impact of the Mombasa SIP since its launch in 1995 to 1999.* Unpublished master's thesis, University of Leeds, Leeds, UK.

Benton, P. (Ed.). (1990). *The Oxford Internship Scheme: Integration and partnership in initial teacher education.* London: Calouste Gulbenkian Foundation.

Black, H., Govinda, R., Kiragu, F., & Derne, M. (1993). *School improvement in the developing world: An evaluation of the Aga Khan Foundation programme.* Geneva: Aga Khan Foundation & Scottish Council for Research in Education (SCRE).

Bray, M. (2000). *Community partnerships in education: Dimensions, variations, and implications.* Commissioned by the Education For All (EFA) Conference, Dakar, Senegal.

Brown, G., Brown, J., & Sumra, S. (1999). *The East African Madrasa Programme: The Madrasa Resource Centres and their community-based preschool programme.* Geneva: Aga Khan Foundation.

Brumfit, A., & Hikmany, H. R. H. (1997). *Secondary English Language Orientation Project* [final evaluation report]. Dar-es-Salaam, Tanzania: Aga Khan Foundation - Tanzania.

Bude, U. (1992). *Field-Based Teacher Development Programme, Northern Areas of Pakistan and Chitral* (evaluation report for the Aga Khan Foundation). Bonn: German Foundation for International Development.

Caillods, F. (Ed.). (1989). *The prospects for educational planning.* Paris: International Institute for Educational Planning.

Capper, J., Nderitu, S. & Ogula, P. (1997). *The School Improvement Programme of the Aga Khan Education Service, Kenya at Kisumu, Western Kenya* [evaluation report]. Geneva: Aga Khan Foundation.

Carasco, J. F., Munene, J. C., Kasente, D. K., & Odada, M. (1996, February). *Factors influencing effectiveness in primary schools: A baseline study.* (Uganda IEQ Project - Phase 1 Research). Kampala: Uganda National Examinations Board (UNEB).

Case, R. (1985). *Intellectual development: Birth to adulthood.* Orlando, FL: Academic Press.

Cohen, D. (1988). *Teaching practice: Plus ça change ...* (Issue Paper 88-3). East Lansing, MI: National Center for Research on Teacher Learning.

Colbert, V., & Arboleda, J. (1990). *Universalization of primary education in Colombia: The New School programme.* Paper presented at the World Conference on Education for All, Jomtien, Thailand.

Corbett, H. D., & Rossman, G. (1989). Three paths to implementing change: A research note. *Curriculum Inquiry, 19,* 163-190.

Corvalan-Vasquez, O. (1988). Trends in technical-vocational and secondary education in Latin America. *International Journal of Educational Development, 8,* 73-98.

Craig, H., Kraft, R., & du Plessis, J. (1998). *Teacher development: Making an impact.* Washington, DC: US Agency for International Development & World Bank (Human Development Network, Effective Schools and Teachers).

Crandall, D., Eiseman, J., & Louis, K. (1986). Strategic planning issues that bear on the success of school improvement efforts. *Educational Administration Quarterly, 22,* 21-53.

Criper, C., & Dodd, W. A. (1984). Report on the teaching of the English language and its use as a medium of education in Tanzania. Dar-es-Salaam, Tanzania: British Council.

Crouch, L. et al. (1997). *Foundations of education reform support.* (Education Reform Support, Volume 2). Washington, DC: US Agency for International Development.

Davies, L. (1996). The management and mismanagement of school effectiveness. In J. D. Turner, (Ed.), *The state and the school: An international perspective* (pp. 91-107). London: Falmer.

Deutsche Gesellschaft für Technische Zusammenarbeit (GTZ). (1995). *Science teaching in government secondary schools: A situation analysis for mainland Tanzania* (Report to the Ministry of Education and Culture, United Republic of Tanzania). Germany: Author.

Djalil, A., & Anderson, L. (1989). The impact of a research-based teacher training program on Indonesian teachers, classrooms, and students. *Teaching and Teacher Education, 5,* 165-178.

Durkin, J. (1999). *Improving your school and planning for its development* (School Improvement and Management Series - Module 1). Dar-es-Salaam, Tanzania: Dar-es-Salaam Primary Schools Project.

Epstein, J. (1995). School/family/community partnerships. *Phi Delta Kappan, 76,* 701-712.

Fagerlind, I. & Sjosted, B. (1990). *Review and prospects of educational planning and management in Europe.* Paris: UNESCO & International Congress on Planning and Management of Educational Development.

Farrell, J. P. (1997). A retrospective on educational planning in comparative education. *Comparative Education Review, 41,* 277-313.

Farrell, J. P. (2000). Why is educational reform so difficult? Similar descriptions, different prescriptions, failed explanations [review essay]. *Curriculum Inquiry, 30,* 83-103.

Farrell, J. P., & Connelly, F. M. (1998). *Final retrospective report: Forum workshops on educational reform in Egypt.* Prepared for UNICEF-Egypt & Egypt Ministry of Education.

Farrell, J. P., & Heyneman, S. (Eds.). (1989). *Textbooks in the developing world: Economic and educational choices.* Washington, DC: World Bank.

Fiske, E. B., & Ladd, H. F. (2000). *When schools compete.* Washington, DC: Brookings Institution Press.

Fullan, M. (1995). The school as a learning organization: Distant dreams. *Theory into Practice, 34,* 230-235.

Fullan, M. (2000). The return of large-scale reform. *Journal of Educational Change, 1,* 5-28.

Fullan, M., Bennett, B., & Rolheiser-Bennett, C. (1990, May). Linking classroom and school improvement. *Educational Leadership, 47,* 13-19.

Fullan, M. with Stiegelbauer, S. (1991). *The new meaning of educational change.* New York: Teachers College Press.

Fullan, M., & Watson, N. (1999, August). *School-based management: Reconceptualizing to improve learning outcomes.* Paper prepared for the Improving Learning Outcomes in the Caribbean seminar sponsored by the World Bank.

Fuller, B. (1991). *Growing up modern: The Western state builds Third-World schools.* New York: Routledge.

Fuller, B., & Clarke, P. (1994). Raising school effects while ignoring culture? Local conditions and the influence of classroom tools, rules, and pedagogy. *Review of Educational Research, 64,* 119-157.

Fuller, B., & Rubinson, R. (1992). *The political construction of education: The state, school expansion and economic change.* New York: Praeger.

Fuller, B. & Snyder, C. (1991). Vocal teachers, silent pupils? Life in Botswana classrooms. *Comparative Education Review, 35,* 274-294.

Gardner, R. (1990). Case study for Chapter 3: The Cianjur Project. In V. D. Rust & P. Dalin (Eds.), *Teachers and teaching in the developing world* (pp. 83-93). New York: Garland.

Gray, J., Hopkins, D., Reynolds, D., Wilcox, B., Farrell, S., & Jesson, D. (1999). *Improving schools – Performance and potential.* Buckingham, UK: Open University Press.

Greenland, J. (Ed.). (1983). *The in-service training of primary school teachers in English-speaking Africa.* London: Macmillan.

Greenland, J., Kanji, G., & Dean, B. (1998). *Training teachers in schools and resource centres in low-income countries.* Paper presented at the World Bank's Human Resource Development Week, Washington, DC.

Guthrie, G. (1990). To the defense of traditional teaching in lesser-developed countries. In V. D. Rust & P. Dalin (Eds.), *Teachers and teaching in the developing world* (pp. 209-232). New York: Garland.

Haiplik, B. (2001). *Educational success stories in the developing world: Community schools and teacher development.* Prepared for Comparative, International and Development Education Centre. Toronto: Ontario Institute for Studies in Education at the University of Toronto.

Hall, E. T. (1986). Unstated features of the cultural context of learning. In Thomas, A., & E. W. Ploman (Eds.). *Learning and development: A global perspective* (pp. 157-176). Toronto: OISE [Ontario Institute for Studies in Education] Press.

Hall, G., & Hord, S. (1987). *Change in schools: Facilitating the process.* Albany, NY: State University of New York Press.

Hardeman, H. (1999). *Methodology sessions for in-service training of primary school teachers: A manual for teacher trainers in mathematics, science and English.* Dar-es-Salaam, Tanzania: Dar-es-Salaam Primary Schools Project.

Hargreaves, A., & Fullan, M. (1998). *What's worth fighting for out there.* Toronto: Ontario Public School Teachers' Federation.

Hargreaves, A., Lieberman, A., Fullan, M. & Hopkins, D. (Eds.). (1998). *International Handbook of Educational Change.* Dordrecht, Netherlands: Kluwer Academic.

Hargreaves, D. (1994). The new professionalism: A synthesis of professional and institutional development. *Teaching and Teacher Education, 10* (4).
Hargreaves, D. & Hopkins, D. (1991). *The empowered school.* London: Cassell.
Hartwell, A. (1995). *Review of Egypt's Community School Project.* Cairo: UNICEF.
Harvey, S. (1999). The impact of coaching in South African primary science InSET. *International Journal of Educational Development, 19,* 191-205.
Heck, S., Stiegelbauer, S., Hall, G., & Loucks, S. (1981). *Measuring innovation configurations: Procedures and applications.* Austin, TX: University of Texas Research and Development Center for Teacher Education.
Heneveld, W., & Craig, H. (1996). *Schools count: World Bank project designs and the quality of primary education in sub-Saharan Africa* (Technical Paper No. 303). Washington, DC: World Bank.
Heneveld, W. & Hasan, P. (1989). *Evaluation of the Sind School Improvement Programme, Pakistan*: Geneva: Aga Khan Education Services.
Hertz-Lazarowitz, R. (1992). Understanding students' interactive behavior: Looking at six mirrors of the classroom. In R. Hertz-Lazarowitz & N. Miller (Eds.), *Interaction in cooperative groups: The anatomy of group learning* (pp. 71-102). New York: Cambridge University Press.
Holbrook, A. & Greenland, J. (1990). *Teachers of the Karakorums. The Field-Based Teacher Development Programme* [videotape and manual]. Geneva: Aga Khan Foundation.
Holliday, A. (1994). *Appropriate methodology and social context.* Cambridge, UK: Cambridge University Press.
Hopkins, D. (2000, November). *Policies for innovation, networking and change.* Paper presented at OECD Conference on Schooling for Tomorrow, Rotterdam, Netherlands.
Hopkins, D. (2001). *School improvement for real.* London: Routledge/Falmer.
Hopkins, D. (2002). *Improving the quality of education for all.* London: David Fulton.
Hopkins, D. & Levin, B. (2000). Government policy and school development, *School Leadership and Management, 20,* 15-30.
Hopkins, D. & Reynolds, D. (in press). The past, present and future of school improvement: Towards the Third Age. *British Educational Research Journal.*
Hornberger, N. H., & Chick, J. K. (1998, March). *Teaching and learning bilingually in post colonial contexts.* Paper presented at Sociolinguistics Symposium 12, Institute of Education, London, UK.
Huberman, M., & Miles, M. (1984). *Innovation up close.* New York: Plenum.
Ibrahim, N. A. (1991). Inservice training in Malaysia for the New Primary Curriculum. In K. Lewin & J. Stuart (Eds.), *Educational innovation in developing countries: Case studies of changemakers* (pp. 95-126). London: MacMillan.
Ingvarson, L., & MacKenzie, D. (1988). Factors affecting the impact of in-service courses for teachers: Implications for policy. *Teaching and Teacher Education, 4,* 139-155.

Johnson, S. (1988). *National assessment: The APU science approach* (for the Assessment of Performance Unit). London: Her Majesty's Stationery Office.

Joyce, B. (1992). Cooperative learning and staff development research: Teaching the method with method. *Cooperative Learning, 12,* 10-13.

Joyce, B., & Showers, B. (1980). Improving in-service training: The messages of research. *Educational Leadership, 37,* 379-385.

Joyce, B. & Showers, B. (1984). Transfer of training: The contribution of coaching. In D. Hopkins & M. Wideen (Eds.), *Alternative perspectives on school improvement.* Lewes, UK: Falmer.

Joyce, B., & Showers, B. (1988). *Student achievement through staff development.* New York: Longman.

Joyce, B. & Showers, B. (1995). *Student achievement through staff development* (2nd ed.). White Plains, New York: Longman.

Joyce, B., Calhoun, E., & Hopkins, D. (1999). *The new structure of school improvement: Inquiring schools and achieving students.* Buckingham, UK: Open University Press.

Kachru, B. (1986). *The alchemy of English: The spread, functions, and models of non-native Englishes.* New York: Pergamon Institute of English.

Kelly, M. J. (1998, December). *Initiatives from Zambia: Multigrade teaching. Initial literacy in a local language. Community schools. Determining learning outcomes for the improvement of quality.* Prepared for the Improving Learning: Perspectives for Primary Education in Rural Africa seminar sponsored by UNESCO & World Bank, Lusaka, Zambia.

Kilosa District Council Education Department. (1999). *Teacher resource centre handbook.* Morogoro, Tanzania: Author.

King, K. (Ed.). (2000). *Special theme on the World Education Forum in Dakar.* (NORRAG News 26). Edinburgh: University of Edinburgh Centre of African Studies.

Knamiller, G. W., Fairhurst, M. G., Gibbs, W., Jain, P., Khatete, D. W., Welford, A. G., & Wiegand, P. A. (2000). *The effectiveness of the teacher resource centre strategy* (Education Research Series Serial No. 34). London: UK Department for International Development.

Knowles, M. (1984). *The adult learner: A neglected species* (3rd ed.). Houston: Gulf.

Kochan, A. (2000, May). *Community projects database: An international list of community projects.* Prepared for Comparative, International and Development Education Centre. Toronto: Ontario Institute for Studies in Education at the University of Toronto.

Kraft, R. J. (1997). *Rural educational reform in the Nueva Escuela Unitaria of Guatemala.* Washington, DC: Academy for Educational Development.

Krashen, S. D. (1981). *Second language acquisition and second language learning.* New York: Pergamon.

Kruse, S., & Louis, K. S. (1995). *Professionalism and community: Perspectives on reforming urban schools.* Thousand Oaks, CA: Corwin.

Kunje, D., & Stuart, J. (1999). Supporting untrained teachers in Malawi. *International Journal of Educational Development, 19,* 157-166.

Leithwood, K., Jantzi, D. & Mascall, B. (1999). *Large scale reform: What works?* Unpublished manuscript, Ontario Institute for Studies in Education, University of Toronto.

Levin, H., & Lockheed, M. (1993). *Effective schools in developing countries.* London: Falmer.

Lewin, K., & Stuart, J. (1991). *Educational innovation in developing countries: Case studies of changemakers.* London: MacMillan.

Lieberman, A. (1996). Networks and reform in American education. *Teachers College Record, 98,* 7-45.

Little, A., Hoppers, W., & Gardner, R. (Eds.). (1994). *Beyond Jomtien: Implementing primary education for all.* London: Macmillan.

Little, J. W. (1987). Teachers as colleagues. In V. Richardson-Koelhler (Ed.), *Educator's handbook: A research perspective* (pp. 491-518). New York: Longman.

Lassibille, G., Tan, J., & Sumra, S. (2000). Expansion of private secondary education: Lessons from recent experience in Tanzania. *Comparative Education Review, 44,* 1-28.

Levin, H. (in press). The Gordian Knot of educational reform. *Curriculum Inquiry.*

Loucks-Horsley, S., Harding, C., Arbuckle, M., Murray, L., Dubea, C., & Williams, M. (1987). *Continuing to learn: A guidebook for teacher development.* Andover, MA: Regional Laboratory for Educational Improvement of the Northeast and Islands & National Staff Development Council.

Louis, K. S., & Miles, M. (1990). *Improving the urban high school: What works and why.* New York: Teachers College Press.

Makau, B. M. (1988). *Computers in Kenya's secondary schools: Case study of innovation in education.* Ottawa: International Development Research Centre.

McEwan, P. (1998). The effectiveness of multigrade schools in Colombia. *International Journal of Educational Development, 18,* 435-452.

McLaughlin, M. (1990). The Rand change agent study revisited: Macro perspectives, micro realities. *Educational Researcher, 19,* 11–16.

Menon, S., & Sandeep, P. (1994). *School Improvement Programme Andhra Pradesh* [evaluation report]. Geneva: Aga Khan Foundation.

Miles, M. (1986). Research findings on the stages of school improvement. Cited in D. Hopkins, *School improvement for real.* London: Routledge/Falmer.

Miwa, C. (2000). *Primary school reform in rural areas of Colombia: An analysis of factors limiting sustainability of the New School Program.* Nagoya, Japan: Nagoya University Graduate School of International Development.

Mosha, H. J., & Welford, A. G. (2000). *The Dar es Salaam Primary Schools' Project* [evaluation report]. Geneva: Aga Khan Foundation.

Newmann, F., King, B., & Young, S. (2000, April). *Professional development that addresses school capacity: Lessons from urban elementary schools.* Paper pre-

sented at the Annual Meeting of the American Educational Research Association, New Orleans.
Nunn, R. (1996). *Describing classroom interaction in intercultural curricular research and development.* Unpublished doctoral dissertation, University of Reading, Reading, UK.
Odada, M. (1997, March). *The Benchmarks of Universal Primary Education in Uganda.* Paper presented to staff and graduate students at the School of Education, Makerere University, Kampala, Uganda.
Olorundare, S. (1990). Discrepancies between official science curriculum and actual classroom practice: The Nigerian experience. *Journal of Education Policy, 5,* 1-19.
Osler, A. (1997). Teachers' biographies and educational development: A Kenyan case study. *International Journal of Educational Development, 17,* 361-371.
Pennycook, A. (1994). *The cultural politics of English as an international language.* New York: Longman.
Peters, L. (1998). Reflections from across the Atlantic: What can we (and the US governors) learn from the UK education reform experience? *Curriculum Inquiry, 28,* 259-262.
Phillipson, R. (1992). *Linguistic imperialism.* Oxford, UK: Oxford University Press.
Pitt, J. (2001). *Civic education and citizenship in Escuela Nueva schools in Colombia.* Paper presented at the annual meeting of the Comparative and International Education Society, Washington, DC.
Prophet, R. (1995). Views from the Botswana junior secondary classroom: Case study of a curriculum intervention. *International Journal of Educational Development, 15,* 127-140.
Psacharopolous, G. et al. (1993). Achievement evaluation of Colombia's Escuela Nueva: Is multigrade the answer? *Comparative Education Review, 37,* 263-276.
Ravitch, D., & Vinovskis, M. (Eds.). *Learning from the past: What history teaches us about school reform.* Baltimore, MD: Johns Hopkins University Press.
Rea-Dickins, P. (1996). *English Language Teachers Support Program (ELTSP): Data analysis and impact assessment* [mimeograph]. Warwick, UK: University of Warwick.
Reimers, F. (Ed.). (2000). *Unequal schools, unequal chances: The challenges to equal opportunity in the Americas.* Cambridge, MA: Harvard University Press.
République de Mali, Ministère de l'éducation de base. (1998, December). *Amélioration de l'apprentissage: Les perspectives pour l'enseignment primaire en milieu rural en Afrique.* Prepared for the Improving Learning: Perspectives for Primary Education in Rural Africa seminar sponsored by UNESCO & World Bank, Lusaka, Zambia.
Reynolds, D. (1995). Turning around ineffective schools: Some evidence and some speculations. In J. Gray et al. (Eds.), *Merging traditions: The future of research on school effectiveness and school improvement.* London: Cassell.
Roberts, J. (1998). *Language teacher education.* New York: Arnold.

Robertson, C. (1984). Formal or nonformal education? Entrepreneurial women in Ghana. *Comparative Education Review, 28,* 639-658.

Robertson, P., & Sammons, P. (1997). *Improving school effectiveness: A project in progress.* Paper presented at the Tenth International Congress for School Effectiveness and Improvement, Memphis, Tennessee.

Rosenholtz, S. (1989). *Teachers' workplace: The social organization of schools.* White Plains, NY: Longman.

Rowell, P., & Prophet, R. (1990). Curriculum-in-action: The 'practical' dimension in Botswana classrooms. *International Journal of Educational Development, 10,* 17-26.

Rubagumya, C. M. (Ed.). (1989). *Language in education in Africa: A Tanzanian perspective.* Philadelphia: Multilingual Matters.

Rust, V. D., & Dalin, P. (1990). *Teachers and teaching in the developing world.* New York: Garland.

Sagor, R. (1992). *How to conduct collaborative action research.* Alexandria, VA: Association for Supervision and Curriculum Development.

Sales, V. (1999). Women teachers and professional development: Gender issues in the training programmes of the Aga Khan Education Service, Northern Areas, Pakistan. *International Journal of Educational Development, 19,* 409-422.

San Miguel B. J. (1996). Programa de mejoriamento de la calidad de la educación para las escuelas multigrados rurales. *Formas y Reformas de la Educación,* 1 (Trimestre 3), 18-24.

Sarkar, S. C. (1994). The BRAC [Bangladesh Rural Advancement Commission] Non-Formal Primary Education Centres in Bangladesh. In S. Shaeffer (Ed.). *Partnerships and participation in basic education* (Volume 2, Case 7). Paris: International Institute for Educational Planning.

Save the Children Fund. (1998, August). *Support to basic education for children in Mtwara Rural District.* Unpublished project memorandum submitted to Department for International Development, London, UK.

Schiefelbein, E. (1991). *In search of the school of the 21st century: Is Colombia's Escuela Nueva the right pathfinder?* Santiago, Chile: UNESCO Regional Office for Education in Latin America and the Caribbean.

Schwille, J. (1982). *Teachers as policy brokers in the content of elementary school mathematics.* East Lansing, MI: Institute for Research on Teaching. (ERIC Document Reproduction Service No. ED 325 953).

Scott, S. (1996). *Education for child garment workers in Bangladesh.* Unpublished master's thesis, Ontario Institute for Studies in Education at the University of Toronto.

Shaeffer, S. (1986). *Participatory approaches to teacher training.* Paper presented at the IMTEC Conference on Improving the Quality of Teachers and Teaching in the Lesser Developed World, Denpassar, Indonesia.

Shaeffer, S. (1990). Participatory approaches to in-service teacher training. In V. D. Rust & P. Dalin (Eds.), *Teachers and teaching in the developing world* (pp. 95-114). New York: Garland.

Siabato, R. C. (1997). *Educación basica primaria en zonas rurales. La Escuela Nueva y su relación con la evaluación del plan de universalización de la educación basica primaria.* Bogota: Ministerio de Educación Nacional.

Simkins, T. & Garrett, V. (1995). *The Education Management Resource Programme* [evaluation report]. Geneva: Aga Khan Foundation.

Siraj-Blatchford, I., Odada, M. & Omagor, M. (1997). *An evaluation of the Kampala Schools Improvement Project.* Geneva: Aga Khan Foundation.

Snyder, C. W. & Ramatsui, P. T. (1990). *Curriculum in the classroom.* Botswana: Ministry of Education.

Stoll, L., Harrington, J., & Myers, K. (1995). *Two British school effectiveness and school improvement action projects.* Paper presented at the Eighth International Congress on School Effectiveness and Improvement.

Stuart, J. (1991). Classroom action research in Africa: A Lesotho case study of curriculum and professional development. In K. Lewin & J. Stuart (Eds.), *Educational innovations in developing countries* (pp. 127-152). London: MacMillan.

Stufflebeam, D. (1983). The CIPP Model for program evaluation. In G. Madaus, M. Scriven, & D. Stufflebeam (Eds.), *Evaluation models: Viewpoints on educational and human services evaluation* (pp. 117-141). The Hague: Kluwer-Nijhoff.

Sweetser, A. T. (1999). *Lessons from the BRAC [Bangladesh Rural Advancement Commission] Non-Formal Primary Education Program.* Washington, DC: Association for Basic Education and Literacy (ABEL).

Swift, D. (2000). *Civil society and the sector wide approach to education reform.* Paper presented to the International Working Group on Education, Florence, Italy.

Swope, J., & Latorre, M. (1998). *Comunidades educativas donde termina el asfalto.* Santiago, Chile: Centro de Investigaciónes y Desarrollo de la Educación (CIDE).

Sylva, K. & Siraj-Blatchford, I. (1996). *Bridging the gap between home and school: Improving achievement in primary schools.* Paris: UNESCO.

Thiagarajan, S. (1990). An integrated approach to primary teacher support and training. In D. W. Chapman & C. A. Carrier (Eds.), *Improving educational quality: A global perspective* (pp. 111-134). Westport, CT: Greenwood.

Thiessen, D., & Anderson, S. E. (1999). *Getting into the habit of change in Ohio schools: The cross-case study of 12 transforming learning communities.* Columbus, Ohio: Ohio Department of Education.

Tyack, D., & Cuban, L. (1995). *Tinkering toward utopia: A century of public school reform.* Cambridge, MA: Harvard University Press.

Uganda National Council for Children (UNCC). (1994). *Equity and vulnerability: A situation analysis of women, adolescents and children in Uganda.* Kampala: Government of Uganda.

UNICEF. (2000). *Defining quality.* Paper presented to the International Working Group on Education, Florence, Italy.

United Republic of Tanzania. (1993). *The Tanzania education system for the 21st Century: Report of the Task Force.* Leeds, UK: University of Leeds Print Services.

United Republic of Tanzania. (1995). *Education and training policy.* Dar-es-Salaam, Tanzania: Ministry of Education and Culture.

United Republic of Tanzania Ministry of Education and Culture. (1998). *Basic statistics in education (BEST).* Dar-es-Salaam, Tanzania: Author.

United States Agency for International Development (USAID). (1996). *Education decentralization in Africa.* Washington, DC: Author.

Vimpany, P. (1998). *Financial management for primary schools in Tanzania* (School Improvement and Management Series Module 3). Dar-es-Salaam, Tanzania: Dar-es-Salaam Primary Schools Project.

Wallace, R. E., Lemahieu, P. G., & Bickel, W. E. (1990). The Pittsburgh experience: Achieving commitment to comprehensive staff development. In Joyce, B. R. (Ed.), *Changing school culture through staff development* (pp. 185-202). Alexandria, VA: Association for Supervision and Curriculum Development.

Wang, M., Haertal, G., & Walberg, H. (1993). Toward a knowledge base for school learning. *Review of Educational Research, 63,* 249-294.

Warwick, D. P. & Reimers, F. (1995). *Hope or despair? Learning in Pakistan's primary schools.* Westport: Praeger.

Welford, G., & Khatete, D. (2000). Teachers' Centres in Kenya. In G. Knamiller et al. (Eds.), *The effectiveness of the teacher resource centre strategy* (Chapter 5). London: UK Department for International Development.

White, M. (2000, May). *Community educators database: An international list of community education specialists.* Prepared for Comparative, International and Development Education Centre. Toronto: Ontario Institute for Studies in Education at the University of Toronto.

Williams, R. (1989). *Report on the FIELOC materials, Aga Khan Mzizima Secondary School Improvement Project* [mimeograph]. Geneva: Aga Khan Foundation.

World Bank. (1990). *Primary Education* [policy paper]. Washington, DC: Author.

Zaalouk, M. (1995). *The children of the Nile. The Community Schools project in Upper Egypt.* Paris: UNICEF.

Zanzibar Ministry of Education. (1996). *Draft education masterplan (ZEMAP).* Zanzibar: Author.

Contributors

Stephen E. Anderson is an Associate Professor in the Department of Theory and Policy Studies in Education at the Ontario Institute for Studies in Education of the University of Toronto (OISE/UT), where he teaches in the graduate and pre-service teacher education programs. He is a member of the International Center for Educational Change at OISE/UT. His research interests include the implementation of education policy and program change, school improvement and teacher development, program evaluation, and innovations in pre-service teacher education. His research and consulting experiences over the past 20 years have been situated in Canada, the United States, Africa (Kenya, Tanzania, Uganda), and Pakistan. Publications of his work have appeared in such journals as School Effectiveness and School Improvement, Curriculum Inquiry, The Journal of Staff Development, and School Organisation, and Educational Leadership.

Ann Brumfit is a Senior Lecturer and Director of the Malaysian Programme B.Ed. at St. Mary's University College, Strawberry Hill, UK. She has extensive education research and consulting experience related to English language education and language policy implementation in many regions of the world, including Tanzania, Zanzibar, Botswana, India, Pakistan, Bangladesh, Malaysia, Kuwait, Qatar, Belgium, Yugoslavia, Poland, Russia, and the United Kingdom. Since 1985 Ms. Brumfit has completed numerous consultancy projects for the British Council, the U.K. Department for International Development, and the Aga Khan Foundation.

Joanne Capper has been working in international development since 1988 in the areas of examination reform, national assessment, teacher training and education technology, primarily in Africa, the Middle East and South Asia. Since 1997, she has worked for the World Bank in education technology planning, and is managing a series of case studies in teacher training and technology in developing countries. She obtained her doctorate in education evaluation and measurement from the University of California at Los Angeles (UCLA). For five years, she worked as an evaluation and measurement specialist at the California State Department of Education. She was Executive Director of the Center for Research into Practice, publishing two monograph series synthesizing research in cognitive science and education technology for educators and policymakers. She has consulted with numerous national and international education policy agencies, including the U.S. Congress, the National Science Foundation, and USAID. Her publications include a book entitled Testing to Learn: Learning to Test on how developing-country policymakers can design examinations to influence improved teaching and learning.

Joe Farrell is a Professor in the Department of Curriculum, Teaching and Learning at the Ontario Institute for Studies in Education of the University of Toronto (OISE/UT), and is director of the OISE/UT Centre for Comparative and International Education. His interests are in planning and policy analysis related to formal and non-formal education of youth and adults, with emphasis on problems of equity in the distribution of opportunities to learn, especially in "developing" regions. More specific areas of work include: diagnosis and design of policies to improve educational quality; educational program and policy evaluation; analysis of the politics of educational policy-making processes; comparative/cross-cultural studies of teacher development and the roles of teachers in different learning systems; and analysis and further development of alternative, action-learning oriented, large-scale systems of primary schooling for severely marginalized children in developing areas. Dr. Farrell's work has been geographic concentrated in Latin America, with major projects in Chile, and others in Uruguay, Colombia, Ecuador, Mexico, Venezuela, Panama, and Costa Rica. He has worked for briefer periods in St. Lucia, Botswana, Kenya, the Philippines, Indonesia, and Thailand. He is currently working on a long-term project supporting the development of a system of small "community schools" for girls in remote rural hamlets in Egypt.

Jeremy Greenland is the Chief Education Officer Aga Khan Foundation (Geneva), and has been actively involved in the conceptualization, mobilization, and evaluation of the all the Aga Khan Foundation teacher development and school improvement projects in East Africa (Kenya, Tanzania, Zanzibar, Uganda), Pakistan, India, and Tajikistan since the mid-d1980's. Prior to joining the Aga Khan Foundation, he was involved in teacher development research and consulting in the developing world, and edited a book on in-service teacher development in Africa (The In-service Training of Primary School Teachers in English-speaking Africa, MacMillan, 1983). Drawing upon the Aga Khan project experiences, he has continued to contribute to theory and policy discussions through conference presentations and publications in professional journals such as The International Journal of Educational Research.

Hamed R. H. Hikmany is Director General Zanzibar Investment Promotion Agency (Ministry of Finance, Zanzibar). He holds a PhD in Economics of Education from the University of Wales College of Cardiff. His career includes experience as a teacher and school administrator in Zanzibar, as a planning officer for the Department of Administration and Planning (Ministry of Education, Zanzibar), as Head of Training (Ministry of State for Planning, Zanzibar), and as Commissioner for the Department of External Finance and International Cooperation. He has served as a policy advisor and consultant to numerous government and internationally supported social and education sector initiatives, and is the author of several reports associated with those assignments.

David Hopkins is Professor of Education at the University of Nottingham and has served as Head of the School and Dean of the Faculty of Education. Prior to coming to Nottingham in 1996, he was for ten years a tutor at the University of Cambridge Institute of Education. David Hopkins is also Chair of the Leicester City Partnership Board and a member of the Governing Council of the National Council for School Leadership. His professional interests are in the areas of teacher and school development, educational change, teacher education, and policy implementation and evaluation. He is a long time consultant to the OECD and the Department for International Development on issues of school improvement and teacher quality, and has lectured and consulted in some twenty countries on these themes. David Hopkins has published over thirty books on educational issues. Among the most recent are recent are: School Improvement for Real (Routledge/Falmer, 2001); A Teacher's Guide to Classroom Research (2nd Edition, Open University Press, 2002); Models of Learning - Tools for Teaching (with Bruce Joyce and Emily Calhoun, 2nd Edition, Open University Press, 2002); and Improving the Quality of Education for All (Fulton, 2002). He is also an International Mountain Guide and has climbed in many of the world's great mountain ranges.

Herme Joseph Mosha is a Professor at the Faculty of Education of the University of Dar-es-Salaam, and a specialist in educational planning and management, program design and evaluation. Prior to his university career, he held teachers college, adult, secondary and primary teaching positions in the Tanzanian school system. He was the first and founder Dean of the Faculty of Education at the University of Dar-es-Salaam, Chairman of the National Task Force on the Education System for the 21st Century and architect of the University of Dar-es-Salaam's internally initiated Institutional Transformation Program in African universities. His research and publications focus on policy, planning and administration issues and practices at all levels of education in Tanzania and other African education systems. Dr. Mosha has served as a consultant to the Tanzanian Ministry of Education and Culture and to numerous internationally sponsored education projects in Africa for such agencies as NORAD, the British Council, the World Bank, DANIDA, UNESCO, UNICEF, ECA and the Aga Khan Foundation.

Shelom Nderitu is a Nairobi-based Education Consultant and Trainer. In this capacity she has worked on numerous training, curriculum development, and research projects in Kenya and Southern Sudan. These include school-based training of teachers for the improvement of teaching and learning mathematics in primary schools, writing of curriculum for both formal and non-formal education (e.g., basic education of street children), and research and evaluation studies of education projects for the Kenya Ministry of Education and for NGO's such as the Aga Khan Foundation and World Bank. Prior to her current career as an Education Consultant, Ms. Nderitu worked at all levels of the Kenyan education system as a National Senior Inspector of Schools, an Education Officer with the Ministry of Education Planning and Development Department, a Curriculum Specialist (Primary Mathematics) with the Kenya Institute of Education, a teachers college lecturer, and a secondary school teacher. She is co-author of three sets of Primary Mathematics Course textbooks and teachers guides in Kenya and in Uganda.

Matthew Odada is an Associate Professor of Social Studies and History at the School of Education, Makerere University, Uganda. He has forty years professional experience in Ugandan education as a secondary school teacher, headmaster, teachers' college tutor and principal, university lecturer and associate professor. He has served as a resource person and consultant on numerous national curriculum writing and education reform projects in Uganda, and has been active locally on PTA executives in both primary and secondary public schools. Dr. Odada has written articles and monographs on the conditions and quality of teaching and learning in Ugandan schools, and on the teaching of social studies in the African context. He participated in research for Uganda's Improving Education Quality (IEQ) project in the 1990's.

Paul Ogula
Director of Research and Evaluation
Kenya Institute of Education
Nairobi, Kenya

Martin Omagor is an education officer with the Inspectorate of the Ministry of Education and Sports, Government of Uganda. He holds an advanced degree in educational psychology from Makerere University, Uganda, and a Bachelor's Degree in Education from the Institute of Teacher Education Kyambogo, Uganda, where he also taught as a lecturer in the 1980's. Mr. Omagor has teaching and teacher education experience in the areas of special education and of arts education within the Ugandan school system.

Iram Siraj-Blatchford lives in Cambridge and is a Professor of Early Childhood Education at the Institute of Education University of London and Head of Child Development and Learning. Her research interests include early childhood and primary curriculum and policy development. She has authored and edited 12 books, including A Curriculum Development Handbook for Early Childhood Educators (Trentham Books, 1998), Bridging the Gap Between Home and School: Improving Achievement in Primary Schools (UNESCO, 1996), Doing Early Childhood Research (Allen & Unwin, 2001), and Educating the Whole Child: Cross-curricular Skills, Themes, and Dimensions (Open University Press, 1995). She is currently researching pedagogy and pre-school effectiveness for the UK Department of Education.

Suleman Sumra is an Associate Professor in the Department of Educational Foundations in the Faculty of Education, University of Dar-es-Salaam where he teaches both in the undergraduate and post-graduate programs. His research interests include policy formulation and implementation, and the role of donor agencies in shaping educational agendas and innovations in primary education. He has provided consulting services for various organizations in Tanzania. Publications of his work have appeared in such journals as the Alberta Journal of Educational Research, Comparative Education Review, and Papers in Education and Development.

Geoff Welford is an independent education consultant, and former Director of International Education in the School of Education at the University of Leeds, where he also taught in the School's graduate and pre-service teacher education programs. He began his career has a secondary school teacher, and later became Deputy Director of the Assessment of Performance Unit's science team at Leeds developing national tests of pupil's practical performance and understanding in Biology. His research and consulting experience embraces many countries, particularly in Sub-Saharan Africa, and include: assessment policy and practice in school science; indigenous science and technology; non-formal approaches to universal basic education; and the acquisition of classroom management skills among beginning teachers. His work has been published in such journals as the International Journal of Science Education, Evaluation and Research in Education, and Research Papers in Education.

Index

D

E

T